T0133135

PROCESSING OF FRUITS AND VEGETABLES

From Farm to Fork

Innovations in Agricultural and Biological Engineering

PROCESSING OF FRUITS AND VEGETABLES

From Farm to Fork

Edited by
Khursheed Alam Khan, MTech
Megh R. Goyal, PhD, PE
Abhimannyu A. Kalne, MTech

APPLE ACADEMIC PRESS

Apple Academic Press Inc.
3333 Mistwell Crescent
Oakville, ON L6L 0A2 Canada

Apple Academic Press Inc.
1265 Goldenrod Circle NE
Palm Bay, Florida 32905 USA

© 2020 by Apple Academic Press, Inc.

First issued in paperback 2021

Exclusive worldwide distribution by CRC Press, a member of Taylor & Francis Group

No claim to original U.S. Government works

ISBN 13: 978-1-77463-403-5 (pbk)
ISBN 13: 978-1-77188-708-3 (hbk)

Library and Archives Canada Cataloguing in Publication

Title: Processing of fruits and vegetables : from farm to fork / edited by Khursheed A. Khan, MTech, Megh R. Goyal, PhD, PE, Abhimannyu A. Kalne, MTech

Names: Khan, Khursheed A. (Khursheed Alam), editor. | Goyal, Megh Raj, editor. | Kalne, Abhimannyu A., editor.

Series: Innovations in agricultural and biological engineering.

Description: Series statement: Innovations in agricultural and biological engineering | Includes bibliographical references and index.

Identifiers: Canadiana (print) 20189068485 | Canadiana (ebook) 20189068493 | ISBN 9781771887083 (hardcover)| ISBN 9780429505775 (PDF)

Subjects: LCSH: Fruit—Processing. | LCSH: Vegetables—Processing. | LCSH: Food industry and trade—Technological innovations.

Classification: LCC TP440 .P76 2019 | DDC 664/.8—dc23

Library of Congress Cataloging-in-Publication Data

Names: Khan, Khursheed A. (Khursheed Alam), editor. | Goyal, Megh Raj, editor. | Kalne, Abhimannyu A. (Abhimannyu Arun), editor.

Title: Processing of fruits and vegetables : from farm to fork / editors, Khursheed A. Khan, Megh R. Goyal, Abhimannyu A. Kalne.

Description: Toronto ; New Jersey : Apple Academic Press, [2019] | Series: Innovations in agricultural and biological engineering | Includes bibliographical references and index.

Identifiers: LCCN 2018058138 (print) | LCCN 2018058650 (ebook) | ISBN 9780429505775 (ebook) | ISBN 9781771887083 (hardcover : alk. paper)

Subjects: | MESH: Food Technology | Fruit--enzymology | Vegetables--enzymology | Food-Processing Industry--methods | Antioxidants | Vitamins

Classification: LCC SB319.95 (ebook) | LCC SB319.95 (print) | NLM WA 703 | DDC 635.028/6--dc23

LC record available at https://lccn.loc.gov/2018058138

Apple Academic Press also publishes its books in a variety of electronic formats. Some content that appears in print may not be available in electronic format. For information about Apple Academic Press products, visit our website at **www.appleacademicpress.com** and the CRC Press website at **www.crcpress.com**

OTHER BOOKS ON AGRICULTURAL & BIOLOGICAL ENGINEERING BY APPLE ACADEMIC PRESS, INC.

Management of Drip/Trickle or Micro Irrigation
Megh R. Goyal, PhD, PE, Senior Editor-in-Chief

Evapotranspiration: Principles and Applications for Water Management
Megh R. Goyal, PhD, PE, and Eric W. Harmsen, Editors

Book Series: Research Advances in Sustainable Micro Irrigation
Senior Editor-in-Chief: Megh R. Goyal, PhD, PE

Volume 1: Sustainable Micro Irrigation: Principles and Practices
Volume 2: Sustainable Practices in Surface and Subsurface Micro Irrigation
Volume 3: Sustainable Micro Irrigation Management for Trees and Vines
Volume 4: Management, Performance, and Applications of Micro Irrigation Systems
Volume 5: Applications of Furrow and Micro Irrigation in Arid and Semi-Arid Regions
Volume 6: Best Management Practices for Drip Irrigated Crops
Volume 7: Closed Circuit Micro Irrigation Design: Theory and Applications
Volume 8: Wastewater Management for Irrigation: Principles and Practices
Volume 9: Water and Fertigation Management in Micro Irrigation
Volume 10: Innovation in Micro Irrigation Technology

Book Series: Innovations and Challenges in Micro Irrigation
Senior Editor-in-Chief: Megh R. Goyal, PhD, PE

Volume 1: Principles and Management of Clogging in Micro Irrigation
Volume 2: Sustainable Micro Irrigation Design Systems for Agricultural Crops: Methods and Practices
Volume 3: Performance Evaluation of Micro Irrigation Management: Principles and Practices

Volume 3: Potential of Solar Energy and Emerging Technologies in Sustainable Micro Irrigation

Volume 4: Potential of Solar Energy and Emerging Technologies in Sustainable Micro Irrigation

Volume 5: Micro Irrigation Management: Technological Advances and Their Applications

Volume 6: Micro Irrigation Scheduling and Practices

Volume 7: Micro Irrigation Scheduling and Practices

Volume 8: Engineering Interventions in Sustainable Trickle Irrigation: Water Requirements, Uniformity, Fertigation, and Crop Performance

Volume 9: Management Strategies for Water Use Efficiency and Micro Irrigated Crops: Principles, Practices, and Performance

Book Series: Innovations in Agricultural & Biological Engineering
Senior Editor-in-Chief: Megh R. Goyal, PhD, PE

- Dairy Engineering: Advanced Technologies and Their Applications
- Developing Technologies in Food Science: Status, Applications, and Challenges
- Emerging Technologies in Agricultural Engineering
- Engineering Interventions in Agricultural Processing
- Engineering Interventions in Foods and Plants
- Engineering Practices for Agricultural Production and Water Conservation: An Interdisciplinary Approach
- Engineering Practices for Management of Soil Salinity: Agricultural, Physiological, and Adaptive Approaches
- Engineering Practices for Milk Products: Dairyceuticals, Novel Technologies, and Quality
- Evapotranspiration
- Flood Assessment: Modeling and Parameterization
- Food Engineering: Emerging Issues, Modeling, and Applications
- Food Process Engineering: Emerging Trends in Research and Their Applications
- Food Technology: Applied Research and Production Techniques
- Modeling Methods and Practices in Soil and Water Engineering
- Nanotechnology and Nanomaterial Applications in Food, Health and Biomedical Sciences

- Nanotechnology Applications in Dairy Science: Packaging, Processing, and Preservation
- Novel Dairy Processing Technologies: Techniques, Management, and Energy Conservation
- Processing of Fruits and Vegetables: From Farm to Fork
- Processing Technologies for Milk and Milk Products: Methods, Applications, and Energy Usage
- Scientific and Technical Terms in Bioengineering and Biological Engineering
- Soil and Water Engineering: Principles and Applications of Modeling
- Soil Salinity Management in Agriculture: Technological Advances and Applications
- State-of-the-Art Technologies in Food Science: Human Health, Emerging Issues and Specialty Topics
- Sustainable Biological Systems for Agriculture: Emerging Issues in Nanotechnology, Biofertilizers, Wastewater, and Farm Machines
- Technological Interventions in Dairy Science: Innovative Approaches in Processing, Preservation, and Analysis of Milk Products
- Technological Interventions in Management of Irrigated Agriculture
- Technological Interventions in the Processing of Fruits and Vegetables
- Technological Processes for Marine Foods, From Water to Fork: Bioactive Compounds, Industrial Applications, and Genomics

ABOUT THE BOOK SERIES: INNOVATIONS IN AGRICULTURAL & BIOLOGICAL ENGINEERING

Under the book series titled *Innovations in Agricultural and Biological Engineering*, Apple Academic Press Inc. (AAP) is publishing volumes in the specialty areas defined by a American Society of Agricultural and Biological Engineers (<asabe.org>) over a span of 8–10 years.

The mission of this series is to provide knowledge and techniques for agricultural and biological engineers (ABEs). The series offers high-quality reference and academic content in ABE that is accessible to academicians, researchers, scientists, university faculty, university-level students, and professionals around the world.

ABEs ensure that the world has the necessities of life including safe and plentiful food, clean air and water, renewable fuel and energy, safe working conditions, and a healthy environment by employing knowledge and expertise of sciences, both pure and applied, and engineering principles. Biological engineering applies engineering practices to problems and opportunities presented by living things and the natural environment in agriculture. ABE embraces a variety of following specialty areas (asabe. org): Aquacultural Engineering, Biological Engineering, Energy, Farm Machinery and Power Engineering, Food and Process Engineering, Forest Engineering, Information & Electrical Technologies Engineering, Natural Resources, Nursery and Greenhouse Engineering, Safety and Health, and Structures and Environment.

For this book series, we welcome chapters on the following specialty areas (but not limited to):

1. Academia to industry to end-user loop in agricultural engineering
2. Agricultural mechanization
3. Aquaculture engineering
4. Biological engineering in agriculture
5. Biotechnology applications in agricultural engineering
6. Energy source engineering
7. Food and bioprocess engineering

8. Forest engineering
9. Hill land agriculture
10. Human factors in engineering
11. Information and electrical technologies
12. Irrigation and drainage engineering
13. Nanotechnology applications in agricultural engineering
14. Natural resources engineering
15. Nursery and greenhouse engineering
16. Potential of phytochemicals from agricultural and wild plants for human health
17. Power systems and machinery design
18. GPS and remote sensing potential in agricultural engineering
19. Robot engineering in agriculture
20. Simulation and computer modeling
21. Smart engineering applications in agriculture
22. Soil and water engineering
23. Structures and environment engineering
24. Waste management and recycling
25. Any other focus area

For more information on this series, readers may contact:

Megh R. Goyal, PhD, PE
Book Series Senior Editor-in-Chief
Innovations in Agricultural and
Biological Engineering
goyalmegh@gmail.com

ABOUT THE LEAD EDITOR
KHURSHEED ALAM KHAN

Khursheed Alam Khan acquired his bachelor's degree in Agricultural Engineering from Allahabad Agricultural Institute, University of Allahabad, and master's degree in Postharvest Engineering and Technology from Aligarh Muslim University, Aligarh, India while securing the university's prestigious award "University Medal" for his outstanding performance among faculty students of the university.

He is presently working as In-charge of the Department of Agricultural Engineering at the College of Horticulture, Mandsaur of Rajmata Vijayaraje Scindia Agriculture University (RVSKVV), Gwalior, India. He has acquired proficiency in the field of postharvest processing through his experience in the food industry and by research and teaching at national and international levels for more than 13 years.

He has gained immense experience in production and quality control of processed food products, such as tomato ketchup, sauces, salad dressings, and mayonnaise during his service in the food industry (MNC) in the Middle East. He has been awarded a certificate in *"Essential HAACP Practices"* from The Royal Society for the Promotion of Health, UK.

He has taught several courses at Awassa College of Agriculture at Debub University (now Hawassa University) of Ministry of Education, Government of Ethiopia. He has also prepared the *"Training Manual on Food Processing"* for technical and vocational program in food processing, run by the Ministry of Education, Federal Democratic Republic of Ethiopia under UNDP (Capacity Building Program). He made a fruitful contribution to the field of research in agricultural process engineering and has published more than 25 technical papers and review articles in different national and international journals/conference papers. He has written several book chapters. He is also a reviewer for the *International Food Research Journal*, Malaysia.

ABOUT THE SENIOR EDITOR-IN-CHIEF: MEGH R. GOYAL

Megh R. Goyal, PhD, PE, is a Retired Professor in Agricultural and Biomedical Engineering from the General Engineering Department in the College of Engineering at the University of Puerto Rico–Mayaguez Campus; and Senior Acquisitions Editor and Senior Technical Editor-in-Chief in Agriculture and Biomedical Engineering for Apple Academic Press, Inc. He has worked as a Soil Conservation Inspector and as a Research Assistant at Haryana Agricultural University and Ohio State University.

During his professional career of 45 years, Dr. Goyal has received many prestigious awards and honors. He was the first agricultural engineer to receive the professional license in Agricultural Engineering in 1986 from the College of Engineers and Surveyors of Puerto Rico. In 2005, he was proclaimed as "Father of Irrigation Engineering in Puerto Rico for the Twentieth Century" by the American Society of Agricultural and Biological Engineers (ASABE), Puerto Rico Section, for his pioneering work on micro irrigation, evapotranspiration, agroclimatology, and soil and water engineering. The Water Technology Centre of Tamil Nadu Agricultural University in Coimbatore, India, recognized Dr. Goyal as one of the experts "who rendered meritorious service for the development of micro irrigation sector in India" by bestowing the Award of Outstanding Contribution in Micro Irrigation. This award was presented to Dr. Goyal during the inaugural session of the National Congress on "New Challenges and Advances in Sustainable Micro Irrigation" on March 1, 2017, held at Tamil Nadu Agricultural University. Dr. Goyal received the Netafim Award for Advancements in Microirrigation: 2018 from the American Society of Agricultural Engineers at the ASABE International Meeting in August 2018.

A prolific author and editor, he has written more than 200 journal articles and textbooks and has edited over 62 books. He is the editor of three book series published by Apple Academic Press: Innovations in Agricultural & Biological Engineering, Innovations and Challenges in Micro Irrigation, and Research Advances in Sustainable Micro Irrigation. He is also instrumental in the development of the new book series Innovations in Plant Science for Better Health: From Soil to Fork.

Dr. Goyal received his BSc degree in engineering from Punjab Agricultural University, Ludhiana, India; his MSc and PhD degrees from Ohio State University, Columbus; and his Master of Divinity degree from Puerto Rico Evangelical Seminary, Hato Rey, Puerto Rico, USA.

ABOUT THE EDITOR:
ABHIMANNYU A. KALNE

Abhimannyu A. Kalne, MTech, is presently working as an Assistant Professor at Swami Vivekanand College of Agricultural Engineering and Technology & Research Station, Indira Gandhi Agricultural University, Raipur, India. Formerly he worked on the Technology Mission on Citrus project of the Government of India at the National Research Centre for Citrus (NRCC), Nagpur, India, and on the World Bank-aided National Agricultural Innovation Project (Component 4: Basic and Strategic Research in the Frontier Areas of Agricultural Science) at the Central Institute of Agricultural Engineering (CIAE), Bhopal, as a Senior Research Fellow and Research Associate. He has gained expertise in postharvest management and processing of citrus fruits and nondestructive quality evaluation techniques for fruits and vegetables. He has also worked on a research project on the application of machine vision for distinguishing crop varieties, funded by the National Fund for Basic, Strategic and Frontier Application Research in Agriculture (ICAR), India. He is presently engaged in teaching of undergraduate- and postgraduate-level courses on Agricultural Processing and Food Engineering. On the research side, he is involved in processing and value addition of medicinal and aromatic plants and non-timber forest produce at the Centre of Excellence.

He was awarded a state-level Young Scientist Award in science and engineering for the postgraduate thesis work on banana chips quality, a national-level poster presentation award on dehydration technique for preparation of aloe vera gel powder, and a national-level best conference paper award for value addition of aloe vera gel. He has published more than seven research papers in national and international journals and three book chapters.

He received his bachelor's degree in Agricultural Engineering from Dr. Panjabrao Deshmukh Agricultural University, Akola (India) and his master's degree in Agricultural Processing and Food Engineering from Indira Gandhi Agricultural University, Raipur, India through ICAR's Junior Research Fellowship Examination.

CONTENTS

CONTRIBUTORS

Singathirulan Balasubramanian, PhD
Dean, College of Fisheries Engineering, Tamil Nadu Fisheries University, Nagapattinam 611001, Tamil Nadu, India. E-mail: balaciphet@gmail.com

Kumari S. Banga, MTech
PhD Research Scholar, ICAR-Central Institute of Agricultural Engineering, Bhopal 462038, Madhya Pradesh, India. E-mail: sheetal.0415@gmail.com

Khalid Bashir, MTech
PhD Research Scholar, Department of Basic and Applied Sciences, National Institute of Food Technology Entrepreneurship and Management, Kundli 131028, Sonipat, Haryana, India. E-mail: kbnaik25@gmail.com

Khan Chand, MTech
Assistant Professor, Department of Postharvest Process and Food Engineering, College of Technology, G.B. Pant University of Agriculture and Technology (GBPUAT), Pantnagar, Udham Singh Nagar 263145, Uttarakhand, India. E-mail: kcphpfe@gmail.com

Kiran Dabas, MTech
Senior Analyst, Chemical Department, TUV SUD South Asia India Pvt. Limited, Gurugram 122016, Haryana, India. E-mail: kirandabas18@gmail.com

Bazilla Gayas, MTech
PhD Candidate, Department of Processing and Food Engineering, Punjab Agricultural University (PAU), Ludhiana 141004, Punjab, India. E-mail: bazilagayass@gmail.com

Abhimannyu Arun Kalne, MTech
Assistant Professor, Department of Agricultural Processing and Food Engineering, Swami Vivekanand College of Agricultural Engineering and Technology and Research Station, Indira Gandhi Agricultural University, Raipur 492012, India. E-mail: abhikalne7@gmail.com

Adinath Kate, PhD
Scientist, Agro-Produce Processing Division, ICAR-Central Institute of Agricultural Engineering, Nabibagh, Berasia Road, Bhopal 462038, Madhya Pradesh, India. E-mail: eradikate02@gmail.com

Khursheed Alam Khan, MTech
Assistant Professor (Agricultural Engineering), Department of Agricultural Engineering, College of Horticulture, Rajmata Vijayaraje Scindia Agricultural University (RVSKVV), Mandsaur 458001, Madhya Pradesh, India. E-mail: khan_undp@yahoo.ca

Sunil Kumar, MTech
PhD Research Scholar, ICAR-Central Institute of Agricultural Engineering, Bhopal 462038, Madhya Pradesh, India. E-mail: sunilciae@gmail.com

Umesh C. Lohani, PhD
Assistant Professor, Department of Postharvest Process and Food Engineering, College of Technology, G.B. Pant University of Agriculture and Technology (GBPUAT), Pantnagar, Udham Singh Nagar 263145, Uttarakhand, India. E-mail: ulohani@gmail.com

Sivashankari Manickam, PhD
Senior Research Fellow, Indian Institute of Food Processing Technology, Ministry of Food Processing Industries, Government of India, Pudukkottai Road, Thanjavur 613005, Tamil Nadu, India. E-mail: sivashankari.m@gmail.com

Debabandya Mohapatra, PhD
Senior Scientist, Agro-Produce Processing Division, ICAR-Central Institute of Agricultural Engineering, Nabibagh, Berasia Road, Bhopal 462038, Madhya Pradesh, India. E-mail: debabandya@gmail.com

Beena Munaza, MTech
PhD Candidate, Division of Food Science and Technology, Sher-e-kashmir University of Agricultural Sciences and Technology, Jammu 180009, India. E-mail: beenamunaza@gmail.com

Akash Pare, PhD
Assistant Professor, Academics and Human Resource Development, Indian Institute of Food Processing Technology, Ministry of Food Processing Industries, Government of India, Pudukkottai Road, Thanjavur 613005, Tamil Nadu, India. E-mail: akashpare@iicpt.edu.in

Mahendrabhai Babulal Patel, MTech
Assistant Professor, Post-harvest Technology Department, College of Horticulture, S.D. Agricultural University, Jagudan 382710, Gujarat, India. E-mail: mb_patel65@yahoo.co.in

Monica Premi, PhD
Assistant Professor, Department of Food Technology, Jaipur National University, Jagatpura, Jaipur 302017, Rajasthan, India. E-mail: befriendlymonica@gmail.com

Savita Rani, MTech
PhD Research Scholar, Department of Food Science and Technology, National Institute of Food Technology Entrepreneurship and Management, Kundli 131028, Sonipat, Haryana, India. E-mail: savitabalhara88@gmail.com

Chandrakala Ravichandran, MTech
PhD Research Scholar, Department of Food Science and Technology, National institute of Food Technology Entrepreneurship and Management, Plot No.: 97, Sector-56, HSIIDC Industrial estate, Kundli 131028, Sonepat, Haryana, India. E-mail: rchandrakala.phd@gmail.com

Anwesa Sarkar, PhD
Senior Research Fellow, Department of Postharvest Process and Food Engineering, College of Technology, G.B. Pant University of Agriculture and Technology (GBPUAT), Pantnagar, Udham Singh Nagar 263145, Uttarakhand, India. E-mail: anwesa29@gmail.com

Navin Chandra Shahi, PhD
Professor, Department of Postharvest Process and Food Engineering, College of Technology, G.B. Pant University of Agriculture and Technology (GBPUAT), Pantnagar, Udham Singh Nagar 263145, Uttarakhand, India. E-mail: ncshahi2008@gmail.com

Gagandeep Kaur Sidhu, PhD
Senior Research Engineer, Department of Processing and Food Engineering, Punjab Agricultural University (PAU), Ludhiana 141004, Punjab, India. E-mail: gagandeep@pau.edu

Anupama Singh, PhD
Professor, Department of Postharvest Process and Food Engineering, College of Technology, G.B. Pant University of Agriculture and Technology (GBPUAT), Pantnagar, Udham Singh Nagar 263145, Uttarakhand, India. E-mail: asingh3@gmail.com

Tanya Luva Swer, MSc
PhD Research Scholar, Department of Food Science and Technology, National Institute of Food Technology Entrepreneurship and Management, Kundli 131028, Sonipat, Haryana, India. E-mail: tanyaswer@gmail.com

Ajita Tiwari
Department of Agricultural Engineering, Triguna Sen School of Technology, Assam University, Silchar 788011, India. E-mail: ajitatiwari@gmail.com

Kamalchandra R. Trivedi, MTech
Senior Research Fellow, Post-harvest Technology, Centre for Research on Seed Spices, S.D. Agricultural University, Jagudan 382710, Gujarat, India. E-mail: trivedi_kamal@rediffmail.com

Ashutosh Upadhyay, PhD
Associate professor, Department of Food Science and Technology, National Institute of Food Technology Entrepreneurship and Management, Plot No.: 97, Sector-56, HSIIDC Industrial estate, Kundli 131028, Sonepat, Haryana, India. E-mail: ashutosh.niftem@gmail.com

ABBREVIATIONS

AA	ascorbic acid
ABTS	2,2'-azino-bis (3-ethylbenzothiazoline-6-sulphonic acid)
AFD	atmospheric freeze drying
AMS	α-amylase
APEDA	Agricultural and Processed Foods Export Development Authority (India)
ASHRAE	American Society of Heating, Refrigerating and Air-Conditioning Engineers
ASME	American Society of Mechanical Engineers
a_w	water activity
BET	Brunauer–Emmett–Teller
BIS	Bureau of Indian Standards
CA	controlled atmosphere
CCD	charge-coupled device
CD	cabinet drying
CFB	corrugated fiberboard
CIPHET	Central Institute of Post harvest Engineering and Technology
CFU	colony-forming unit
CMOS	complementary metal oxide semiconductor
COP	coefficient of performance
CP	conducting polymers
CSM	charge simulation methods
CT	computed tomography
CV	coefficient of variation
D	decimal reduction time (s)
d.b.	moisture content on dry basis
DCA	dynamic controlled atmosphere
DHF	dihydrofolate
DNA	deoxyribonucleic acid
DPPH	1,1-Diphenyl-2-picrylhydrazyl
DRI	daily reference intake
DRP	deformation relaxation phenomenon
e-nose	electronic nose
e-tongue	electronic tongue

e-vision	electronic vision
EC	European Communities
ECM	electrical capacitance method
EQ	external quality
FAO	Food and Agriculture Organization
FAV	fruits and vegetables
FD	freeze drying
FDA	Food and Drug Administration
FFA	free fatty acid
FIFO	first in first out
FIR–VD	far-infrared vacuum drying
FMCG	fast-moving consumer goods
FPO	Food Product Order
FRAP	ferric reducing antioxidant power
FSSAI	Food Safety and Standard Authority of India
GRAS	generally recognized as safe
GSH	glutathione
HACCP	Hazard Analysis and Critical Control Points
HACD	hot air convective drying
HDM	hydrodynamic mechanism
HHP	high hydrostatic pressure
HMF	hydroxymethyl furfural
HMT	homocystein methyl trasferase
HPP	high-pressure processing
IARW	International Association of Refrigerated Warehouses
IFT	Institute of Food Technologists
IIHR	Indian Institute of Horticultural Research
INR	Indian Rupees
IQ	internal quality
IRD	infrared drying
ISL	Instrumentation and Sensing Laboratory
ISO	International Organization for Standardization
KMS	potassium metabisulfite
LCTF	liquid crystal tunable filter
LDL	low-density lipoprotein
LED	light emitting diode
LPG	liquefied petroleum gas
LPS	lipase
LPSSD	low-pressure superheated steam drying
MAE	mean absolute error

MAE	microwave-assisted extraction
MAP	modified atmosphere packaging
MAPE	mean absolute percentage error
MC	moisture content
MeJa	methyl jasmonate
MOS	metal oxide semiconductors
MOSFET	metal oxide semiconductors field effect transistors
MRI	magnetic resonance imaging
MVP	moderate vacuum packaging
MW	microwave
MW-VF	microwave vacuum frying
MWVD	microwave vacuum drying
NADPH	nicotinamide adenine dinucleotide phosphate
NCCD	National Centre for Cold-chain Development
NG	no growth
NGOs	nongovernmental organizations
NIR	near infrared radiance
NMR	nuclear magnetic resonance
OD	osmotic dehydration
ORAC	oxygen radical absorbance capacity
ORS	optical ring sensor
PCM	phase changing material
PA	polyacrylamide
PAL	phenylalanine ammonia lyase
PEF	pulsed electric field
PEFE	pulse electric field extraction
PG	polygalacturonase
pH	negative log of hydrogen ion concentration
PHF	potential hazardous food
PL	pulsed light
PME	pectin methylesterase
POD	peroxidase
PPOs	polyphenol oxidases
PV	photovoltaic
QCM	quartz crystal microbalance
RFID tags	radio-frequency identification tags
RMSE	root mean squared error
RMSPE	root mean squared percentage error
ROS	reactive oxygen species
RRR	reduce, reuse, and recycle

RTE products	ready to eat products
SAW	surface acoustic wave
SC-CO$_2$	supercritical-carbon dioxide
SFE	supercritical fluid extraction
SFMAE	solvent-free microwave-assisted extraction
SSC	soluble solids content
SSF	simultaneous saccharification and fermentation
TBA	thiobarbituric acid
TCS	temperature control for safety
THF	tetrahydrofolate
TMP	trimithopium
TOF	time of flight
TSS	total soluble solid
UAE	ultrasound-assisted extraction
UNDP	United Nations Development Programme
UNEP	United Nations Environment Programme
USDA	United States Department of Agriculture
UV	ultraviolet
VC	vacuum cooling
VD	vacuum drying
VF	vacuum frying
VHF	vacuum hermetic fumigation
VI	vacuum impregnation
VIS	visible infrared spectrum
VO	vacuum osmotic
VOD	vacuum osmotic dehydration
VP	vacuum packaging
VSD	vacuum spray drying
w.b.	moisture content on wet basis
WHO	World Health Organization
WVPD	water vapor pressure deficit

FOREWORD 1 BY NAWAB ALI

Fruits and vegetables, also known as the elixir of life, are vital for human health as they are the major source of valuable vitamins, minerals, dietary fibers, and phytochemicals that make the human diet a balanced one and keep the body physically and mentally fit and fine. However, careful handling of horticultural produce is important to avoid bruises, cuts, scratches, and abrasion injuries. Harvesting at the right maturity, grading, and curing will increase the storability and marketing of fresh fruits and vegetables. Although the quality requirements of horticultural crops are generally the same for fresh consumption as for processing, standards for raw material quality for making processed products may be less stringent because the produce may ultimately be peeled, sliced, pureed, or even juiced. Some of the important factors for the raw materials are variety, maturity, and quality specifications.

Although it is desirable to consume horticultural products when they are fresh (that is, when their color, texture, flavor, and nutritional value are optimal), it is not generally possible for everyone to maintain gardens and/ or greenhouses that could supply a year-round desired variety of horticultural produce. Besides, many fruits and vegetables are seasonal and region specific and, therefore, cannot be produced and supplied around the year in a fresh form. Some horticultural crops are only produced during a short specific season, and once that time span is over, fresh produce is no longer available. In such a situation, it is suggested that people may consume seasonal and regional horticultural produce, as far as possible, to derive the maximum nutritional and economic benefits leading toward better health and living. However, for these and some other reasons, processing and later consumption of fruits and vegetables is often a preferred alternative to fresh consumption. Horticultural crops play an unique role in food and nutritional security of India and its economy. Horticulture, being a labor-intensive activity, provides more employment opportunities and remunerations. It also has a better export potential.

This book, *Processing of Fruits and Vegetables: From Farm to Fork*, describes in detail the various aspects of fruits and vegetable processing, products, technology, sensors for physical and biochemical properties of fruits, vegetables, and their products, challenges and solutions of fruits- and vegetables-processing industry, and a few other related aspects, such as the effects of processing on nutritional contents and its bioavailability, economic utilization of bio-wastes, and processing by-products.

It is expected that the book would be useful to students, scientists, teachers, policymakers, research administrators, industry professionals, and all those who are involved in horticultural-based processed products and marketing and utilization. I take this opportunity to congratulate the editors of this much-needed book, especially for the students of food science, engineering, and technology.

Nawab Ali, PhD
Former Deputy Director General (Engineering)
Indian Council of Agricultural Research (ICAR)
House No.: SDX-40, Minal Residency,
J.K. Road, Bhopal 462023, Madhya Pradesh, India
Phone: +91-755-2590592
Mobile: +91-7898842501
E-mail: alinawab11@gmail.com

FOREWORD 2 BY PITAM CHANDRA

Fruits and vegetables constitute an essential component of diet to ensure the adequacy of human nutrition. We are fortunate that India today produces about 170 million tons of vegetables and 90 million tons of fruits. Considering the growth of the horticulture sector in the recent past, it is not difficult to realize that we would be reaching the level of nutritional adequacy expected from fruits and vegetables in very near future. The disturbing issues, however, are the extent of postharvest losses and very low levels of processing in the fruit and vegetable sector. India is losing about billion rupees (60.00 Rs. = 1.00 USD) worth of revenue annually due to postharvest losses. Further, value addition through processing would increase the revenue generation from the horticulture sector very significantly.

The agriculture and food processing segments of the Indian economy are alive to the twin issues of postharvest losses and food processing, and some important policy interventions have been made to improve the situation. Research and development activities are directed toward developing effective technologies to ensure additional farm income and enhancement of food processing. The importance of education and skill development could, however, not be underestimated to meet the targets. The acute shortage of skilled human resource in the country is a serious bottleneck in this endeavor. While the skill development programs need to be expanded, the relevant education and training literature need to be developed on different aspects of production and processing of fruits and vegetables.

It is in this background that efforts of editors to write this book on *Processing of Fruits and Vegetables: From Farm to Fork* are appreciated. It is important to realize that there could not be a single comprehensive book to do justice with the subject matter and to meet the aspirations of all the stakeholders. There is need for a whole range of publications instead. The present book is the result of long years of research and teaching experience of authors at the grass-root level. The organization of the contents is nonconventional; it is more oriented toward meeting the needs of skills. The editors have tried their best to be up to date in the treatment of subject matter.

I greatly appreciate the initiative and the efforts of my colleagues, Khursheed Alam Khan, Megh R. Goyal, and Abhimannyu A. Kalne to edit this book, and I hope that the book would be well received by students, teachers, and budding entrepreneurs.

Pitam Chandra, PhD
Former Assistant Director General (Process Engineering) at ICAR Former Director, ICAR—Central Institute of Agricultural Engineering, Bhopal; and now Professor (Food Engineering)
National Institute of Food Technology Entrepreneurship and Management (Deemed University) Plot No. 97, Sector 56
HSIIDC, Industrial Estate, Kundli,
Sonipat 131028, India, Mobile: +91-8199901306
E-mail: pc1952@gmail.com

FOREWORD 3 BY N. C. PATEL

The issue of food wastage is central to India's efforts in combating hunger and improving food security. While focus has been on improving production, reducing food supply chain losses remains a relatively unaddressed problem till very recently. It is hard to put a figure on how much food is lost and wasted in India due to lack of inadequate infrastructure; however, a 2011 report by a UN body, FAO, puts wastage in fruits and vegetables as high as 45% of produce (postharvest to distribution) for developing Asian countries including India.

Supply chain management plays an integral role in keeping business costs minimum and profitability as high as possible. Although India has many positives in the fruit and vegetable production and marketing sector, the country lacks an efficient supply chain for the distribution of fruits and vegetables. It has been found that 30–40% of fruits and vegetables are wasted due to postharvest losses, leading to low availability for consumers and the need for import. There is a lack of basic as well as specialized infrastructure including the modern technologies backed by relevant research. Also, there is a missing link between production, research systems, and consumers. Today, consumers expect protection from hazards occurring along the entire food chain. Providing adequate protection to the consumers by merely sampling and analyzing the final products is not possible; hence the emphasis is on introduction of preventive measures at all stages of the food production and distribution chain. This calls for a determined, innovative, inclusive, and participative approach from all the stakeholders that are involved in food chain, from farm to fork.

I am happy that the book *Processing of Fruits and Vegetables: From Farm to Fork* addressing the above-mentioned issues through technologies and research is being published. The book includes all the aspects of fruits and vegetables processing and highlights the novel processing technologies, enzyme profiles of fruits and vegetables, engineering interventions, and challenges faced by this sector along with solutions.

I am sure the book would be very useful not only to the students, faculty, and teachers concerned to this field but would also help food industry professionals for their respective academic, scientific, and professional pursuits. The compilation has a flow, with some chronological order with regards to the main topics and subtopics, which should help the readers in conceiving the essence of the overall focus of the book.

I congratulate the contributing authors for the much-needed compilation and wish them all success for reaching the stakeholders so as to bring in desired academic and professional interventions.

N. C. Patel, PhD
President—Indian Society of Agricultural Engineers (ISAE) and
Vice Chancellor
Anand Agricultural University
Anand 388110, Gujarat, India
Phone:+91-2692-261273 (office)
E-mail: vc@aau.in

PREFACE 1

Competent personnel in academia, research, and food industries in the field of fruits and vegetables processing are required in the coming decades of this century. To ensure food security and food safety in the future, there is an imperative need for abundant and reliable literature, which will be extremely helpful to students, researchers, and professional engaged in the fruits- and vegetables-processing industry. This book is a genuine endeavor toward the same, and an effort has been made to portray some of the very important advanced technologies used in the processing of fruits and vegetables.

This book consists of four main parts, namely, *processing, antioxidant/ enzyme profiles of fruits and vegetables* (role of antioxidants, enzymes role in processing, use of solar energy in processing, and techniques used in making processed products from fruits and vegetables); *novel processing technologies: methods and applications* (ultraviolet light, pulsed light technology, hurdle technology, and vacuum technology); *engineering interventions in fruits and vegetables* (nondestructive size determination methods and sensors for physical and biochemical properties); and *challenges and solutions: the food industry outlook* in waste reduction, negative effects of processing, and effects of processing on vitamins of fruits and vegetables.

I am highly thankful to the authors of the individual chapters for their sincere effort in writing the assigned book chapter. All the chapters impart the basics of the subject in addition to discussing the latest application of technology in improving the processing used in fruits and vegetables. This book will contribute effectively to ensuring the development of competent technical manpower to support academia, research, and industry.

I sincerely acknowledge the support received from my wife (Dr. Ruby Khan) and I am thankful to my son (Ariz Alam Khan) for giving me time to work at home after office hours. I am very thankful to my late father (Fateh Mohammed Khan), who always guided me in every aspect of life and am thankful to my mother (M. Nisha) for her true love and support.

I am grateful to all acquaintances, relatives, and colleagues who have faith in me and for the support I received from Rajmata Vijayaraje Scindia Agriculture University, Gwalior, India, specifically from the Honorable Vice-Chancellor, Dr. S. K. Rao; Dean, Faculty of Agriculture Dr. Mridula Billore, Director of Instructions; and Dean, College of Horticulture, Mandsaur.

I personally owe gratitude to my colleagues: Dr. Tarun Kapur, Dr. Nachiket Kotwaliwale (Principal Scientist, Central Institute of Agricultural Engineering, Bhopal, India), Er. Abhimannyu Kalne (Assistant Professor, Indira Gandhi Agricultural University, Raipur, India), who helped me in various ways.

I also thank Apple Academic Press for giving an opportunity to publish this book and I am highly thankful to Professor Megh R. Goyal for his unselfish support and guidance during this book project.

I shall, of course, appreciate constructive criticism and suggestions from readers.

—Khursheed Alam Khan
Lead Editor

PREFACE 2

To be healthy, it is our moral responsibility,
Towards Almighty God, ourselves, and our family;
Eating fruits and vegetables makes us healthy,
Believe and have a faith;
Reduction of food waste can reduce the world hunger and
can make our planet eco-friendly.
—Megh R. Goyal

In 1955, S. M. Henderson and R. L. Perry published their first classical book, *Agricultural Process Engineering*. In their book, they define agricultural processing as "*any processing activity that is or can be done on the farm or by local enterprises in which the farmer has an active interest ... any farm or local activity that maintains or raises the quality or changes the form of a farm product may be considered as processing. Agricultural processing activities may include cleaning/sorting/grading/treating/drying/grinding or mixing/milling/canning/packing/dressing/freezing/conditioning/and transportation, etc. Specific farm products may involve a specific activity(ies).*" This focus has evolved over the years as new technologies have become available.

Each one of us eats a processed food daily. In today's era of technological interventions in agriculture, processing of agricultural produce has become a necessity of our daily living. I am not an exception as I have seen and tasted all kinds of processed food (except meat and fish, as I am a vegetarian). I enjoy eating raw fruits and raw or cooked vegetables. Although I am not a professional cook, I feel proud of my culinary skills as I am able to develop recipes, cook, or process the vegetables and fruits (it is outside my profession of engineering), and am able to prepare jams, jellies, and marmalades from fruits and vegetables (this is the way of getting me on a TV show). I am not a professional cook; however, I love to cook. My wife has prohibited me to enter into her kitchen as it has been too expensive for her (kitchen has been on fire three times, as I leave the food unattended most of the times). Of course, our children love my food.

Apple Academic Press Inc. published my first book, *Management of Drip/Trickle or Micro Irrigation*, and a 10-volume set under the book series *Research Advances in Sustainable Micro Irrigation*, in addition to other books in the focus areas of agricultural and biological engineering.

The book series, *Plant Science for Better Health: From Soil to Fork* is to introduce processing of fruits and vegetables. This book volume, complements my previous volume titled *Technological Interventions in the Processing of Fruits & Vegetables* edited by Rachna Sehrawat, Khursheed A. Khan, Megh R. Goyal, and Prodyut K. Paul.

At the 49th annual meeting of the Indian Society of Agricultural Engineers at Punjab Agricultural University during February 22–25, 2015, a group of agricultural and biological engineers (ABEs) convinced me that there is a dire need to publish book volumes on the focus areas of ABE. This is how the idea was born on book series, titled *Innovations in Agricultural and Biological Engineering*.

The contributions by all the cooperating authors to this book volume have been most valuable in the compilation. Their names are mentioned in each chapter and in the list of contributors. This book would not have been written without the valuable cooperation of these investigators, many of whom are renowned scientists who have worked in the field of ABE throughout their professional careers. The book volume would not have become reality without the esteemed and dedicated professional editing by my fellow colleagues, namely, Khursheed A. Khan and Abhimannyu A. Kalne. They are staunch supporters of my profession. Their contributions to the content and quality of this book volume has been invaluable. I owe them my gratitude.

I thank the editorial staff, Sandy Jones Sickels, Vice President, and Ashish Kumar, Publisher and President at Apple Academic Press, Inc., for making every effort to publish the book when the diminishing water and food resources are a major issue worldwide. Special thanks are due to the AAP Production Staff also.

I express my deep admiration to my wife (Subhadra Devi Goyal) for her understanding, collaboration, and spiritual support during the preparation of this book.

I dedicate this book to all people who want to have health as their first priority among the stressful situations in this world.

As an educator, there is a piece of advice to one and all in the world: *"Permit that our almighty God, our Creator and excellent Teacher, allow us to process agricultural products wisely without contaminating our planet.*" I invite my community in agricultural engineering to contribute book chapters to the book series by getting married to my profession. I am in total love with my profession by length, width, height, and depth.

—**Megh R. Goyal, PhD, PE**
Senior Editor-in-Chief

PREFACE 3

Processing of fruits and vegetables is one of the major sectors in the food-processing industry, which has gained increasing recognition with time. Although fresh fruits and vegetables have their special importance in diet from the nutritional point of view, it is a well-accepted truth that demand of processed fruit and vegetable products has been booming in recent years, due to increasing urbanization. Processing of fruits and vegetables is being done in many ways and at many levels. Some processed products could be easily developed at home with little knowledge and effort, whereas high-end processed products need knowledge of food processing and technology and specialized machines/equipment for the unit operations. This book has some chapters on easy-to-use processing methods, and some chapters are written on advanced techniques in fruit/vegetable processing.

This book is written for postgraduate students and manufacturers of fruit/vegetable processed products. The book is also equally important to other readers interested in this field. The content of this book is designed with the expectation that it will fulfill the needs of targeted users.

The first section of book covers knowledge of antioxidants and enzymes in fruit/vegetables, application of solar energy in fruit and vegetable processing, and manufacturing of jam, jelly, and marmalade from fruit/vegetables, as these processing methods are easy to apply and have enormous potential in product development. The second section covers some established technologies, such as application of ultraviolet light and pulsed light, use of vacuum technology, and hurdle technology in many aspects of fruit/vegetable processing. For rapid measurements of size parameters and biochemical properties, potential of nondestructive technologies and electronic sensors are explored in the third section of the book. Finally, in the fourth and last sections, emphasis has been given on some real challenges, such as how to reduce waste in the fruit/vegetable processing sector, negative effects of processing, and effects on vitamins during processing, along with possible solutions and precautions.

This book is not intended to be a complete book on processing of fruits and vegetables. It provides a theoretical basis for selected processes and operations so that readers can gain a strong understanding of the concepts. The last section of this book is designed to engage readers with further in-depth discussions about challenging subjects in processing of fruits and vegetables.

—**Abhimannyu A. Kalne**
Editor

BOOK ENDORSEMENTS

The book *Processing of Fruits and Vegetables: From Farm to Fork* contains information on the latest developments in the field, and the authors have discussed in detail the methods to preserve quality of food during processing operations, specifically during the newer food processing methods. The book also covers the application of pulsed light, ultraviolet radiations, and vacuum technology among others in food processing applications. Also the desirable and negative effects of different food processing methods have been discussed, which will be of interest to practitioners. I am sure the book will fill the void of information in the field and meet the need of students and researchers.

—**Sanjaya K. Dash, PhD**
Professor and Head
Department of Agricultural Processing and Food Engineering
College of Agricultural Engineering and Technology
Orissa University of Agriculture and Technology
Bhubaneswar 751003, Odisha, India
Mobile: +91-94372 05952

This book volume has well conceived all the important areas of fruits and vegetables processing. In the second part, inclusion of the topic of extraction technique will be useful in this era of organic search everywhere. The third part, on engineering intervention will be definitely very useful for realizing the future potential of this sector. The fourth part highlights the negative effects of processing which is almost not even thought of by many contributing authors yet is very much important for responsible processing of fruits and vegetables, which are naturally loaded with phytochemicals. Overall the idea of writing this book is well conceived and will be very useful for stakeholders involved in this sector.

—**Nagarajan Manimehalai, PhD**
Associate Professor and Head
Department of Fish Process Engineering, College of Fisheries Engineering
Tamil Nadu Fisheries University
Nagapattinam 611001, Tamil Nadu, India

PART I

Processing, Antioxidant/Enzyme Profiles of Fruits and Vegetables

CHAPTER 1

ANTIOXIDANTS IN FRUITS AND VEGETABLES: ROLE IN THE PREVENTION OF DEGENERATIVE DISEASES

MONICA PREMI and KHURSHEED ALAM KHAN

ABSTRACT

Antioxidants present in fruits and vegetables play a vital role in preventing or decreasing the risk of degenerative diseases. Free radicals are most unstable and harmful molecules that damage the body cells and are responsible for degenerative diseases. Antioxidants act as defense mechanism of our body against free radicals which prevent cellular damage and thus fight with cancer, variety of diseases, and are abundantly available in various fruits and vegetables. Antioxidants are nonnutritive compounds that protect the cells from harmful free radicals. Fruits and vegetables contain various antioxidant compounds such as flavonoids, isoflavones, flavones, anthocyanins, catechins, and isocatechins rather than from vitamins C, E, or β-carotene.

1.1 INTRODUCTION

In current scenario, consumers have become much more aware about their health and food habits. The linkage between food habits and their effects on health tends consumers to buy functional foods, which are capable to provide additional physiological function besides basic nutritional requirement. Diet enriched in fruits and vegetables is widely recommended for health-promoting properties as it reduces the risk of diseases.[14] Antioxidants present in fruits and vegetables play a vital role in preventing or decreasing the risk of degenerative diseases (Fig. 1.1). Fruits and vegetables contain natural antioxidants molecules, which normally occur as vitamins, minerals, phytochemical, and other compounds.

Free radicals are most unstable and harmful molecules that damage our body cells[2] and are formed in body during metabolism, cigarette smoking, and ultraviolet (UV) light exposure, etc. Therefore, these free radicals often interact with other molecules within body and can develop many chronic diseases due to oxidation.

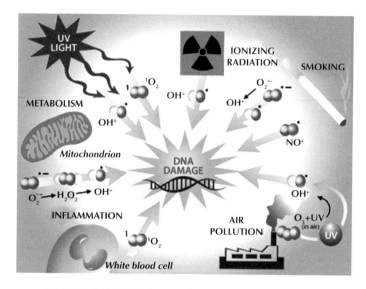

IMPLICATED DISEASE STATES

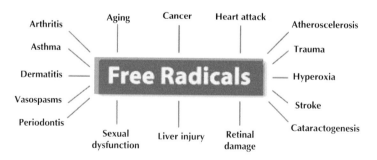

FIGURE 1.1 (See color insert.) Factors responsible (top) for production of free radicals that cause various diseases (bottom).

Source: http://altered-states.net/barry/newsletter708/.

This chapter focuses on different types of antioxidants and their possible sources, factors responsible for affecting the efficiency of antioxidants, and health benefits of antioxidants found in fruits and vegetables.

1.2   REACTIVE OXYGEN SPECIES

Reactive oxygen species (ROS) are oxygen-containing molecules that are very reactive, which includes free radicals (Table 1.1). ROS causes cellular damage by reacting with nucleic acids, membrane lipids, proteins and enzymes, and other small molecules.[20]

TABLE 1.1   Types of Reactive Oxygen Species (ROS) and Its Neutralizing Antioxidants.

ROS	Antioxidants for neutralization
Hydrogen peroxide	Vitamin E, vitamin C, and β-carotene
Hydroxyl radical	Flavonoids and vitamin C
Lipid peroxides	Glutathione, peroxidase, β-carotene, flavonoids, and vitamin E
Superoxide radical	Flavonoids, vitamin C, and glutathione

ROS, reactive oxygen species.

Free radicals are molecules that are electrically charged, which have an unpaired electron to neutralize themselves by capturing the electrons from other substances. Free radicals attack DNA of healthier cells, thus causing them to lose its specific function and structure. Excess amount of ROS leads to the development of unhealthy conditions (Fig. 1.1 and Table 1.2) such as[22]

- accelerated (Table 1.4),
- accelerating the risk of heart disease,
- cancer,
- damage to brain nerves cell leads to Parkinson's and Alzheimer's disease,
- damage to the healthy cells,
- deterioration of eye lens leads to blindness,
- inflammation of joints (arthritis),
- liver disease, and
- oxidation of bad cholesterol (low-density lipoproteins).

TABLE 1.2   Examples of Diseases Associated with Oxidative Stress.

- *Blood:* atherosclerosis, leukemia, thalassemia
- *Brain:* Parkinson's, Alzheimer's, schizophrenia
- *Eyes:* cataract
- *Heart:* heart attack, atherosclerosis, myocardial infarction
- *Kidney:* urolithiasis
- *Liver and pancreas:* Wilson's disease
- *Lungs:* asthma, allergy
- *Reproductive systems:* infertility
- *Skin and hair:* gray hairs, melanin abnormalities

1.3   EFFECT OF ANTIOXIDANTS: PROTECTION AGAINST VARIOUS DEGENERATIVE DISEASES

Antioxidants are natural occurring nonnutritive compounds that safeguard the healthy cells from free radicals. It is also known as "free radical scavengers." Antioxidants are nutrients that are present in low concentration in fruits and vegetables, which provide protective covering against various degenerative diseases by neutralizing the harmful free radicals.[1] Due to this protective effect, fruits and vegetables consumption are increased in diseased patients to delay the rate of lipid oxidation reactions.

Antioxidant activity of fruits or vegetables is mainly due to compounds such as β-carotene, flavonoids, vitamins C, isoflavones, anthocyanins, vitamin E, and catechins.[13] These bioactive compounds normally safeguard the cells against damage by free radicals oxidation.[24] Majority of nutrients are antioxidants, which are present in the food in the form of vitamins, minerals,[10] polyphenols; and carotenoids such as vitamin A, vitamin E, vitamin C, anthocyanins, β-carotene, lycopene, lutein, ellagic acid, catechins, and resveratrol.[15]

Antioxidants scavenge the harmful free radicals from the body[5] such as purple-blue color of blueberries, blackberries, and grapes; the dark red color of tomatoes and cherries; the orange color of carrots; and the yellow color of saffron and mango.

1.4 CLASSIFICATION OF ANTIOXIDANTS

Antioxidants can be classified on the basis of origin, chemical composition, solubility, etc. (Table 1.3).

TABLE 1.3 Classification of Antioxidants.

Basis	Antioxidant
Origin[3]	• *Endogenous (endo: inside):* if formed inside the body • *Exogenous (exo: outside):* acquired through diet

Exogenous antioxidant Endogenous antioxidant

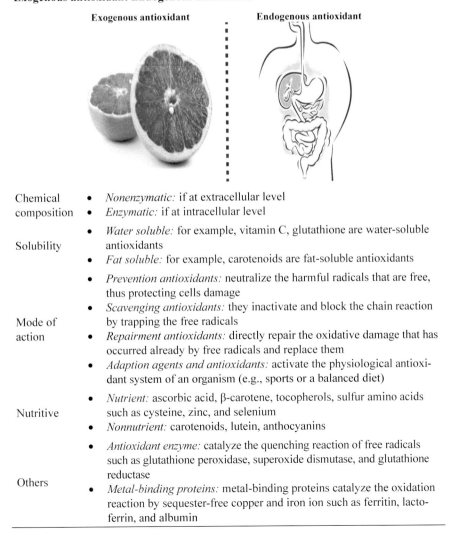

Chemical composition	• *Nonenzymatic:* if at extracellular level • *Enzymatic:* if at intracellular level
Solubility	• *Water soluble:* for example, vitamin C, glutathione are water-soluble antioxidants • *Fat soluble:* for example, carotenoids are fat-soluble antioxidants
Mode of action	• *Prevention antioxidants:* neutralize the harmful radicals that are free, thus protecting cells damage • *Scavenging antioxidants:* they inactivate and block the chain reaction by trapping the free radicals • *Repairment antioxidants:* directly repair the oxidative damage that has occurred already by free radicals and replace them • *Adaption agents and antioxidants:* activate the physiological antioxidant system of an organism (e.g., sports or a balanced diet)
Nutritive	• *Nutrient:* ascorbic acid, β-carotene, tocopherols, sulfur amino acids such as cysteine, zinc, and selenium • *Nonnutrient:* carotenoids, lutein, anthocyanins
Others	• *Antioxidant enzyme:* catalyze the quenching reaction of free radicals such as glutathione peroxidase, superoxide dismutase, and glutathione reductase • *Metal-binding proteins:* metal-binding proteins catalyze the oxidation reaction by sequester-free copper and iron ion such as ferritin, lactoferrin, and albumin

1.5   SOURCES OF ANTIOXIDANTS

Fresh fruits and vegetables are known for providing significant amount of antioxidants to human body, which are required to prevent oxidation of other molecules. Figure 1.2 indicates the sources of antioxidants from fruit and vegetables.

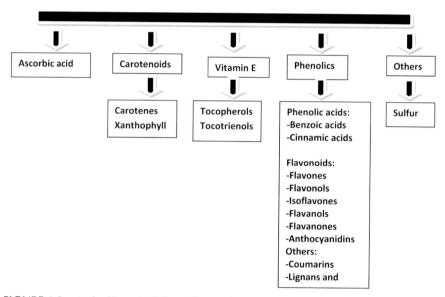

FIGURE 1.2   Antioxidants in fruits and vegetables.

1.5.1   TOCOPHEROL (VITAMIN E)

It is a fat-soluble vitamin and the strongest antioxidant found in green leafy vegetables, oils, nuts, and whole grain cereals. Vitamin E (Fig. 1.4, top) basically donates electron or hydrogen to free radicals unpaired electron to make cells stable (Fig. 1.3). It protects the DNA in cells by harmful free radicals and helps to prevent degenerative diseases.[4] It helps to maintain the healthy blood vessels and also improves the skin condition. Figure 1.4 presents selected examples of foods: sources of vitamin E.

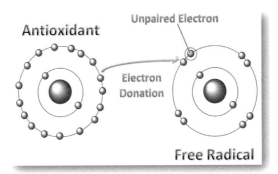

FIGURE 1.3 **(See color insert.)** Mechanism of antioxidant activity by vitamin E.

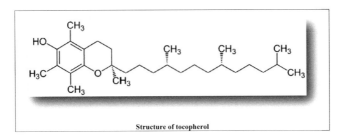

FIGURE 1.4 **(See color insert.)** Structure of vitamin E (top) and selected examples of foods: sources of vitamin E.

1.5.2 ASCORBIC ACID (VITAMIN C)

Vitamin C (Fig. 1.5, top) is a water-soluble antioxidant. It has dual functions as it protect the cells from damage and also helps body to absorb vitamin E. It acts as a protective covering for cigarette smokers because vitamin C helps to fight against toxins and free radicals developed from smoke. It promotes the iron absorption from foods and controls the blood cholesterol. It boosts the immune system and protects against health-related problems. It also protects the skin from the UV damage and infection. Citrus fruits, strawberries, broccoli, potato, and tomato are the good source of vitamin C (Fig. 1.5).

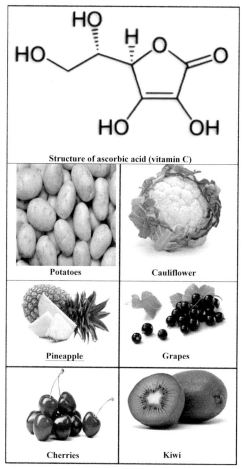

FIGURE 1.5 (See color insert.) Structure of vitamin C (top); selected examples of foods: sources of vitamin C.

1.5.3   VITAMIN A (β-CAROTENE)

Vitamin A (Fig. 1.6, top) is fat soluble, which functions as antioxidant. It ($C_{40}H_{56}$) lowers the cholesterol, improves immunity, helps to repair and replacement and damaged tissue, delays. In Table 1.4, phytonutrients present in fruits and vegetables prevents calcification and prevents certain forms of cancer and eye-related problems. Vitamin A (1,3,3-trimethyl-2-[(1*E*,3*E*,5*E* ,7*E*,9*E*,11*E*,13*E*,15*E*,17*E*)-3,7,12,16-tetramethyl-18-(2,6,6-trimethylcy-clohexen-1-yl)octadeca-1,3,5,7,9,11,13,15,17-nonaenyl]cyclohexene)   is mainly available in red- and orange-colored fruits and vegetables (Fig. 1.6).

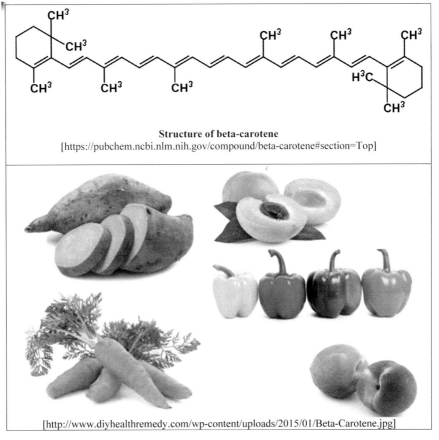

Structure of beta-carotene
[https://pubchem.ncbi.nlm.nih.gov/compound/beta-carotene#section=Top]

[http://www.diyhealthremedy.com/wp-content/uploads/2015/01/Beta-Carotene.jpg]

FIGURE 1.6   (See color insert.) Structure of vitamin A (top); selected examples of foods: sources of vitamin A (β-carotene).

1.5.4   SELENIUM

It is an important mineral antioxidant, which is involved in synthesis of powerful enzymes such as glutathione peroxidase (POD), which basically mops up the enzymes and deactivate it. It helps in maintaining healthy hair, eyes along with skin. It reduces the anxiety and mental fatigue, promotes the normal liver function, increases male potency, decreases the heart-related diseases and reduces the possibilities of cancer. Figure 1.7 indicates selected examples of foods that are sources of selenium.

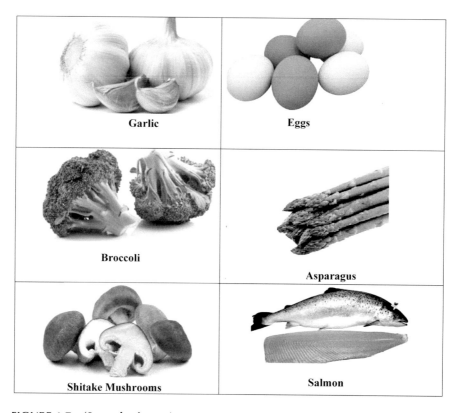

FIGURE 1.7   (See color insert.) Selected examples of foods: sources of selenium.

1.5.5   PHYTONUTRIENTS IN FRUITS AND VEGETABLES

These are naturally occurring plant chemicals, which are nonnutritive but act as antioxidants. Types of phytonutrients present in fruits and vegetables are carotenoids, flavonoids, glutathione, indoles, isoflavones, polyphenols, and saponins (Table 1.4 and Fig. 1.2). Table 1.5 indicates sources and functional benefits of bioactive components. Table 1.6 indicates examples of antioxidant vitamins and minerals.[10]

TABLE 1.4 Phytonutrients Present in Fruits and Vegetables.

Vitamin C, terpenoids	α- and β-carotene	Anthocyanins and resveratrol	Lycopene	Lutein	Glucosinolate	Allyl sulfides
Color						
Yellow-orange	Orange	Purple-red	Deep red	Green-yellow	Green	White-green
Orange, peaches, tangerines, papayas, nectarines, pineapple, yellow grapefruit	Carrots, mangos, apricots, canta-loupes, pumpkin, sweet potatoes, winter squash	Red grapes, fresh plums, cranberries, raspberries, blackberries, strawberries, beetroot, prunes, grapes, red apple	Tomato juice, pink grapefruit, watermelon	Spinach, avocado, mustard greens, green peas, cucumber, honeydew, melon	Broc-coli, Brussels, sprouts, cabbage, kale, Swiss chard	Garlic, chives, onions, celery, asparagus, cauli-flower, mushrooms

TABLE 1.5 Sources and Functional Benefits of Bioactive Components.

Components/class	Sources	Functional benefits
Carotenoids		
Zeaxanthin and lutein	Eggs, kale, citrus fruits, spinach, and corn	Helps in maintaining healthy vision
β-carotene	Carrots	Act as antioxidant defenses and neutralizes free radicals to prevent damage
Lycopene	Tomatoes and tomato-related products	Helps in maintaining health related to prostate
Flavonoids		
Anthocyanidins	Red grapes, berries, cherries	Helps in maintaining brain function, also act as antioxidant defenses
Flavanols: procyani-dins, catechins, and epicatechins	Grapes, apples, tea, choco-late, cocoa	Helps in maintaining health-related to heart
Flavanones	Citrus fruits	Act as antioxidant defenses and neutralizes free radicals to prevent damage

TABLE 1.5 *(Continued)*

Flavonols	Tea, onions, broccoli, apples	Act as antioxidant defenses and neutralizes free radicals to prevent damage
Proanthocyanidins	Wine, cranberries, straw-berries, peanuts, apples, grapes	Helps in maintaining health-related to urinary tract and heart
Isothiocyanates		
Sulforaphane	Cabbage, cauliflower, Brussels sprouts, broccoli	Act as antioxidant defense and enhance the detoxification of unde-sirable compounds
Phenols		
Ferulic acid and caf-feic acid	Citrus fruits, apples, pears	Act as antioxidant defense and maintains the health of eyes and heart
Thiols (sulfides)		
Allyl methyl trisulfide and diallyl sulfide	Garlic, leeks, onions	Helps in maintaining immune system efficiently and detoxify the undesirable compounds
Dithiolthiones	Cruciferous vegetables such as collards, cabbage, broccoli	Helps in maintaining immune sys-tem efficiently

TABLE 1.6 Examples of Antioxidant Vitamins and Minerals.

Vitamins	**Daily reference intake (DRI)**[a]	**Antioxidant activity**	**Sources**
Selenium	20–55 µg/day	Helps prevent cellular damage from free radicals	Brazil nuts, meats, tuna, plant foods
Vitamin A	300–900 µg/day	Protects cells from free radicals	Liver, dairy products, fish
Vitamin C	15–90 mg/day	Protects cells from free radicals	Bell peppers, citrus fruits
Vitamin E	6–15 mg/day	Protects cells from free radicals, helps with immune function and DNA repair	Oils, fortified cereals, sunflow-er seeds, mixed nuts

[a]Here, DRIs provided are a range for Americans aged 2–70.

1.5.5.1   POLYPHENOL

These antioxidants (resveratrol and flavonoids) are easily characterized by the presence of functional group "phenol." These compounds in human health help to combat oxidative stress. For example, fruits such as pears, cantaloupe, cranberries, raspberries, apples, blackberries, cherries, grapes, plums, and strawberries; and vegetables such as onion, broccoli, parsley, cabbage, and celery are rich in polyphenol antioxidants. Alternative sources are green tea, chocolate, red wine, olive oil, grains, and bee pollen.

1.5.5.2   GLUTATHIONE

It protects the cells from free radicals. It is produced in the human body by the formation of prime amino acids that are glycine, cysteine, and glutamic acid. Naturally occurring highest amounts of glutathione food sources are peach, asparagus, watermelon, avocado, cantaloupe, grapefruit, spinach, squash, potato, zucchini, strawberries, and broccoli. Food sources for increasing and maintaining glutathione levels in human body are meat, fish, and sulfur-containing amino acids. GSH is the only active supplemental form of glutathione available.

1.5.5.3   FLAVONOIDS

These compounds are effective when different types are combined together. It promotes the cellular health, normal tissue growth, and renewal and antioxidant activity. It reduces the oxidative stress in combined effect of vitamin C for particularly water-based cell portion and may effectively slow down aging effects. Richest sources of flavonoids are grape fruit, kale, berries, cranberries, beets, black and red grapes, lemons, oranges, and green tea.

1.5.5.4   INDOLES

These are antioxidant compounds and normally produced by the breakdown of the glucosinolate glucobrassicin. These are basically anticancer food compounds. It helps to prevent the breast and colon cancer and help to prevent cancer by deactivation process. It is easily available in the form of dietary supplements. Good sources of indoles are cruciferous vegetables such as kale, turnip, cabbage, broccoli, Brussels sprouts, drumstick, and cauliflower.

1.5.5.5 SAPONINS

These are plant steroids that consist of polycyclic aglycones attached to sugar side chains. It is very effective in lowering plasma cholesterol by inhibiting its absorption from small intestine. These compounds have antioxidant property and prevent the liver cancer by toxic compounds produced by free radicals. Commercially available supplement is "ginseng," it effectively calms the mental condition. Saponin containing foods are soybean, spinach, onion, garlic, pea, eggplant, and tomato.[12]

1.5.5.6 PEROXIDASE

It is an enzyme that consists of a protein complex with group "hematin," which catalyzes the oxidation reaction. POD sources in food are soybean, turnip, horseradish root, and mango fruit.

1.6 FACTORS AFFECTING THE ANTIOXIDANT EFFICIENCY

Antioxidant efficiency is largely affected by genetic and environmental factors, which are listed below[11]:

- Genetic: species, variety
- Environmental

Preharvest: radiation; stress during development (water, fertility, pathogens, etc.).
Harvest: maturity, handling
Postharvest: storage, postharvest treatments (UV, ozone), processing

1.6.1 GENETIC FACTORS

a. *Species* is the prime factor in determining the prevalence of different antioxidants.

b. *Cultivar:* Antioxidant efficiency is significantly affected by the cultivar. Nelson and coworkers[19] observed variations from 19 to 71 mg of ascorbic acid per 100 g in six varieties of strawberry. Similar differences among varieties have been observed for phenolic compounds.[25]

c. *In transgenic plants,* there is an increase in the levels of other antioxidants such as phenolic compounds. Muir et al.[18] observed that tomato transformation with a Petunia gene for chalcone isomerase increased the concentration almost 80 times in flavonols present in peel.

1.6.2 ENVIRONMENTAL FACTORS

1.6.2.1 RADIATION

Due to radiation, changes had been associated with phenolic compounds, ascorbic acid, and carotenoids. Lee and Kader[16] found that fruits have higher levels of vitamin C and phenolics in sun-exposed regions than the shaded ones. In tomato, the total phenols content was increased twofolds in plants exposed to higher radiance.

1.6.2.2 CULTURAL PRACTICES

These also have significant effect on antioxidant activity. Strawberry grown in bed with plastic mulch had higher antioxidant capacity than fruits grown in beds without plastic mulch.[26] Toivonen et al.[23] found that vitamin C content is inversely related with rainfall.

1.6.2.3 MATURITY AT HARVEST

The stage of maturity affects the fruits antioxidant capacity.[21] Particularly in tomato and pepper, total antioxidant capacity increases because of the vitamin C and carotenoids accumulation. In the case of carotenoids, in some products (e.g., pepper, tomato, and mango) the concentration increases during development.[9]

1.6.2.4 WOUNDING

It causes alterations in the levels of antioxidants. According to Loaiza Velarde et al.,[17] there is alteration in synthesis and degradation of phenolics due to wounding. From the perspective of molecular level, wounding induces de novo synthesis of phenylalanine ammonia lyase, a key enzyme in phenyl propanoid metabolism.[8] Besides having its role on phenolic biosynthesis, it also affects the degradation. First, in response to wounding, there is an increase in enzymes activity associated with phenolics oxidation such as polyphenol oxidases (PPOs) and PODs. Due to wounding, there is cell disruption that allows direct contact between preexisting phenolic degrading enzymes, which lead to the production of hydrogen peroxide favoring PPO activity. This promotes the oxidation of phenolics, which then polymerizes and leads to the formation of brown-colored pigments that may ultimately cause quality reduction.

1.6.2.5 STORAGE

Effect of storage on antioxidants is related to the role of ethylene in the ripening process in many cases. In some cases, specific antioxidants are induced by the ethylene. In carrots, ethylene stimulated the accumulation of an isocoumarin (6-methoxymellein). In berries, it has been observed that ambient conditions with high levels of oxygen (60% and 100%) result in increased antioxidant capacity by favoring anthocyanins and other phenolics accumulation.[27]

1.6.3 OTHER FACTORS

a. Antioxidant activity of fruits can be increased by the manipulation in the *metabolism of products* by the application of postharvest treatments.[14] In response to stress conditions (infection by microorganisms or wounding, UV irradiation or the exposure of the products to ozone-enriched atmospheres), phenolic compounds are synthesized. In strawberry, UV treatments increased the level of phenolic compounds and the antiradical capacity.[6]

b. **Processing** affects the level and bioavailability of antioxidants.[7] In some cases, processing increased the availability of antioxidants. Lycopene bioavailability increases in heat-treated tomato. In fresh carrots, β-carotene is present in the *trans* form, while canning causes significant formation of *cis* isomers. Carotenoids are in general susceptible to oxidation. Ascorbic acid is one of the antioxidants that are easily susceptible to degradation. Blanching or even freezing and thawing can cause losses up to 25%. Finally, processing can also cause losses of phenolic antioxidants. For instance, peeling or cutting reduces quercetin levels by only 1%, but cooking in water may reduce the content of this component by 75%. Factors that cause changes in antioxidant level during processing are:

- stability to pH and processing,
- reduction potential should be low,
- oxidation potential should be high,
- solubility in oil should be high, and
- activation energy of antioxidants to donate hydrogen should be low.

1.7   HEALTH BENEFITS OF ANTIOXIDANTS

Since the damage caused by free radicals are not localized, as they deteriorate the overall health and well-being of the person. To neutralize, it is necessary to consume fruits and vegetables rich in antioxidants. Benefits of consuming antioxidant-rich fruits and vegetables are summarized below:

- Improves nervous system functioning
- Improves quality of sleep
- Improves reproductive function
- Maintains good dental health
- Maintains health vision
- Offers protection against digestive disorders
- Protects liver
- Reduces obesity
- Supports immune system and improves defense power of body
- Supports kidney function
- Supports respiratory system

1.8   SUMMARY

In this chapter, authors have discussed causes that lead to development of stress in human body; the sources, benefits, and classification of various antioxidants; and factors affecting antioxidant efficiency. Free radicals are produced naturally in form of by-products either due to chemical reaction in body cells or due to exposure to toxins. These are also produced when the consumed food is converted into energy, during exposure to sunlight, smoking cigarette, and irradiation. Formation of such free radicals can cause damage to important cellular components, such as DNA or cell membrane and cell function may disturb; harmful molecules may damage the body cell and are responsible for degenerative disease. Diet that is richer in fruits and vegetables is widely recommended to reduce the risk of disease as it contains the antioxidants that are very beneficial.

Sufficient intake of antioxidants is beneficial to prevent the damage caused by free radical formation, which is possible by consuming variety of fruits and vegetables. Fresh berries, green leafy vegetable, vitamin C-rich fruits, vitamin E-rich seed/nuts, and other antioxidant-rich foods such as pomegranates, red grapes, plums, cherries are good sources of variety of antioxidants. Fruits and vegetables contain various antioxidant compounds such as flavonoids, isoflavones, flavones, anthocyanins, catechins, and isocatechins.

KEYWORDS

- **antioxidant**
- **ascorbic acid**
- **carotene**
- **flavonoid**
- **free radicals**
- **lycopene**
- **polyphenol**

REFERENCES

1. Ames, B. M.; Shigena, M. K.; Hagen, T. M. Oxidants, Antioxidants and the Degenerative Diseases of Aging. *Proc. Nat. Acad. Sci. U.S.A.* **1993,** *90*, 7915–7922.
2. Anonymous. *Lifelong Vitality Program*, 2016. www.slideplayer.com (accessed July 17, 2017).
3. Anonymous. *Free Radicals and Antioxidants*, 2016. www.micaleofaleo.com (accessed Aug 24, 2017).

4. Anonymous. *Antioxidant Protocol*, 2016. www.syromonoed.com (accessed Aug 31, 2017).

5. Antolovich, M.; Prenzler, P. D.; Patsalides, E.; McDonald, S.; Robards, K. Methods for Testing Antioxidant Activity. *Analyst* **2002**, *127*, 183–198.

6. Ayala-Zavala, J. F.; Wang, S. Y.; Wang, C. Y.; Gonzalez-Aguilar, G. A. Effect of Storage Temperatures on Antioxidant Capacity and Aroma Compounds in Strawberry Fruit. *LWT—Food Sci. Technol.* **2004**, *37*, 687–695.

7. Bernhardt, S.; Schlich, E. Impact of Different Cooking Methods on Food Quality: Retention of Lipophilic Vitamins in Fresh and Frozen Vegetables. *J. Food Eng.* **2006**, *77*, 327–333.

8. Choi, Y. J.; Tomás-Barberán, F. A.; Saltveit, M. E. Wound-induced Phenolic Accumulation and Browning in Lettuce (*Lactuca sativa* L.) Leaf Tissue Is Reduced by Exposure to *n*-Alcohols. *Postharv. Biol. Technol.* **2005**, *37*, 47–55.

9. DeAzevedo, C. H.; Rodriguez-Amaya, D. B. Carotenoid Composition of Kale as Influenced by Maturity, Season and Minimal Processing. *J. Sci. Food Agric.* **2005**, *85*, 591–597.

10. Food and Nutrition Board, Institute of Medicine, National Academics. *DRI Reports & National Institutes of Health Office of Dietary Supplements*, 2011. www.iom.edu (accessed Aug 24, 2017).

11. Florkowski, W. J.; Shewfelt, R. L.; Brueckner, B.; Prussia, S. E. *Postharvest Handling: A Systems Approach*, 2nd ed.; Elsevier, Academic Press: San Diego, CA, 2009; pp 89–90.

12. International Food Information Council Foundation. *Media Guide on Food Safety and Nutrition*; 2004–2006; p 210.

13. Kahkonen, M. P.; Hopia, A. I.; Vuorela, H. J.; Rauha, J. P.; Pihlaja, K.; Kujala, T. S.; Heinonen, M. Antioxidant Activity of Plant Extracts Containing Phenolic Compounds. *J. Agric. Food Chem.* **1999**, *47* (10), 3954–3962.

14. Kalt, W.; Forney, C. F.; Martin, A.; Prior, R. L. Antioxidant Capacity, Vitamin C, Phenolics, and Anthocyanins After Fresh Storage of Small Fruits. *J. Agric. Food Chem.* **1999**, *47*, 4638–4644.

15. Larson, R. A. The Antioxidants of Higher Plants. *Phytochemistry* **1988**, *27*, 969–972.

16. Lee, S. K.; Kader, A. A. Preharvest and Postharvest Factors Influencing Vitamin C Content of Horticultural Crops. *Postharv. Biol. Technol.* **2000**, *20*, 207–220.

17. Loaiza-Velarde, J. G.; Tomás-Barberán, F. A.; Saltveit, M. E. Effect of Intensity and Duration of Heat-shock Treatments on Wound-induced Phenolic Metabolism in Iceberg Lettuce. *J. Am. Soc. Hortic. Sci.* **1997**, *122*, 873–877.

18. Muir, S. R.; Collins, G. J.; Robinson, S.; Hughes, S.; Bovy, A.; Ric De Vos, C. H.; van Tunen, A. J.; Verhoeyen, M. E. Overexpression of Petunia Chalcone Isomerase in Tomato Results in Fruit Containing Increased Levels of Flavonols. *Nat. Biotechnol.* **2001**, *19*, 470–474.

19. Nelson, J. W.; Barritt, B. H.; Wolford, E. R. Influence of Location and Cultivar on Color and Chemical Composition of Strawberry Fruit. *Washington Agric. Exp. Stat., Tech. Bull.* **1972**, *74*, 1–7.

20. Percival, M. Antioxidants. *Clin. Nutr. Insight* **1998**, *31*, 1–4.

21. Prior, R. L.; Cao, G.; Martin, A.; Sofi c, E.; McEwen, J.; O'Brien, C.; Lischner, N.; Ehlenfeldt, M.; Kalt, W.; Krewer, G.; Mainland, C. M. Antioxidant Capacity as

Influenced by Total Phenolic and Anthocyanin Content, Maturity, and Variety of *Vaccinium* Species. *J. Agric. Food Chem.* **1998,** *46,* 2686–2693.

22. Rahman, T.; Hosen, I.; Islam, M. M. T.; Shekhar, H. U. Oxidative Stress and Human Health. *Adv. Biosci. Biotechnol.* **2012,** *3,* 997–1019.

23. Toivonen, P. M. A.; Zebarth, B. J.; Bowen, P. A. Effect of Nitrogen Fertilization on Head Size, Vitamin C Content and Storage Life of Broccoli (*Brassica oleracea* var. *italica*). *Can. J. Pl. Sci.* **1994,** *74,* 607–610.

24. Wada, L.; Ou, B. Antioxidant Activity and Phenolic Content of Oregon Caneberries. *J. Agric. Food Chem.* **2002,** *50* (12), 3495–3500.

25. Wang, S. Y.; Lin, H. S. Antioxidant Activity in Fruits and Leaves of Blackberry, Raspberry, and Strawberry Varies with Cultivar and Developmental Stage. *J. Agric. Food Chem.* **2000,** *48,* 140–146.

26. Wang, S. Y.; Zheng, W.; Galletta, G. J. Cultural System Affects Fruit Quality and Antioxidant Capacity in Strawberries. *J. Agric. Food Chem.* **2002,** *50,* 6534–6542.

27. Zheng, Y.; Wang, C. Y.; Wang, S. Y.; Zheng, W. Effect of High-oxygen Atmospheres on Blueberry Phenolics, Anthocyanins, and Antioxidant Capacity. *J. Agric. Food Chem.* **2003,** *51,* 7162–7169.

CHAPTER 2

EFFECTS OF ENZYMES ON PROCESSING OF FRUITS AND VEGETABLES

SIVASHANKARI MANICKAM and AKASH PARE

ABSTRACT

Enzymes are in general present naturally in fruits and vegetables. There are many factors that affect the activity of these enzymes. Knowing the effect of each factor will be helpful in controlling the chemical reactions. Due to the low availability of enzymes that occur naturally, there is a need to produce larger amounts at the industrial level. These industrially produced enzymes have varied applications in the processing of fruits and vegetables. Also enzymes are not always beneficial, because they also cause some deleterious effects like causing undesirable color, flavor, and texture changes in the fruits and vegetables. One of the major problems caused is enzymatic browning. Also there are techniques available to inactivate the action of these browning causing enzymes. Therefore, this chapter explores the information available on important properties of enzymes in fruits and vegetables technology, enzyme activity, various factors affecting the enzyme activity, available sources of enzymes, their effects in fruits and vegetables, their applications in the processing of fruits and vegetables, their deleterious effects on the quality attributes of fruits and vegetables, different techniques to inactivate these enzymes and finally their applications.

2.1 INTRODUCTION

Enzymes are special proteins found in plant cells. They change the rate of chemical reactions either by promoting chemical changes or by controlling chemical activities.[28] This process is called catalysis. They are also involved in reducing the amount of time and energy required to complete a chemical process. Plant substances normally consist of two types of enzymes, namely hydrolytic enzymes and oxidoreductases. They are named after the compound that they change. Among these, some of these enzymes are considered beneficial and some others are considered detrimental. The substances upon which enzymes work when performing catalysis are known as substrates, and each enzyme will only fit and act upon a specific set of substrates.[35] The metabolic changes observed in any physical characteristics of fruits (such as softness or color change) may be a result of changes in the activity of enzymes.[12]

Some enzymes are naturally present in fresh fruits and vegetables. In preparation, processing, or preservation of fruits and vegetables, we want to preserve color, flavor, texture, and nutritive value. As a general rule, less heat processing produces fewer undesirable pigment and texture changes in fruits and vegetables. Many changes occur resulting in the loss of anti-oxidant qualities,[20] during harvesting, preparation, and storage of fruits and vegetables, due to the presence of antioxidant compounds and the activity of the naturally present enzymes. The activities of these enzymes are also responsible for the development of objectionable color and flavor changes during processing and storage. Inactivation of enzymes prevents discolor-ation and development of unpleasant taste during storage. Colors caused by the presence of chlorophylls or carotenoids are also protected from enzymatic degradation.[4] Commercial enzyme preparations are also used as processing aids in the processing of fruits and vegetables to improve the product quality.

This chapter explores the information available on important properties of enzymes in fruits and vegetables technology, enzyme activity, various factors affecting the enzyme activity, available sources of enzymes, their effects in fruits and vegetables, their applications in the processing of fruits and vegetables their deleterious effects on the quality attributes of fruits and vegetables, different techniques to inactivate these enzymes and finally their applications.

2.2 IMPORTANT PROPERTIES OF ENZYMES IN FRUITS AND VEGETABLES TECHNOLOGY

Enzyme categories (like hydrolases, amylase, cellulase, chlorophylase, invertase, lipase, and tannase) and oxidoreductases [ascorbinase, catalase, peroxidase (POD), polyphenol oxidase (PPO), and tyrosinase] play impor-tant roles in the processing of fruits and vegetables. Some of the important properties of enzymes are listed below[16]:

- Enzymes present in the living fruits and vegetables controls the reactions associated with ripening.
- They continue the ripening process even after harvest and if it is not controlled or destroyed by any means may even lead to spoilage of fruits and vegetables due to over ripening.
- They also cause various color, flavor, textural, and nutritional changes in fruits and vegetables due to their involvement in enormous number of biochemical reactions.
- Blanching is normally done not only to kill the surface microorganisms but also to inactivate the enzymes so as to increase the shelf life of the produce.

The optimal temperatures for enzymes are around 50°C, where they exhibit maximum activity. Deactivation of the enzyme will occur, if it is exposed to any temperature beyond this range. Apart from temperature, pH, water activity, and use of chemicals are other aspects that are useful in controlling the enzyme activity.

2.3 ENZYME ACTIVITY

The biological catalysts require an optimal reaction conditions for their activity and this varies with each enzyme. As mentioned earlier, they just take part in increasing or decreasing the rate of the chemical or biochemical reactions without being actually consumed by the reaction. They can be derived from anywhere as they are found everywhere in the nature: in human bodies, throughout the environment, and almost in all living organisms.[2] The optimum conditions for their maximum activity are usually based on the following parameters:

- source of the enzyme from where it is obtained;
- the part or the location of the source where the enzyme functions; and
- the environmental conditions of the source where it survives.

The activity of the enzymes follows a lock-and-key principle. The location, where the enzyme is going to act, is called as a substrate. And, each substrate will have a specific site on its surface in the shape of a pocket and the chemical molecules responsible for the reaction to occur, goes, and fit into those pocket-like structures in a similar fashion as the key fits the lock (Fig. 2.1). These pockets will hold the locked chemical molecules in position so as to initiate the reaction. By this way, enzymes catalyze the reactions. Once the reaction is over, these enzymes will release the formed product(s) and get ready to catalyze the next reaction. It is summarized as below:

$$\text{Step 1: } S + E \rightarrow E - S; \text{ Step 2: } E - S \rightarrow P + E \qquad (2.1)$$

where E is an enzyme molecule, S is the substrate molecule or molecules, E – S is the *enzyme–substrate complex; and* P is the formation of the product due to catalysis.

The following are some major criteria for the selection of enzymes for any specific purpose like processing or preservation of any food commodity:

- The enzyme with good specificity is essential.
- Choosing an enzyme with high enzyme activity will limit the usage of enzyme for any purpose so as to avoid any detrimental effects like taste or texture change that might happen to the product as a result of over usage.

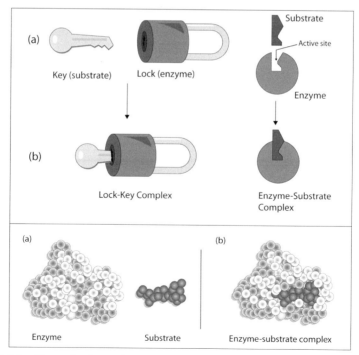

FIGURE 2.1   (See color insert.) Enzyme–substrate reaction (lock-and-key model).
Source: https://saylordotorg.github.io/text_the-basics-of-general-organic-and-biological-chemistry/s21-06-enzyme-action.html

2.4 FACTORS AFFECTING THE ENZYME ACTIVITY

Enzyme activity is affected by many factors such as temperature, pH, and water activity,[29] inhibitors, and activators.[35]

- *Temperature* plays a role in the stability of enzymes and in the speed of the reaction. Also, enzyme activity increases with increase in the temperature. However, the enzymes will become unstable once the temperature reaches beyond 60–70°C and the denaturation of the enzyme will take place resulting in reduced enzyme activity.[35]
- Another important factor is the *pH*. Each enzyme has an optimal pH for its maximum activity. For example, the optimal pH for lipoxygenase in soybean is 7–9, while that for polygalacturonase (PG) in tomatoes is 4.[35]
- *Water activity* is also an important factor that affects the enzyme activity. Some moisture content is required for the enzyme activity to mobilize and solubilize the reactants. Enzyme activity increases with increased water activity. Enzyme activity decreases with decreased water activity. For example, phospholipase has no activity on lecithin at water activity below 0.35.[35]

2.5 SOURCES OF ENZYMES

Enzymes have several advantages for industrial application of fruits and vegetables processing due to their functions, such as requirement of low temperature for its activity thereby lowering the energy requirement for the process, by-product formation is less, improved quality of the product, nontoxic if used correctly, can be biologically degraded, and also can be immobilized to reuse, has increased stability, and can be easily separated from the environment.[3] Some enzymes can be used as processing aids for fruits and vegetables.

Enzymes available naturally yield very less amount of enzymes that is insufficient for industrial use. Usually for industrial uses, large quantities of enzymes are required to speed up or slow down the chemical reactions. Enzymes are normally present in naturally occurring microorganisms like bacteria, yeast, and fungi. To obtain pure and unlimited source of enzymes, these microorganisms can be multiplied by the fermentation process, which is a biotechnological technique.

2.6 APPLICATIONS OF ENZYMES IN THE PROCESSING OF FRUITS AND VEGETABLES

Preservation of fruits and vegetables can be done by slowing down the natural processes of decay caused mainly by enzymes.[22] Enzymes are very beneficial to fruit juice industry. Pectinases and amylases are two main group of enzymes used in fruit juice industry.[30] In case of fruit juice extractions, most processes do not produce satisfactory quantity and quality of juices. However, enzymatic juice extractions are found to result in higher yields by breaking down the complex molecules of plant tissues to simpler molecules.[30] The enzymes that are normally used for juice extractions are: pectinases, cellulases, hemicellulases, etc. Apart from increasing the yield, it also increases the release of various phenolics and other important nutritional components in the juice.[17] Enzymes also improve the juice clarification, improve the quality of juice in terms of reduced viscosity, decreased turbidity and improved filterability, prevent darkening of juices, and debittering of citrus fruit juices.[17]

2.7 EFFECTS OF ENZYMES DURING STORAGE/POST HARVEST HANDLING AND PROCESSING OF FRUITS AND VEGETABLES

The main oxidative reactions in fruits and vegetables are enzymatic browning, which happens in the presence of oxygen, when the substrate located in vacuoles and enzymes located in the cytoplasm come in contact as a result of cutting or loss of firmness of fruits and vegetables. Two oxidoreductases enzymes, namely PPO and PODs are involved in this. Therefore, inactivation of these enzymes is compulsory in both fruits and vegetables before further processing. Various methods are developed to prevent this reaction either by inactivating PPO or by avoiding the enzyme and the substrate to come to contact either by adding antioxidants or by maintaining the structural integrity of the food.[3] Those include different types of blanching, such as water blanching or steam blanching or microwave blanching, freezing, modified atmosphere packaging, or edible coating.[3]

During processing, enzymes normally affect the color, texture, and flavor of the processed fruits and vegetables. Table 2.1 indicates enzymes that cause the color, flavor, odor, and textural changes in the fruits and vegetables during processing.

TABLE 2.1 List of Enzymes Affecting the Color, Flavor, Odor, and Textural Properties of Fruits and Vegetables.

Enzymes responsible	Reactions taking place	Changes occurred	References
Lipoxygenase	Hydro peroxides and free radicals are produced as a result of oxidation of unsaturated fatty acids catalyzed by lipoxygenase	*Affects color* Bleaching of chlorophylls and carotenoids in unblanched stored vegetables resulting in loss of color	[23]
Pectin methyl esterase and polygalacturonase, pectinase	Conversion of pectin polymers to pectic acid	*Affects texture* Decrease in textural properties like hardness or firmness resulting in softening of fruits and vegetables of high pectin content such as tomato, apple, banana, etc., due to over ripening	[21]
Peroxidase, lipase, and lipoxygenase	Oxidation of unsaturated fatty acids and other compounds	*Affects flavor* Results in the flavor change due to the formation of compounds such as *cis*-3-hexenal, *cis*-3-hexenol, *trans*-2-hexenol, and 2-methylbutanal in lettuce, corn, broccoli, green beans, cauliflower, and peas	[1,3,9,17]
Polyphenol oxidases	Oxidation of anthocyanins and many other flavor compounds	*Affects color* Results in undesirable browning in cut fruits and vegetables such as apples, bananas, potatoes, and lettuce	[25]

TABLE 2.2 Enzymes Responsible for Quality Deterioration in Unblanched Fruits and Vegetables.

Enzymes responsible for deterioration	Deterioration caused
Ascorbic acid oxidase	Affects the vitamin availability
LPS, peroxidase	Causes flavor and odor changes
Lipoxygenase	Affects flavor and odor development in some vegetables
Pectinases	Fruit softening
Peroxidase	Causes tissue softening in vegetables
Phenol oxidases	Discoloration in fruits and vegetables
Polygalacturonase	Causes over ripening in fruits
Polyphenolase	Causes certain fruits and vegetables to turn brown when cut
Polyphenol oxidase	Enzymatic browning, pigment degradation in fruits
AMS	Degradation of starch

AMS, α-amylases; LPS, lipase.

Even after maturity, the enzymes that are naturally occurring in fruits and vegetables continue to react with the substrate, causing changes in the quality and nutritional characteristics. Some of the enzymes that are primarily responsible for the quality deterioration of unbleached fruits and vegetables are listed in Table 2.2.

2.8 ENZYMATIC BROWNING

Reactions taking place during drying of fruits and vegetables can result in qualitative loss of the product, particularly loss of nutrients and other deteriorations caused by browning reactions. These reactions are evident when the fruits and vegetables are subjected to processing or mechanical injury. In case of fruits and vegetables, these browning reactions are not considered desirable as it results in off-flavors and colors unlike in case of coffee, beer, and bread toasting where it is desirable. Therefore, it is essential to know the mechanisms and inhibition methods of browning reactions. Browning also results in the lowering of nutritive value of fruits and vegetables.[11]

Phenolase or polyphenol oxidases are the group of enzymes that cause enzymatic browning in many fruits and vegetables like apples, bananas, and potatoes, when the tissue is bruised, diseased, cut, peeled, or exposed to any number of abnormal conditions. This results in the rapid darkening of injured fruit tissue on exposure to air, due to the conversion of phenolic compounds to brown melanins.

2.9 CONTROLLING OF ENZYMATIC BROWNING REACTIONS IN FRUITS AND VEGETABLES BY INACTIVATING THE ENZYMES

Browning reduces the shelf life of a product, lowers its commercial value, and in some cases, results in the complete rejection of a product.[14–16] To prevent any browning reactions in fruits or vegetables, phenolase, and all other enzymes can be inactivated by exposing the fruits and vegetables to high temperature for a sufficient length of time. However, the heating time and temperature should be controlled to avoid any unfavorable texture or flavor changes.[11] Use of inhibitory agents is one of the most commonly used methods to hinder browning reactions. These compounds affect the enzymes that are responsible and prevent the formation of coloring compounds.[7]

To inhibit the activity of phenolase, salts like sulfur dioxide, sulfates like sodium sulfate, sodium bisulfate, and sodium metabisulfate are widely used in industries. Gaseous sulfur dioxide or dilute aqueous solutions of the sulfites can also be used either by dipping or spraying.[33] Gases will penetrate at a faster rate into the fruits or vegetables but the aqueous solutions are easier to handle.

Another widely used method for controlling enzymatic browning reactions is use of acids, such as citric, malic, phosphoric, and ascorbic acids. The general action of these salts is to lower the pH thereby decreasing the rate of browning reaction. The optimum pH of phenolase lies in the range of 6–7; and virtually there is no enzymic activity for pH below 3.[19] Cut fruits such as peaches are often immersed in the dilute solutions of citric acids prior to processing to inhibit the enzymatic browning activity. They act not only by lowering the pH but also by chelating with the copper moiety of the enzyme.[19] Ascorbic acid is also been used, which acts by reducing the o-quinones formed by phenolase to the original o-dihydroxyphenolic compounds, which in turn prevents the formation of brown substances.[19]

In the last decade, power ultrasound has emerged as an alternative processing option to conventional thermal approaches for pasteurization and sterilization of food products. While sonication alone is not sufficient for inactivation of various spoilage and harmful enzymes present in food, ultrasound in combination with mild heat treatment and/or pressure has shown possible for both enzyme and pathogen inactivation. Numerous studies have investigated ultrasound for inactivating enzymes such as pectinmethylesterase, PPOs and PODs that are responsible for deterioration of fruit and vegetable juice.

2.10 DISCOLORATIONS OF RAW AND PROCESSED FRUITS AND VEGETABLES

Each enzyme acts in a particular manner on a particular chemical component or closely related group of compounds. It is essential to know the different types of enzymes that are present in a particular fruit or vegetable, the chemical substance called the substrate on which these enzymes act, and also the factors affecting its activity. Browning of fruits and vegetables normally occurs as a result of the reaction between the oxidizing enzymes, oxygen, and a substrate. The factors such as pH, temperature, and concentration of the substrate influence the rate and intensity of browning reaction.[16] In case of fruits, the enzymes responsible for discoloration may be destroyed or inactivated by using fruits containing low oxidizing enzyme, or by decreasing the pH, or by lowering the substrate concentration, or by creating anaerobic environment, or by using antioxidants, or by subjecting to mild heat.[16]

In general, heat treatment is well recognized to destroy the enzymes. However, it is necessary to find specific heat conditions under which there will be minimal damage to other organoleptic attributes of the product. These conditions would be different for different products. Another way to lower the enzyme activity is to lower the storage temperature, but lowering the temperature even below −370°F will not destroy the enzymes. Therefore in case of vegetables, blanching is done prior to further processing such as drying or freezing to prevent the production of undesirable off-flavors during storage.

2.11   THERMAL INACTIVATION OF ENZYMES BY BLANCHING

Enzymes play a major role in the color change of fruits and vegetables. Some are positive changes as in case of ripened tomatoes, strawberries, apples, bananas, etc., where the color changes from green to red or yellow. On the other hand, some color changes are considered undesirable as in case of bruised or sliced potatoes, peaches, or apples, where the browning takes place as a result of oxidation of PPO enzyme. Also, colors bleaching of some green vegetables are considered undesirable.[34]

Therefore, blanching which is nothing but the minimal heat treatment is normally given to the cut fruits and vegetables to inactivate the enzymes and to kill the surface microorganisms so that it can be stored for several months at frozen state.[19] The minimum heat is suggested to be applied to preserve the fresh quality attributes (color, flavor, taste, texture, aroma, and nutrition) of the blanched fruits and vegetables. POD is universally considered as an indicator enzyme to check for blanching adequacy, as it is the most heat-stable enzyme that is in most of the fruits and vegetables.[34]

2.12   ENZYMES AS PROCESSING AIDS

Not all enzyme activities are undesirable for fruits and vegetables processing, as some are beneficial too. Clarification is desirable in some fruit products. For example, for fruit juices like apple and wines, the pectic enzymes produced by certain molds are used as a clarifying agent.[16] To improve the product quality or to increase the efficiency of operations such as peeling, extraction of juice, clarification, production of value-added products, enzymes can also be used as processing aids. Some of the enzymes that are used in the fruits and vegetables processing industries for various purposes are listed in Table 2.3.

TABLE 2.3   Enzymes as Processing Aids in Fruits and Vegetables Processing.

Enzyme category	Functions	Advantages	References
Pectolytic enzymes	Pectin degradation	Decrease of juice viscosity, increased yield, reduced amount of pectin in juice	[3,19,24]
Pectic enzymes including pectinlyase, pectin methylesterase, endo- and exo-poly-galacturonases, pectin acetylesterase, rham-nogalacturonase, endo- and exo-arabinases	Pectin degradation	Extraction and clari-fication of fruits and vegetables juices	[13]
Pectin methyl esterase	Removes methoxyl group from pectin	Clarification of juice	[2]
Pectate lyases	Nonhydrolytic break-down of pectates or pectinates	Clarification of juice	[32]
Cellulase	Degradation of cel-lulose into glucose	Extraction and clari-fication of fruits and vegetable juices for production of nectars and purees	[5,6]
Hemicellulases	Hydrolyzes hemicellulose	Clarification of fruit juices	[27]
Amylases	Breaks down starch to sugars	Clarification of fruit juices	[8]
Limonoate dehydroge-nase, naringinase	Catalyses the oxida-tion of A-ring lactone, converts naringin to naringenin	Removes bitterness in juices	[11]

2.13   SUMMARY

Enzymes are useful in the processing industries for fruits and vegetables. The major applications are for fruit juice industry. At industrial level, usage of these enzymes results in fruit juices with higher yield and with improved physical quality characteristics such as clarity, viscosity, filterability, color, etc. Enzymatic processing of fruits and vegetables improves the nutritional content and the organoleptic properties by reducing the bitterness and by preventing the darkening of juices.

KEYWORDS

- **enzymatic browning**
- **hydrolytic enzymes**
- **processing**
- **enzyme activity**
- **blanching**

REFERENCES

1. Azarnia, S.; Boye, J. I.; Warkentin, T.; Malcolmson, L. Changes in Volatile Flavour Compounds in Field Pea Cultivars as Affected by Storage Conditions. *Int. J. Food Sci. Technol.* **2011,** *46* (11), 2408–2419.
2. Baker, R. A.; Bruemmer, J. H. Pectinase Stabilization of Orange Juice Cloud. *J. Agric. Food Chem.* **1972,** *20* (6), 1169–1173.
3. Barrett, D. M.; Beaulieu, J. C.; Shewfelt, R. Color, Flavor, Texture, and Nutritional Quality of Fresh-cut Fruits and Vegetables: Desirable Levels, Instrumental and Sensory Measurement, and the Effects of Processing. *Crit. Rev. Food Sci. Nutr.* **2010,** *50* (5), 369–389.
4. Bayindirli, A. *Enzymes in Fruit and Vegetable Processing: Chemistry and Engineering Applications*; CRC Press: Boca Raton, FL, 2010; p 402.
5. Béguin, P.; Aubert, J. P. The Biological Degradation of Cellulose. *FEMS Microbiol. Rev.* **1994,** *13* (1), 25–58.
6. Bhat, M. K.; Bhat, S. Cellulose Degrading Enzymes and Their Potential Industrial Applications. *Biotechnol. Adv.* **1997,** *15* (3), 583–620.
7. Biegańska-Marecik, R.; Czapski, J. The Effect of Selected Compounds as Inhibitors of Enzymatic Browning and Softening of Minimally Processed Apples. *Acta Sci. Polon. Technol. Aliment.* **2007,** *6* (3), 37–49.
8. Couto, S. R.; Sanromán, M. A. Application of Solid-State Fermentation to Food Industry: A Review. *J. Food Eng.* **2006,** *76* (3), 291–302.
9. Dauthy, M. E. Fruit and Vegetable Processing (Chapter 3). In *Deterioration Factors and Their Control, Agricultural Services Bulletin No. 119*; FAO: Rome, 1995; pp 32–39.
10. Deza-Durand, K. M.; Petersen, M. A. The Effect of Cutting Direction on Aroma Compounds and Respiration Rate of Fresh-cut Iceberg Lettuce (*Lactuca sativa* L.). *Postharv. Biol. Technol.* **2011,** *61* (1), 83–90.

11. Eskin, N. M.; Shahidi, F. Biochemistry of Foods (Chapter 2). In *Biochemical Changes in Raw Foods: Fruits and Vegetables*; Academic Press: London, 2012; pp 49–54.

12. Fox, P. F., Ed. *Food Enzymology*; Elsevier Applied Science: London, 1991; pp 371–398.

13. Galante, Y. M.; De Conti, A.; Monteverdi, R. Application of Trichoderma Enzymes in Food and Feed Industries. In *Trichoderma and Gliocladium: Enzymes, Biological Control and Commercial Applications*; CRC Press: Boca Raton, 1998; Vol. 2, pp 327–342.

14. Ioannou, I. Prevention of Enzymatic Browning in Fruit and Vegetables. *Eur. Sci. J.* **2013,** *9* (30), 310–341.

15. James, C. S., Ed. *Analytical Chemistry of Foods*; Springer Science & Business Media: New York, NY, 2013; pp 13–31.

16. Joslyn, M. Role of Enzymes in the Commercial Processing of Fruits and Vegetables. *Calif. Agric.* **1947,** *1* (2), 4.

17. Kumar, S. Role of Enzymes in Fruit Juice Processing and Its Quality Enhancement. *Health (Bundelkhand University, Jhansi, India)* **2010,** *3,* 4.

18. Kohman, E. F. Enzymes and the Storage of Perishables. *Food Ind.* **1936,** *8,* 287–288.

19. Lee, W. C.; Yusof, S.; Hamid, N. S. A.; Baharin, B. S. Optimizing Conditions for Enzymatic Clarification of Banana Juice Using Response Surface Methodology (RSM). *J. Food Eng.* **2006,** *73* (1), 55–63.

20. Lindley, M. G. The Impact of Food Processing on Antioxidants in Vegetable Oils, Fruits and Vegetables. *Trends Food Sci. Technol.* **1998,** *9* (8), 336–340.

21. Madhavi, D. L.; Salunkhe, D. K. Tomato. In *Handbook of Vegetable Science and Technology: Production, Composition, Storage, and Processing*; Salunkhe, D. K., Kadam, S. S., Eds.; Marcel Dekker: New York, NY, 1998; pp 171–202.

22. Moline, H. E.; Buta, J. G.; Newman, I. M. Prevention of Browning of Banana Slices Using Natural Products and their Derivatives. *J. Food Qual.* **1999,** *22* (5), 499–511.

23. Oliw, E. H. Plant and Fungal Lipoxygenases. *Prostagland. Other Lipid Mediat.* **2002,** *68,* 313–323.

24. Płocharski, W.; Szymczak, J.; Markowski, J. Effects of Zymatish Apple Mash Treatment on Juice Extraction and Pectin Amount of Substance (*Auswirkungenzymatish Apfelmaische Behandlung auf Saftsbeute und Pektin Stoffmenge*). *Flussiges Obst* **1998,** *6,* 325–330.

25. Ruenroengklin, N.; Sun, J.; Shi, J.; Xue, S. J.; Jiang, Y. Role of Endogenous and Exogenous Phenolics in Litchi Anthocyanin Degradation Caused by Polyphenol Oxidase. *Food Chem.* **2009,** *115* (4), 1253–1256.

26. Sapers, G. M.; Miller, R. L. Enzymatic Browning Control in Potato with Ascorbic Acid-2-Phosphates. *J. Food Sci.* **1992,** *57* (5), 1132–1135.

27. Shallom, D.; Shoham, Y. Microbial Hemicellulases. *Curr. Opin. Microbiol.* **2003,** *6* (3), 219–228.

28. Srilakshmi, B. Food Science (Chapter 8). In *Vegetables and Fruits*; New Age International: New Delhi, 2003; pp 171–211.

29. Sumonsiri, N.; Barringer, S. A. Fruits and Vegetables: Processing Technologies and Applications (Chapter 16). In *Food Processing: Principles and Applications*; Clark, S., Jung, S., Lamsal, B., Eds.; John Wiley & Sons: Hoboken, NJ, US; 2014; pp 363–381.

30. Tapre, A. R.; Jain, R. K. Pectinases: Enzymes for Fruit Processing Industry. *Int. Food Res. J.* **2014,** *21* (2), 447–453.

31. Voragen, A. G. J.; Schols, H. A.; Beldman, G. Tailor-Made Enzymes in Fruit Juice Processing. *Fluessiges Obst (Germany, FR)* **1992,** *7*, 98–102.
32. Watada, A. E.; Ko, N. P.; Minott, D. A. Factors Affecting Quality of Fresh-Cut Horticultural Products. *Postharv. Biol. Technol.* **1996,** *9* (2), 115–125.
33. Whitaker, J. R. Importance of Enzymes to Value-Added Quality of Foods. *Food Struct.* **1992,** *11* (3), 2.
34. Whitaker, J. R. *Principles of Enzymology for the Food Sciences*; CRC Press: Boca Raton, FL, 1993; p 61.

CHAPTER 3

USE OF SOLAR ENERGY IN PROCESSING OF FRUITS AND VEGETABLES

MAHENDRABHAI BABULAL PATEL, KAMALCHANDRA R. TRIVEDI, and KHURSHEED ALAM KHAN

ABSTRACT

Because of free of cost and unlimited supply, solar energy holds good scope and can be used either as main or auxiliary source of power for the food-processing units. A major part of research on applications of solar energy in processing has been devoted to various dryers designed for drying of fruits/vegetables/grains. These dryers are generally classified into passive and active with both types containing subtypes of indirect, direct, and hybrid dryers. They are designed and fabricated according to local climate conditions and availability of materials. Solar energy can also be used, in a concentrated form, in solar furnaces. Solar energy has also good potential in being used for greenhouse operations: both in active form—use of solar PV for greenhouse lighting—and in passive form—for greenhouse ventilation. Similarly, solar energy can be used for lighting and ventilation of processing units. The use of solar cookers for cooking food is also widespread and large numbers of solar cooker's design are to meet the requirement of local climate conditions. Apart from these applications, solar energy has also been used for storage of fruits/vegetables in cold storage.

3.1 INTRODUCTION

Although intake of fruits/vegetables is preferred in fresh form by people, yet increase in population and changing lifestyle has increased the trend toward consumption of processed food products. Postharvest practices are often followed for maintaining the quality of harvested fruits/vegetables and to lower the storage and nutritional losses among them. Processing of fruit and vegetable sector comprises various unit operations which are performed according to the type of processed products to be made. For example, in preparation of juice, fruits/vegetables have to go through crushing operation while to obtain the product in powdered form, drying, and grinding operations are followed.

At the time of harvest, the moisture content in fruits/vegetables is around 80% and water activity is also high.[12] Thus, due to biochemical reactions such as enzymatic activity and respiration, they are prone to physiological, physicochemical, or microbial spoilage. To prevent these food items from spoilage, certain measures are taken to lower the water activity to prevent spoilage and extend shelf life. Drying is an age-old practice where benefit of solar energy is being harnessed from centuries to preserve these fruits/vegetables by lowering their water content.

The manufacturing activities which constitute the fruit and vegetable processing industry can be enlisted as frozen fruits, vegetables, their juices and specialty foods obtained from them; dried, pickled, and canned fruits and vegetables and specialty food obtained from them including minimally processed fruits and vegetables.[33] The various unit operations employed to obtain the aforementioned processed products can generally include:

- sorting/grading;
- washing/cleaning/disinfection;
- peeling/deseeding/coring/trimming/cutting;
- extraction of juice/pulp (dewatering and packaging in case of fresh-cut fruits and vegetables);
- filtration/clarification;
- heating/cooling;
- bottle washing; bottle filling/canning; and
- sterilization and packaging.

In preparation of dried products, disinfected fruits/vegetables are cut, dried, and packed.

All of these unit operations require input of energy, for example, drying is an energy guzzling process; heating and pasteurization processes require generation of steam (sometimes superheated); large amount of electricity is required for operation of electric motors and air compressors; heat exchangers, sterilizers, and boilers operate at elevated temperatures; and in freezing and cooling operations, extremely low temperatures are required.

Thus, for performing all these operations large amount of electricity is needed, which in turn results in high operating cost. Sources of energy for such operations include mostly the conventional fuels; however, with their shortage and the eventual hike in their prices, more emphasis is now given to the application of renewable energy sources in performing these operations.[8]

Solar energy is one of the best sustainable energy source which can supply energy in constant and concentrated form to food processing facilities.[33] Assimilation of renewable energy is imperative to replace the electricity power which is currently being used in food processing sector.[22] Application of solar energy in performing processing operations at small-scale level can reduce the processing costs, which can result in the availability of processed products to larger part of population in developing countries. Applying solar energy and adopting sustainable practices for processing of fruits and vegetables has a wide scope, especially in developing countries, which have plenty of sunshine hours around the year.

This chapter explores the various avenues in which solar energy can be applied in different unit operations for processing of fruits and vegetables. This chapter also discusses about the drying principles, application of solar energy for drying, different types of solar dryers, and their uses for drying of fruits and vegetables.

3.2 USE OF SOLAR ENERGY IN DRYING OF FRUITS AND VEGETABLES PROCESSING

Solar energy can be widely used in the processing of fruit/vegetable processing in many ways. Further at remote locations where availability of electricity is still a major hurdle, supply by renewable energy sources can speed up the processing tasks. Solar energy can be used at such places to establish processing facilities. For example, use of solar energy in drying of fruit/vegetable commodities can reduce the processing cost. Numerous researchers have carried out research studies for solar drying of fruits/vegetables to develop the drying procedure and to optimize the drying parameters for particular fruit/vegetables.

Baradey et al.[9] conducted an experiment with natural and forced convection mode to obtain 5-, 10-, and 15-mm thick solar-dried apple, orange, and mango slices to observe the effect of thickness of fruit slices on drying rate. The results revealed that forced convection solar dryer was better than natural convection dryer in terms of moisture removal. Also highest weight loss occurred in slices of thickness 5 mm for all three types of fruits.

Bano et al.[8] has described different types of dryers primarily operated by solar radiation or in hybrid mode, such as solar natural dryer (greenhouse type), semiartificial solar dryers (where, air preheated in a collector is made to flow through a bed of material to be dried), indirect-type solar dryer (solar energy heats a packing material and air passing through this heated material is then used to dry food material), and hybrid dryers (air is heated by both "sun and an auxiliary" source like biomass, natural gas, or oil). It was observed that the quality of final product is inferior as compared to freeze-dried product, but still for energy-deprived farmers in developing countries, the use of a good solar dryer is beneficial.

Hegde et al.[13] designed and fabricated an indirect and active-type solar dryer for drying of banana fruit. It has a flat plate with three insulation layers, a chamber for drying the material, a fan, and mechanism to regulate the air flow. The experiment was conducted in two ways: two types of air flow—one flow was sent between the plate and glass (top flow) and the other between the plate and insulation (bottom flow); and three variations in air velocity—from 0 to 3 m/s. It was found that the bottom flow provided a temperature 2.5°C higher than top flow for the same input of energy resulting in a heat loss of 201.9 W from top flow and 139 W in bottom flow, thereby saving 62 W energy. On the other hand, banana slices dried at air velocity of 1 m/s were superior to those dried at 0.5 and 2 m/s.

Krishna et al.[18] fabricated a multipass multirack solar dryer for fruits. The experimental setup consisted of a solar collector consisting aluminum sheet as absorber plate, granite as heat storage medium, a drying chamber consisting four perforated trays to accommodate grapes for drying, thermocol was used an insulation for both collector and drying chamber. The collector area was 1.6 m². A glass sheet was used to cover the collector to maximize incoming heat absorption. The multipass flow is arranged so that three passes are below the trays and the other three are above the trays. Grapes were loaded in the trays for the experiment. To utilize the full capacity of the hot air, the grapes were so arranged in the trays that the bottom of the trays is not visible. The results showed that 32 h were required to reduce the moisture content of grapes to a safe storage level of 18.6% (wet basis, w.b.), whereas in case of open sun drying, it took 38 h.

Kumar et al.[19] have reported the results of extending the shelf life of vegetables by passive cooling. A solar passive cool chamber was constructed, working on the principle of direct evaporative cooling. The chamber had two walls—inner and outer. The space between the walls was filled with jute bags soaked with water. The base of the chamber was filled with coarse sand

and provided with a channel system to attain reduction of temperature and maintenance of humidity in the cooling area. Clear glass was used in the south facing wall to increase passive cooling. The chamber was covered with shade net. Pipes were arranged in between the outer and inner walls for supply of water in droplet form on the jute bags. A tank of 200-L capacity was used for storage of water. In total 1 kg of tomatoes, 1 kg of radish, 1 kg of cabbage, 500 g of peas, and 250 g of coriander leaves were put in different plastic baskets and loaded in the chamber. The same amount was kept at room temperature in a room as control. Temperature and relative humidity were measured simultaneously for the chamber and the room. It was reported that the temperature within the chamber, as compared to ambient condition, is reduced by 15–19°C when humidity is maintained above 94% level. Significant increase in shelf life of vegetables was observed. The shelf life of tomatoes, radish, cabbage, peas, and coriander leaves in cool chamber was 12, 7, 9, 9, and 5 days, respectively, whereas those stored at room temperature was 7, 3, 5, 4, and 2 days, respectively.

Ringeisen et al.[26] determined the effect of a concave solar concentrator on the working of a typical solar air heater. Two solar crop dryers, both identical, were fabricated: one acting as control and the other was fitted with the concentrator. Tomato slices (moisture content 94% w.b.) were placed on the racks provided in both dryers. They were considered sufficiently dried when the moisture content reaches up to 10% (w.b.). Results indicated that due to the use of concentrator, the drying time was reduced by 21% in comparison to the control conditions. At the same time, the use of concentrator did not affect the biochemical properties of tomatoes.

Janjai[17] developed a greenhouse-type drier for establishing its viability for drying of fruits and vegetables on commercial scale. The dryer has dimensions of 8 m width, 20 m length, and 3.5 m height. It is a hybrid dryer; in that both solar radiation and an LPG burner is used for heating incoming air—solar radiation when there is adequate sunlight and LPG burner when the atmosphere is cloudy. Moreover, fans are installed on the opposite to incoming air to ventilate the dryer. The fans are operated by installed solar PV modules. To simulate the procedure at commercial-level, 1000 kg of tomatoes were taken for drying. Prior to drying, the tomatoes were subjected to predrying treatments for color preservation: blanching and soaking in sugar solution. Sensors to measure internal temperature, relative humidity, solar radiation, air velocity, etc. were placed inside the dryer. Moisture content of samples was taken every 3-h interval and compared with similar samples dried in open sun. The drying continued till the required moisture content (17% w.b.) was obtained, which was similar to that of samples available in market. The study found that the time taken to reduce the moisture content (w.b.) of tomatoes from 54% to 17% was 4 days, whereas in the same time period, the moisture content of open sun-dried tomatoes could be brought down to 29%. The study provides a good example of scaling-up of drying technology of fruits and vegetables at commercial level.

Rajeshwari and Ramalingam[24] constructed a box-type solar dryer with dimensions of 30 cm length, 15 cm width, and 45 cm height. It consists of a blackened absorber plate at the base and a glass cover of thickness 5 mm at top. In between are fixed two wire-mesh screens—one on top of another and 10 am apart—on which the food to be dried can be placed. A small hole of diameter 3 cm is provided on the top of dryer to remove the moisture content. Experiments were conducted for drying of potato slices, chilly, and grapes. The average system drying efficiency was estimated 69.6% during the observation period. Depending on intensity of radiation received on earth surface, the dryer was able to remove 80% of moisture on dry basis in 1-day period.

Rathore[25] studied technical feasibility of different solar dryers for drying of chili for domestic, farm, and industrial use. All three dryers have low initial and operating costs as compared to a conventional dryer. To boost efficiency of these dryers, various measures are taken to reduce the heat losses to surrounding. The dryers have the capacity to accommodate 5, 25, and 500 kg of raw material. Domestic solar dryer is having provision of variable inclination with side and back insulation whereas the farm and industrial dryers are having provision of UV-protected polythene as glazing surface. During no-load testing, the maximum temperature inside the domestic solar dryer, farm solar dryer, and solar tunnel dryer was 43, 42, and 42°C, respectively, after 14 h as compared to maximum ambient temperature of 26°C. During full-load testing, the maximum inside temperature in all dryers was 41.8, 39, and 42°C, respectively, at 14 h. The performance of solar dryer is excellent in terms of drying time as compared to a conventional dryer. The cost-economics of drying Aonla in all three dryers is given in Appendix A.[25]

Palaniappan[23] briefly outlines the work done by Planters Energy Network (PEN), an NGO working in southern India as a link between planters and energy scientists. It has done work in solar hot air technology mainly for drying and dehydration of products like tealeaves, spices, leather, fruits and vegetables, latex rubber, pulses and paddy, salt, ceramic, fish, etc. In case of fruits and vegetables, owing to absence of infrastructural facilities and inadequate sunlight in the extreme Northern India, the PEN has installed eight large solar drying units in the region. The roof of the processing house acts as the base for solar collector and heated air is passed to the recirculation drier placed inside the building. Fruits such as apricots and apples are dried in this manner.

Bala et al.[6] has conducted an experiment to dry pineapple slices in a solar tunnel dryer. The dryer consisted of a flat plate air heater, a tunnel-type dryer unit, and a fan to facilitate the required flow over the material. All these components are series connected. The collector and the dryer are covered with plastic. For good absorbance, black paint is used in collector. The material is placed in thin layer on a plastic-net in the tunnel drier. To reduce the heat loss, glass wool is used. Ten-millimeter thick pineapple slices were treated with sulfur dioxide. Pineapple slices were then spread on plastic-net in a thin layer. To compare the tunnel dryer with sun drying, pineapple slices as control samples were placed in single layer on trays inside the dryer on a raised platform. The moisture content of pineapple treated with sulfur reduced from 87.32% (w.b.) to 14.13% (w.b.) in the solar tunnel drier in 3 days. In traditional method, a similar sample was dried to 21.52% (w.b.) in same period. Proximate analysis was carried out for both fresh and dried pineapple and the solar-dried pineapple contained higher amount of protein and vitamin C.

3.3 THEORETICAL ASPECTS OF SOLAR ENERGY APPLICATION IN FRUIT–VEGETABLE PROCESSING

It is essential to process the fruits and vegetables after harvesting to preserve them for long duration without the quality loss. To achieve this, many technologies (such as dehydration, freezing, caning, etc.) for preservation have been developed and are used extensively. Literature available regarding the application of solar/renewable energy in processing of food items especially fruits and vegetables, suggest that its major use is in drying of foods.[11] Other applications are concerned with industrial processing such as heating and refrigeration using solar energy. Hence, this theory will outline some of the theoretical aspects of solar energy application in fruit–vegetable processing.

3.3.1 DRYING

It is suited for those countries, which have poorly established processing facilities. It is a cheap and practical means of preservation, which helps in reducing losses and also compensates when there is a shortage in supply. In drying, the moisture in the material is removed to reach a predetermined level and thus drying is an operation which requires large amount of energy. Apart from extending the storage life, drying also enhances the quality, makes handling easier, and simplifies further processing. Hence, it is one of the oldest methods of preserving food.[20] In drying, the moisture in the material is vaporized and a mechanism is to be made for the removal of vapor so that proper drying takes place. Thus, it is a process which involves the transfer of both heat and mass. Moisture once removed, prevents the growth of microbes, which otherwise may cause spoilage of foods. Removal of moisture also leads to reduction in bulk, minimizing costs related to package and transportation.[32]

Most of the energy required during drying is that which is used in conversion of water to its vapor (2258 kJ/kg at 101.3 kPa). Water which is present in free moisture or bound forms tend to affect the drying rate directly. Moisture content of a food product is expressed—as a fraction—either on wet basis (w.b.) or dry basis (d.b.) as shown below:

$$\text{Moisture content on wet basis, } M_w = \frac{m_w}{m_w + m_d} \tag{3.1}$$

$$\text{Moisture content on dry basis, } M_d = \frac{m_w}{m_d} \tag{3.2}$$

where m_w is the mass of water contained in the sample (g); m_d is the mass of dried sample (g).

Drying consists of two processes: (1) heat transferred to liquid and (2) mass transferred in the form of liquid or vapor in solid and as vapor from the surface. In all dryers, except dielectric and microwave drying, heat flows to the external surface first and then in the solid. Drying of food items, in principle, occurs in two modes:

a. The *constant-rate period*, in which the moisture moving from within the solid to outer surface is equal to the moisture removed from the outer surface to air as vapor. Drying occurs when the vapor diffuses from the surface to the environment and as the rates of mass and heat transfer are equal, the temperature of the surface tends to remain constant.

b. *Falling-rate period*, in which the rate of vapor removed from surface decreases as less mass transfer is taking place from inside the solid to outer surface.

The rate of drying depends on many factors such as type and weight of the product. Thus, fruits, vegetables are better to be dried in thin layers, but grains are dried in deep beds.

Solar drying has been used since ancient times to dry agricultural products. Drying of agricultural products in closed structures by forced air is helpful in reducing losses and low quality of products, which are otherwise obtained in traditional open-sun drying methods.[16]

Drying systems are classified according to their operating temperature: ranges from high temperature to low temperature dryers. They are also classified according to their source of heating: fossil fuel and solar-energy dryers. Generally, dryers operating at high temperature are fossil fuel powered while the dryers operating at low temperature are either fossil fuel or solar-energy powered.[11]

The techniques of drying are divided into following general types based on the way the material is heated[34]:

- In open-air drying, the process takes place when material is exposed to the sun and wind on the ground.
- In direct sun-drying, the food is enclosed in a container and the sunlight falls directly on the material. The material is heated by greenhouse effect apart from sun's rays.
- Indirect sun driers first heat the air in a solar collector and the hot air is sent to the material. So, the food is not exposed to direct sunlight.
- The features of direct and indirect type dryers are combined in a mixed mode dryer; air is heated in a separate collector and then further heated by direct sunlight.
- Solar energy is combined with a fossil or biomass based fuel in a hybrid drier.[34]
- When drying is done by air which is heated and circulated by buoyancy, it is known as Passive solar dryer; this type of dryer is devoid of any mechanical systems, for example, Cabinet dryer and greenhouse dryers. These types of dryers are primitive, and can be constructed with materials available locally and are easily operated at off-grid sites. Small batches of fruits and vegetables can be dried in the passive dryers.[14]

- Active solar dryers incorporate external means for circulation of air such as fans, for transporting the heated air to the drying chamber, a mechanical system is employed. They are mainly used in commercial drying, having conventional fossil-fuel system so that the glut in sunlight can be offset and drying can commence uninhibited.

Both passive and active solar dryers are further classified as indirect, direct, and hybrid types. Indirect-type dryers employ natural air convection for drying. The trays are arranged vertically with some space between to increase the dryer's capacity. In direct-type passive dryers, when certain fruits and vegetables are exposed to direct sunlight, it tends to enhance the color. Types of dryers in this category are the cabinet and greenhouse dryers. The structural makeup of a hybrid-type passive dryer is similar to indirect and direct-type dryers and is lined with glazed walls on the inside to facilitate the sunlight impinging directly on the material to be dried.

The indirect-type active solar dryers have a separate collector and drying unit. These are made up of four components, namely, a solar heater, drying unit, a fan, and piping accessories. Higher temperatures can be obtained with controlled flow of air due to separated air heating unit.

As all drying parameters are variable and dependent on the type and amount of material to be dried, there is no single design that fits all.[3] Due to nonlinearity of the involved processes, a thorough knowing of heat and mass transfer and material characteristics is required. As a result, scaling of a solar dryer is difficult. The important aspect to be considered for design of solar dryer is the working temperature for drying a material. Calculations on volume of air required, mass of moisture to be removed in a definite time period, and amount of energy required are to be determined. For dryers working in active mode, studies regarding materials of construction, size of chamber, volume of trays, size of piping, air-flow rate of fan, etc. are taken under consideration. Other design parameters considered are location, local climate, intensity of insolation, etc. In designing a solar dryer where an air heater is joined with a drying unit, the following design parameters are considered:

- Amount of material to be dried and size of drying cabinet.
- Capacity of dryer.
- Materials used for fabrication.
- Method for loading/unloading.
- Method of passing the heated air through the raw material.
- The way the moist air will be removed.

The quantity of air that will be required for drying a particular mass of product has to be calculated. This is done either by the help of a psychrometric chart or by the energy-balance equation. Psychrometric chart is used for the determination of thermodynamic properties of moist air.

The energy contained in the air passing through the material must be equal to the energy needed to remove moisture (the energy-balance equation). Thus, the removal of water by evaporation will require an amount of heat equal to the latent heat of evaporation of water plus sufficient air to remove the vapor. Hence, the main point in solar dryer is to determine optimum temperature, T_p, air flow, m_a to remove vapor, and m_w. Therefore,

$$m_w L = \left(T_f - T_i\right) m_a C_p \qquad (3.3)$$

where m_w is the mass of vapor; L is the latent heat of vaporization; m_a is the mass of air supplied; C_p is the specific heat of air; T_f and T_i = final and initial temperature of air. The volume of air is determined as

$$V_{air} = \frac{m_a RT}{P} \qquad (3.4)$$

Working fluid temperature and flow rate of heated air will determine the quality of the product.[3] Velocity of air, humidity, and final moisture content also play vital role. High drying temperature can damage the product. Low drying temperature increases the drying time, which can cause microbial infection. If the relative humidity of air is less, it will lead to increase the drying rate and reduce the time taken for drying. Materials that have higher moisture content have less drying times as the moisture tends to flow from the inside to the surface more readily than in a material with a lower moisture content.[3]

An example is provided here regarding the design and sizing of a solar dryer based on phase-change material.[1] The system is designed to dry 5 kg of raw material up to a particular moisture content level. Based on this, the amount of energy required to remove the said moisture is determined. It is then used to determine the size of various parts of the dryer and the amount of phase-change material that will be needed. The raw material to be dried in this case is taken as cassava root. The average active solar energy available is taken as 833 W/m². The dryer consists of solar air heater, blower, drying chamber, and PCM chamber. Therefore,

Material to be dried = cassava roots
Quantity (M) = 5 kg

Initial moisture content $= 62\%$ (w.b.)
Required final moisture content $= 14\%$ (w.b.)
Therefore, mass of moisture to be removed from the raw material (M_m)

$$M_m = \frac{M\left(\text{initial moisture content} - \text{final moisture content}\right)}{100 - \text{final moisture content}} \qquad (3.5)$$

where M_m is the mass of water that needs to be removed from cassava in kg; M is the mass of raw material in kg.

In eq 3.5, the initial and final moisture content values are to be put in fractional form, that is, 0.62 for initial and 0.14 for final. Thus, the amount of water to be removed is 2.8 kg. Therefore, amount of heat required to remove 2.8 kg of moisture is given as

$$Q_R = \left(M_m \times h_{fg}\right) + \left(M_m \times h_f\right) \qquad (3.6)$$

where h_{fg} is the latent heat of evaporation of water (2358.40 kJ/kg at 60°C, as per steam-table); h_f is the enthalpy of water (251.16 kJ/kg at 60°C, as per steam-table).

Therefore, $Q_R = (2.8 \times 2358.40) + (2.8 \times 251.16)$ kJ $= 7306.76$ kJ

Therefore, assuming that the sunlight is available for 12 h a day, the power required is

$$= (7306.76 \times 1000 \text{ J} / (12 \times 60 \times 60 \text{ s}) = 161.138 \text{ W}$$

Therefore, putting the respective values in eq 3.6, the power required to remove 2.8 kg of moisture will be 169.138 W. The equation used to determine the area of collector for delivering Q_R amount of heat is given below:

$$Q_u = A_c\left[I_t\left(\tau\alpha\right) - UL\left(T_p - T_a\right)\right]FR \qquad (3.7)$$

where A_c is the collector area (m²); I_t is the solar irradiance (taken here, 833 W/m²); τ is the transmittance percentage of light transmitted by glass cover (taken here, 0.5); α is the absorptance: percentage of light absorbed by the glass cover (taken here, 0.5); UL is the overall loss coefficient (taken here, 5 W/m² C); T_p is the average temperature at upper surface of the absorber (taken here, 40°C); T_a is the average atmospheric temperature (taken here, 30°C); FR is the collector heat removal factor depending on material (taken here, 0.92).

Therefore, $Q_u = A_c \times [833 \times (0.5 \times 0.5) - 5 (40 - 30)] \times 0.92 = A_c \times 145.5$
For the design purpose, $Q_u = Q_R$. Therefore,
$A_c = (169.138/145.5) = 1.16$ m²

3.3.1.1 THE QUANTITY OF PCM (HERE, PARAFFIN WAX)

Latent heat capacity of paraffin wax = 220 kJ/kg

Required amount of energy = 7306.76 kJ

Therefore, amount of PCM required = (7306.76/220) = 33.2 kg.

For extending the working of dryer during night time for 12 h = (33.2/12) = 2.76 kg. Thus, approximately 3 kg of PCM is required.

The efficiency of solar dryers working on natural circulation can be hampered drastically when the ambient humidity is very high during the wet season. This leads to moist air getting entrained in the dryer chamber during night, which can further lead to reabsorption of moisture by the products kept for drying. Moreover, in some products like grapes, direct exposure to sunlight is important for their ripening as direct sunlight decomposes residual chlorophyll during dehydration. However, insolation in the drying chamber can, on the one hand, cause an asymmetry in temperature distribution, and on the other hand, shorten the life of dryer components in the presence of UV light. Thus, in such cases, the duration of drying has to be critically monitored and regulated constantly. Some models of solar dryers, which have been developed and are working successfully (Table 3.1). Data on solar dehydration of selected foods are presented in Appendix B.

TABLE 3.1 Some Successful Solar Dryers Working in India.

Description	SPRERI forced circulation solar dryer[10]	Solar cabinet dryer, CUSAT[29]	Solar dryer, CUSAT[4,5]	Solar dryer, Pondicherry University[4,5]	Solar dryer, PAU[28]
Characteristics	Modular air heaters integrated to drying chamber	Collector at top and drying chamber at bottom	Dryer integrated with solar air heater	Dryer with wire-mesh air heater	Natural convection, low cost
Surface area	2 m²	1.27 m²	46 m²	6 m²	–
Outlet temperature from air heater	70°C	–	–	–	–
Inside temperature	50–60°C	–	–	–	75°C
Capacity	100 kg/day	4 kg/batch	200–250 kg	30 kg	–
Mechanical systems	–	Two axial fans; 20 W each	–	0.5 hp blower; 500 m³/h	–
Performance	100 kg onion flakes dried per day	Bitter gourd dried from 95% to 5% m.c. in 6 h	200 kg apple dried in 8 h	–	–

3.3.2 SOLAR FURNACES

A solar furnace can reach higher temperatures (approx. 3800°C) than furnaces working on conventional fuels and also avoids the contamination of products which can occur in the conventional ones. A solar furnace creates a concentrated beam that has intensity quite higher than the initial beam. A surface exposed to such high intensity beam gets heated rapidly. To achieve deposition of vapor and metallization of ceramics, rapid heating of surface is essential. Concentrated insolation such as this can be applied especially in bonding of metals on ceramic surfaces to fabricate delicate electronic components. Production of fullerenes in a solar furnace uses less energy than other technologies. Solar-pumped lasers make use of concentrated solar radiation.

3.3.3 GREENHOUSE HEATING AND COOLING

Greenhouses are structures, which help maintain suitable conditions for cultivation of plants, especially off-season crops.[31] Multispan greenhouses are most common form but the quonset-type greenhouses, which utilize light-weight plastic cover material, are also constructed.[22] The principle of these structures is that short-wave radiation transmitted through a plastic cover is absorbed by internal surfaces; the longer wavelength radiation emitted by these surfaces cannot penetrate the cover and thus increases the temperature inside. Thus, depending on the type of cover (polyethylene or glass), the microclimate of a greenhouse undergoes cyclical heating and cooling during the daytime. This facilitates the cultivation of off-season crops in a greenhouse. Thus, greenhouse heating and cooling is a major energy consumer in greenhouse operations. Greenhouse heating should be mainly achieved by direct insolation. However, during times of insufficient sunlight, supplementary heating systems can be used, for example, an air heater can heat air and can be coupled with a sensible heat storage system.[22] This system is particularly useful where daytime temperatures are quite high and subsequent nights are cold. In another system, the ground can be used for storage of sensible heat by supplying warm air from the greenhouse which can then be used to warm-up the greenhouse at night.

For greenhouse cooling, the most common method is providing ventilation in temperate regions—along the ridge, at the sides or both—to provide wind and buoyancy induced ventilation.[22] Cooling can also be achieved by evaporative cooling either by wetting the inside air or the ground surface or the cover. Fan-and-pad systems can also be installed in greenhouses where a fan is installed in one side of the greenhouse and pads are installed on the opposite ends. Water is continuously circulated through the pads.

3.3.4   HEATING AND VENTILATION OF AGRICULTURAL BUILDINGS

Solar energy can be used for unit operations in processing premises by installing solar air heaters on the roof of the building. The principal attraction in roof-installed solar air heaters is small amount of start-up investment which ensures availability of low cost heated air for different unit operations such as drying. These types of collectors enable collection and distribution of solar heat properly. Solar energy can also be used in active form through installation of solar PV panels on the roofs for providing lighting for the processing unit.

The space-heating systems working on solar option are classified as active and passive type. An active method is one which utilizes a pump or a blower to circulate the fluids involved in the space-heating system. In contrast, space heating which gives comfortable working conditions can be achieved by employing passive methods, for example, space heating by Trombe wall.[30]

3.3.5   SOLAR COOKING

The use of solar energy in cooking of food is widespread and well-documented. Cooking of food by employing solar energy has been advocated extensively in rural areas whereby the dependence on wood and other harmful sources can be avoided.[27] Designs of the cookers vary depending on the type of food cooked. In cookers of focusing type, a device concentrates the incoming solar radiation on to an area at which the vessel can be placed. In this type of cookers, the heat loss due to convection is substantial and the cooker is able to utilize only the direct insolation. A hot-box cooker contains a box which is insulated and painted black on the inside and consists of double glazing covers. Continuous adjustment of the cooker to face the sun is not necessary as in case of focusing-type units. In most hot-box cookers, temperature achieved is in the range of 50–80°C and hence the numbers of dishes that can be prepared are limited. In indirect cookers, the energy obtained from sun is directly transferred to the cooking unit in the kitchen.

Thus, the necessity of cooking at outdoors can be removed. These types of cookers use either a flat-plate type or focusing-type collector which collects heat energy and transfers it to the cooking unit.

3.3.6   SOLAR REFRIGERATION

Refrigeration is the process in which heat is removed from a definite space at a temperature which is higher than the temperature of the surrounding. Generally, this term is used for food to be stored at temperature lesser than 15°C and above the freezing point.[15] Once fruits/vegetables are harvested, they start deteriorating due to perishable nature and their market value decreases. Decreasing the storage temperature decreases the rate of deterioration reactions, so the shelf life of food is improved. Refrigeration systems used in processing of fruit/vegetable products are often working on a large scale and hence require large input of power and electricity.

To operate the conventional refrigeration system through solar option, one requires converting the sun's energy into electricity or conversion of direct current output available from a solar PV array to an alternating current. An intermittent vapor absorption refrigeration plant works on a 24-h cycle consisting heating and refrigeration which match the diurnal behavior of the sun's movement, that is, heating during the day and "cooling" at night.[2] An adsorbent–refrigerant unit requires the parameters, such as (1) refrigerant whose latent heat of vaporization is high, (2) a working pair whose thermodynamic efficiency is high, (3) the amount of heat desorption is small under the overall operating pressure and temperature conditions, and (4) a low thermal capacity.[22]

Solar refrigeration is used mainly for cold storages for storing fruits/vegetables and their products. The requirement for these systems is more at places where there is absence of grid-connected electricity. In such cases, a cold chain can be extended to these parts possessing of processing units working on solar energy. The NCCD guidelines provide standards for implementation of cold-chain components.[21] The alternate energy options that can be used to operate the equipment of a fruit/vegetable processing unit consist of solar photovoltaic systems, thermal systems, heat storage banks of phase change material, and vapor absorption-based cooling systems. The solar PV systems comprises solar PV panels (to convert solar radiation to electricity), inverters (to convert the electric current generated in solar panel to AC form), battery (for storing of excess electrical energy), and wiring and other miscellaneous components.

A solar thermal system consists of a solar collector (to collect heat energy from incoming solar radiation), storage tank (for storing the heated water), and integrated piping (for supplying the hot water for different unit operations in the processing unit). Thermal banks devices, working on phase change materials, store and release energy as per demand and serve as buffer during erratic electricity supply. A vapor absorption refrigeration system is that which uses a heat source to drive the cooling unit. In this case, the prime refrigerant is not compressed, but turns to vapor by utilizing the heat energy before entering the condensation and expansion phases.

The most commonly used refrigerant–absorbent pairs are water (refrigerant)–lithium bromide (absorbent), and ammonia (refrigerant)–water (absorbent). Water heated in a flat plate collector passes through a generator (heat exchanger) and transfers the heat to a refrigerant–absorbent mixture rich in refrigerant. The boiled-off vapor of refrigerant goes to a condenser to become a high-pressure liquid. It is then throttled to low temperature and pressure by an expansion valve which then goes to an evaporator to absorb heat and result cooling in the space around the coil. From the evaporator, the refrigerant vapor is now withdrawn into a solution mixture weak in refrigerant concentration. This solution mixture is then conveyed to the generator thus completing the cycle. Typical values of coefficient of performance for a solar refrigerator range between 0.5 and 0.8.[30]

A system based on above-mentioned principle has been reported. For designing the system, following values were taken[7]:

- capacity of system = 3 TR
- Temperature of the evaporator = 2°C
- Generator or condenser pressure = 10.7 bar
- Evaporator pressure = 4.7 bar
- Temperature of condenser = 54°C
- Temperature of absorber = 52°C
- Temperature of generator = 120°C
- Concentration of ammonia in refrigerant = 0.98
- Concentration of ammonia in solution = 0.42
- Concentration of ammonia in absorbent = 0.38

The enthalpy–concentration diagram for aqua–ammonia has been used to determine the enthalpy and temperature as the concentrations and corresponding saturation pressure values were known.

Based on these enthalpy values and mass-flow rates, the authors calculated the heat transfer in various components of the system. Although the system has a COP of 0.2, yet it achieves an evaporator temperature of 2°C, which is ideal for storage of most fruits and vegetables.

3.4   SUMMARY

All over the world, the consumption of both fruits and vegetables is widespread and has been so from time immemorial. With an increase in population of the world, the demand and consumption of good quality fruits and vegetables are bound to increase. Moreover, the perishable nature of harvested fruits/vegetables necessitates the adoption of their processing after they are harvested to increase their shelf life. As processing units of fruits/ vegetables are heavy users of electrical power, the rise in fossil fuel prices and electricity derived from them is bound to affect them adversely. In such cases, the search and use of alternative sources of energy to be used in these processing units becomes imperative.

As solar energy is available freely and everywhere on earth, it holds good scope to be used as either a main or auxiliary source of power for the processing units. Of the various unit operations employed in the processing of fruits/vegetables after harvesting, the one using a large amount of energy is drying. Thus, a major part of research on applications of solar energy in processing has been devoted to various dryers designed for drying of fruits/vegetables. A large variety of dryers are manufactured and available which use solar energy for working. These dryers are generally classified in two types; passive and active type with both types containing subtypes of indirect, direct, and hybrid dryers. All the types have their own merits and limitations, so they are designed and fabricated according to local climate conditions and availability of materials. Apart from drying, solar energy can also be used in a concentrated form in solar furnaces. Solar energy has also good potential in being used for greenhouse operations; both in active form (use of solar PV for greenhouse lighting) and in passive form (for greenhouse ventilation). Similarly, solar energy can be used for lighting and ventilation of processing units. The use of solar cookers for cooking food and vegetables is also widespread and large numbers of suitable designs are available for local climate conditions. Apart from these applications, solar energy has also been used in refrigeration, that is, storage of fruits/ vegetables in cold storage. A refrigeration unit working on solar energy will provide ideal storage conditions for fruits/vegetables at places where grid electricity is not available. Thus, it can form a cold chain in which the food product is stored at low temperature right from where it is harvested till it is consumed without affecting its quality.

KEYWORDS

- greenhouse dryer
- photovoltaic
- solar air heater
- solar cooker
- solar dryer
- solar furnace
- solar refrigeration

REFERENCES

1. Aiswarya, M. S. Economic Analysis of Solar Dryer with PCM for Drying Agricultural Products. *Int. J. Sci. Eng.* **2015,** *3*, 124–134.
2. Alghoul, M. A.; Sulaiman, M. Y.; Azmi, B. Z.; Wahab, M. A. Advances in Multi-purpose Solar Adsorption Systems for Domestic Refrigeration and Water Heating. *Appl. Therm. Eng.* **2007,** *27*, 813–822.
3. Aravindh, M. A.; Sreekumar, A. Solar Drying: A Sustainable Way of Food Processing. In *Energy Sustainability Through Green Energy*; Sharma, A., Kar, S. K., Eds.; Springer: New Delhi, India, 2015; pp 27–46.
4. Aravindh, M. A.; Sreekumar, A. An Energy Efficient Solar Drier. *Spice India* **2014,** *27* (5), 10–12.
5. Aravindh, M. A.; Sreekumar, A. Experimental and Economic Analysis of a Solar Matrix Collector for Drying Application. *Curr. Sci.* **2014,** *107* (3), 350–355.
6. Bala, B. K.; Mandol, M. R. A.; Biswas, B. K.; Chowdhury, B. L. D. Solar Drying of Pineapple Using Solar Tunnel Drier. In *4th International Conference on Mechanical Engineering*, 2001, Part I; pp 47–51.
7. Bangotra, A.; Mahajan, A.; Design Analysis of 3 TR Aqua–Ammonia Vapor Absorption Refrigeration System. *Int. J. Eng. Res. Technol.* **2012,** *1* (8), 1–6.
8. Bano, T.; Goyal, N.; Tayal, P. K. Innovative Solar Dryers for Fruits, Vegetables, Herbs and Ayurvedic Medicines Drying. *Int. J. Eng. Res. Gen. Sci.* **2015,** *3* (5), 883–888.
9. Baradey, Y.; Hawlader, M. N. A.; Ismail, A. F.; Hrairi, M.; Rapi, M. I. Solar Drying of Fruits and Vegetables. *Int. J. Rec. Dev. Eng. Technol.* **2016,** *5* (1), 1–6.
10. Chavda, T. V.; Kumar, N. *Indian Renew. Energy Dev. Agency* **2008,** *5* (2), 44–48.
11. Ekechukwu, O. V.; Norton, B. Review of Solar-Energy Drying Systems II: An Overview of Solar Drying Technology. *Energy Convers. Manage.* **1999,** *40*, 615–655.
12. Eswara, A. R.; Ramakrishnarao, M. Solar Energy in Food Processing: A Critical Appraisal. *J. Food Sci. Technol.* **2013,** *50* (2), 209–227.

13. Hegde, V. N.; Hosur, V. S.; Rathod, S. K.; Harsoor, P. A.; Narayana, K. B. Design, Fabrication and Performance Evaluation of a Solar Dryer for Banana. *Energy Sustain. Soc.* **2015,** *5* (23), 1–12.

14. Hughes, B. R.; Oates, M. Performance Investigation of a Passive Solar-assisted Kiln in the United Kingdom. *Solar Energy* **2011,** *85,* 1488–1498.

15. Ibarz, A.; Barbosa-Canovas, G. V. *Unit Operations in Food Engineering*; CRC Press: Boca Raton, FL, 2003; p 550.

16. Jain, D.; Tiwari, G. N. Thermal Aspects of Open Sun Drying of Various Crops. *Energy* **2003,** *28,* 37–54.

17. Janjai, S. A. Greenhouse Type Solar Dryer for Small-Scale Dried Food Industries: Development and Dissemination. *Int. J. Energy Environ.* **2012,** *3* (3), 383–398.

18. Krishna, H. O.; Jihin, C.; Thoufeer, K. V.; Mohammed, D. P. N. M.; Sujith, R. P.; Shoukathali, K.; Muhammed, S. K. Design and Fabrications of a Multi-Pass Multi-rack Solar Dryer. *Int. J. Emerg. Technol. Adv. Eng.* **2015,** *5* (3), 465–473.

19. Kumar, C.; Chaurasia, P. B. L.; Singh, H. K.; Yadav, S. S. Enhancement of Shelf-life of Vegetables by Advance Solar Passive Cool Chamber. *Int. J. Sci. Eng. Technol.* **2015,** *3* (4), 852–858.

20. Mujumdar, A. S. *Handbook of Industrial Drying*; Taylor & Francis Group: London, UK, 2007; pp 4–24.

21. NCDD. *Guidelines and Minimum System Standards for Implementation of Cold-Chain,* 2015; pp 68–74. http://www.nccd.gov.in/PDF/NCCDGuidelines2014-15.pdf (accessed Sept 17, 2016).

22. Norton, B. Industrial and Agricultural Applications of Solar Heat. In *Comprehensive Renewable Energy*; Sayigh, A., Eds.; Elsevier: Amsterdam, 2012; pp 568–593.

23. Palaniappan, C. In *Perspectives of Solar Food Processing in India,* International Solar Food Processing Conference, Indore, India; Solar Processing Network: Indore, India; Jan 14–16, 2009 (Online).

24. Rajeshwari, N.; Ramalingam, A. Low-cost Material Used to Construct a Box-type Solar Dryer. *Arch. Appl. Sci. Res.* **2012,** *4* (3), 1476–1482.

25. Rathore, N. S. In *Low Cost Solution of Solar Drying for Spices Processing at Domestic, Farm and Industrial Level,* Proceedings of National Seminar on Emerging Trends in Spice Processing and Its Impact on Rural Economy; Department of Processing and Food Engineering, College of Tech. and Eng., MPUAT: Udaipur, Rajasthan, India, 2011; pp 24–31.

26. Ringeisen, B.; Barett, D. M.; Stroeve, P. Concentrated Solar Drying of Tomatoes. *Energy Sustain. Dev.* **2014,** *19,* 47–55.

27. Schwarzer, K.; Vierada Silva, M. E. Characterization and Design Method of Solar Cookers. *Solar Energy* **2008,** *82,* 157–165.

28. Singh, P. P.; Singh, S.; Dhaliwal, S. S. Multi-shelf Domestic Solar Dryer. *Energy Convers. Manage.* **2006,** *47* (13–14), 1799–1815.

29. Sreekumar, A. Techno-economic Analysis of a Roof-Integrated Solar Air Heating System for Drying Fruit and Vegetables. *Energy Convers. Manage.* **2010,** *51,* 2230–2238.

30. Sukhatme, S. P.; Nayak, J. P. S*olar Energy: Principles of Thermal Collection and Storage*; Tata McGraw Hill Education Pvt. Ltd.: New Delhi, India, 2010; pp 37–68.

31. Tiwari, G. N. *Greenhouse Technology for Controlled Environment*; Narosa Publishing House: New Delhi, India, 2003; pp 5–27.

32. Van Arsdel, W. B. Food Dehydration. *Food Technol.* **1965,** *19,* 484–487.

33. Wang, L. *Energy Efficiency and Management in Food Processing Facilities*; CRC Press: Boca Raton, FL, 2009; pp 4–39.
34. Wankhade, P. K.; Sapkal, R. S.; Sapkal, V. S. Innovations in Drying for Preservation of Fruits and Vegetables. *Int. J. Sci. Res.* **2013**, *2* (12), 151–153.

APPENDIX A

Cost-Economics of Aonla Dried in Domestic Solar Dryer.

Description	Domestic solar dryer
Initial investment	2000 INR (US$ 33.3)
Purchase price of Aonla	INR 3 (US$ 0.05) per kg
Weight of dried product	1/10th of wet Aonla
Cost of dried product at home	INR 30 (US$ 0.5) per kg
Cost of branded dried product in market	INR 140 (US$ 2.33) per kg
Quantity dried in two drying days	5 kg of Aonla
	0.5 kg of dried Aonla
Saving in two drying days	INR 40 (US$ 0.66) per batch
Saving per drying days	INR 20 (US$ 0.33) per drying day
Payback period	100 drying days

Cost-Economics of Aonla Dried in Farm Solar Dryer and Solar Tunnel Dryer.[25]

Description	Farm solar dryer	Solar tunnel dryer
Initial investment	5000 INR (US$ 83.3)	100,000 INR (US$ 1666.66)
Purchase price of Aonla	INR 3 (US$ 0.05) per kg	INR 3 (US$ 0.05) per kg
Quantity loaded in one batch	20 kg	1000 kg
Total purchase price	60 INR (US$ 1)	3000 INR (US$ 50)
Other investment before drying	20 INR (US$ 0.33)	2000 INR (US$ 33.3)
Weight of dried product in one batch	2 kg	100 kg
Drying time of one batch	2.5 days	2 days
Earning/batch due to drying of Aonla	280 INR (US$ 4.66)	14,000 INR (US$ 233.33)
Savings per batch	200 INR (US$ 3.33)	9000 INR (US$ 150)
Earning/drying day due to drying of Aonla	80 INR (US$ 1.33)	4500 INR (US$ 75)
Payback period	63 days	22 days

APPENDIX B

Data on Solar Dehydration of Selected Foods.

Food type	Drying time (h)	Yield (%)	Ambient temperature (°C)	Cabinet temperature (°C)
Amla	6.5	32	31	50
Bitter gourd	6	11	26	42
Carrot	10	15	31	51
Coconut	5	5	31	50
Coriander leaves	6	12	30	51
Curry leaves	8	35	29	55
Drumstick leaves	5.5	15	29	55
Fenugreek leaves	6	13	27	40
Grapes	25	20	31	53
Green chilies	6	12	40	25
Guava bar (10 mm thick)	35	45	31	48
Mango bar (10 mm thick)	20	45	40	65
Mint leaves	5	17	29	55
Mushrooms	12	15	33	50
Onion	18	17	31	51
Papaya bar (10 mm thick)	20	45	30	51
Pineapple bar (10 mm thick)	20	45	30	51
Potatoes	4	30	31	50
Red chilies	15	34	32	56
Sapota bar (10 mm thick)	20	36	34	42
Sapota slices	8	27	34	49
Spinach leaves	15	8	29	55
Tamarind leaves	12	11	29	55
Tomato	10	10	33	60

CHAPTER 4

MANUFACTURING OF JAMS, JELLIES, AND MARMALADES FROM FRUITS AND VEGETABLES

TANYA LUVA SWER, SAVITA RANI, and KHALID BASHIR

ABSTRACT

Daily consumption of fresh fruits and vegetables is recommended for significantly reducing the risk of chronic diseases and to meet the micronutrient requirements. However, fruits and vegetables have very poor shelf life owing to their high moisture content and are subjected to seasonal and regional availability, which restrict the longer terms utilization of these commodities. To overcome these drawbacks, different processed products such as jams, jellies, and marmalades are made extensively at home as well as on commercial level with different fruits and vegetables sources. Therefore, acquiring the proper knowledge about the manufacturing of different processed products can be extended to value addition and increased variety of products with prolonged shelf life.

4.1 INTRODUCTION

Fruits and vegetables are consumed by human since centuries as a source of food and for therapeutic values. Their therapeutic properties have been attributed to the presence of various phytochemicals and bioactive compounds and dietary nutrients, namely, vitamins, fiber, minerals, etc. Their daily consumption is highly recommended for significantly reducing the risk of many chronic diseases.

India ranks second in the world for fruits and vegetables production.[5] The unique diverse geographical conditions across the regions in the country together with diverse agroclimatic conditions and consistent efforts using science and technology enables availability and increasing production of wide varieties of fruits and vegetables in the country.

The world production of fruits and vegetables is about 608.926 and 1044.38 MT in 2010 to which India's contribution is about 75.121 and 100.405 MT, respectively.[3] This offers the country tremendous potential for export and huge opportunity for investment in the food processing sector. In India, out of the total production of fruits and vegetables, nearly 76% is consumed in fresh form. There are certain limitations in utilization of fresh fruits and vegetables because these commodities are highly perishable. Moreover, their availability is subjected to seasonal and regional factors. Furthermore, the presence of high moisture in fruits and vegetables makes them very susceptible to spoilage.

A large quantity of fruits and vegetables are spoiled every year due to poor postharvest management and supply chain management especially in the developing countries. In India, wastage and postharvest losses of fruits and vegetables accounts to about 20–22%, which is worth 2000 billion Rs. per year. Therefore, there is a need to preserve these produces in the form of value added processed products. However, only 2% of vegetables production and 4% of fruits production are being processed in the country.[1,8] This is in big contrast when compared to several other developing countries in which processing of fruits and vegetables have developed to an extent of 70% in Brazil, 83% in Malaysia, 78% in Philippines, and 30% in Thailand.[5]

Various methods can be adopted for preserving and processing of fruits and vegetables that are made available in the market today. Out of these, the process for manufacturing of jams, jellies, and marmalades using fruits, sugar, pectin, and edible acids is one of the oldest food preserving processes and presents a way of making food stable by increasing the content of soluble solids.

This chapter focuses on the processing of fruits and vegetables into jam, jellies, and marmalade and enlists important features to be ensured during processing to obtain satisfactory products.

4.2 OVERVIEW OF JAM, JELLIES, AND MARMALADES

Jams, jellies, and marmalades are very similar to one another and are made from fruits, preserved by infusing with sugar, and thickened either by boiling till concentrated or by using thickening agents such as pectin. These products differ in the degree of gel attained, manner of preparation of fruits, and ingredient composition. These products are collectively referred to as preserves. Although the exact date is unknown, yet the art of making these preserves probably dates back to centuries ago during the middle ages in the Middle Eastern countries where the sugarcane grew naturally. However, it was only in 1541 when the first book to describe the art of boiling jam and preserving fruits was published by Venedig and Lyon in France with the title "Le Bastimentdecrettes." Soon after, many more books on the art of producing these preserves soon followed.[4] In India, it was only during the World War II, that large quantities of these preserves were imported from the United States, United Kingdom, and Australia. Today, these preserves form an important class of food products and are being manufactured extensively in the food processing industries and prepared even at household level throughout India.[7]

4.3 PRINCIPLE OF PRESERVATION: JAM, JELLIES, AND MARMALADE

The preservation of jam, jellies, and marmalades occurs due to the high concentration of sugar (>65% or 68%), which acts as a preservative. Moreover, the moisture becomes bound, thereby reducing the water activity and hence creating an unfavorable condition for microbial growth.

4.4 JAM

Jam is prepared by boiling the fruit pulp with sugar, acid, and pectin to a suitable consistency firm enough to hold the fruit tissue in position. Jam can be made from fruits and vegetables. Fruits such as apple, aonla, mango, pear, papaya, strawberry, and pineapple are commonly used for jam making. Jam can be made from single fruit or combinations of fruits. Tutti-frutti is the mixed fruit jam prepared from the combination of mango, pear, apple, and pineapple fruits in equal proportion. Jam contains 0.5–0.6% acid and invert sugar should not be less than 40%.[12] For making the jam, usually 45 parts of pulp are used with every 55 parts of sugar. Table 4.1 indicates composition of different ingredients for jam making. FPO specifications for jam are[11]

- minimum percent of total solid in final product 68, and
- minimum percent of fruit juice in final product should be 45%.

TABLE 4.1 Composition of Different Ingredients for Jam Making.

Fruit/Vegetable	Ingredients for 1 kg fruit/vegetable pulp			Reference
	Sugar (kg)	Citric acid (g)	Water (mL)	
Apple	0.75	2.5	150	
Aonla	0.75	1.5	150	
Mango	0.75	2.0	100	
Pear	0.75	2.0	100	
Loquat	0.75	1.5	150	
Papaya	0.75	3.0	100	[11]
Strawberry	0.75	2.0	100	
Goose berry	0.75	1.5	100	
Pineapple	0.75	2.0	150	
Tomato	0.75	2.0	100	
Carrot	0.5	2.0	150	
Apricot	0.60	1.0	100	[12]
Grapes	0.70	1.0	50	
Guava	0.75	2.5	150	
Karonda	0.80	–	100	
Musk melon	0.75	2.5	50	
Plum	0.80	–	150	
Peach	0.80	3.0	100	
Raspberry	0.75	2.0	100	
Sapota	0.75	3.0	150	
Mixed jam (papaya + pineapple pulp in equal amounts or pineapple + guava + mango pulp in equal amount)	0.80	2.5	100	
Tutti-frutti jam (mango + pear + apple + pineapple in equal amount)	0.80	2.5	100	

4.4.1 PREPARATION OF JAM

The step-by-step procedure for preparation of jam is given in Figure 4.1. Jams are prepared using ground fruit/vegetable and it shows generally thick consistency because of high pectin content. Following steps are followed during preparation of jam[10,12]:

a.   *Selection of fruits:* Fruits to be used for jam production should have reached full maturity to possess a rich flavor and be of most desirable texture.

b.   *Preparation of fruit pulp:* Fruits are treated in various ways for pulp preparation such as all berries must be carefully sorted and washed; strawberries must be steamed; peaches, pear, apples, and other fruits with heavy skins must be peeled; while apricot, plums, and other thin-skinned fruits do not require peeling. Apricots, plums, and fresh prunes can be pitted by machine, such as the Elliott pitter.[2] Softening of firm fruits (plums, apricots, peaches, and apple) are done by heating them in 10% water and finally passing through the pulper to separate skin and stones, etc. It is also necessary to remove the SO_2 when the jam is prepared from the pulps, which have been preserved in 1000–1500 ppm SO_2. Therefore, the preserved pulps are boiled with addition of small quantities of water till the SO_2 is not perceived.

c.   *Addition of sugar:* Cane sugar in the ratio of 45:55 (fruit part:sugar) is added during boiling stage. It is also necessary that the finished product should have 30–50% of invert sugar or glucose so that the crystallization of cane sugar can be avoided during storage. However if the reducing/invert sugars exceed 50%, the jam will develop into a heavy honey like syrupy mass.

Ripe firm fruits

[usually a combination of ripe and unripe fruits should be taken as ripe fruits impart good flavor and unripe fruits are rich in pectin necessary for a good set]

↓

Washing

[done to remove dust, leaves, stalks, or any other undesirable portion]

↓

Peeling

[done for the removal of the skin or rind]

↓

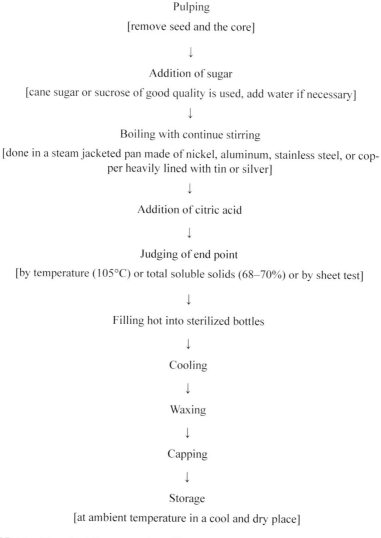

Pulping

[remove seed and the core]

↓

Addition of sugar

[cane sugar or sucrose of good quality is used, add water if necessary]

↓

Boiling with continue stirring

[done in a steam jacketed pan made of nickel, aluminum, stainless steel, or copper heavily lined with tin or silver]

↓

Addition of citric acid

↓

Judging of end point

[by temperature (105°C) or total soluble solids (68–70%) or by sheet test]

↓

Filling hot into sterilized bottles

↓

Cooling

↓

Waxing

↓

Capping

↓

Storage

[at ambient temperature in a cool and dry place]

FIGURE 4.1 Flowchart for preparation of jam.

d. *Addition of acids:* Acid is necessary for setting of pectin to form gel. Some amount of acidity is present in fruits. However, usually additional acid is required for proper setting of gel. The addition of acid reduces the sweetness of the sugar and hence the amount of acid added depends on the required acidity of finished product which in case of jam varies between 0.5% and 0.7%.[9]

e. *Cooking:* The cooking or processing is done to cause intimate mixing of fruit pulp and sugar and partially to concentrate the product by evaporation of excess moisture. Steam-jacketed copper, stainless steel, nickel, or aluminum kettles are commonly used for the processing of jams. However, vacuum evaporators, plate evaporators may also be used sometime for boiling of jam.[2]

f. *Addition of pectin:* Pectin is added to impart thickening and forms gel network, which keep the liquid together leading to setting of the jam. Jam should contain 0.5–1.0% pectin. During jam preparation, 1 part of the pectin is mixed with about 10 parts of sugar to dissolve it uniformly and then is added slowly in the boiling mixture normally after the jam attains 60°Brix.

g. *Addition of colors and essences:* Color and essence are added to improve the esthetic value of the jam to make it more appealing to the consumers. These additives should be dissolved in minimum quantity of water and dispensed dropwise over the almost finished jam with constant stirring.[2]

h. *Determination of the end point of a jam*
   - *Sheet or flake test:* A small portion of jam is withdrawn during boiling using a spoon or wooden ladle and cooled to ambient temperature. It is then allowed to drop. If the product falls off in the form of a sheet or flakes instead of flowing in a continuous stream or syrup, it means that the end point has been reached and the product is ready, otherwise boiling is continued till the sheet is positive.[12]
   - *Total soluble solids:* Total soluble solids (TSSs) of the jam near the endpoint are above 68°Brix. This can be checked by using a hand refractometer. In this method, following steps are followed:
   - First, take a small portion of jam from the pan and allow it to cool to 20°C.
   - Spread one or two drops of jam evenly on the surface of the prism and close it carefully.
   - Hold the refractometer near the source of light and look through the end piece.
   - The line between the dark and light fields will be seen through the viewer. Read the corresponding number on the scale, which is the percentage of sugar in the sample.
   - After the test, sample is removed from the surface of the prism with a piece of tissue paper or wet cotton wool.[10]

- *Boiling test:* Jam is presumed to be ready at a point where the temperature of the boiling mixture reaches 106°C at sea level.
- *Weight test:* At the finish point, the weight of the jam is about 1.5 times the weight of added sugar. For performing this test, the weights of the boiling pan and sugar are to be noted down before starting to cook. The boiling mixture is weighed along with the pan for determination of end point.

i.   *Filling and packaging:* Prepared jam is transferred hot into glass jars. Additionally, 40 ppm of SO$_2$ may be added to ensure the safe preservation of jam during storage.[10–12]

j.   Table 4.2 indicates obstacles in jam preparation and their precautions.[11,12]

TABLE 4.2   Problems in Jam Preparation and the Precautions.

Problem	Causes	Precautions
Crystallization	Jam should have 30–50% invert sugar, if percent of invert sugar is less than 30, then crystallization occurs	Corn syrup or glucose along with cane sugar may be added to avoid this problem
Microbial spoilage	Sometime molds cause spoilage	Add 40 ppm, SO$_2$ in the form of KMS or storage of jam at 80% relative humidity
Premature setting	Low total soluble solids and high pectin content in jam	Add more sugar or by adding a small quantity of sodium bi-carbonate to reduce the acidity and hence prevent coagulation
Sticky jam (gummy jam)	High percentage of total soluble solids	Add pectin or citric acid or both
Surface graining and shrinkage	Evaporation of moisture during storage	Storing in cool place

KMS, potassium metabisulfite

4.5   JELLY

Jelly is a semisolid product obtained by boiling a clear strained solution of pectin containing fruit extract with sugar and acid to a thick consistency. Fruits such as guava, sour apple, plum, karonda, wood apple, loquat, papaya, and gooseberry are generally used for preparation of jelly. Fruits having low pectin content such as apricot, pineapple, strawberry, and raspberry can also be used for jelly making only after addition of sufficient pectin powder. Jelly contains TSSs not less than 65% and acidity 0.5–0.7%. FPO specifications for fruit jelly are[11]

- minimum percentage of TSS in final product 65, and
- minimum percentage of fruit juice in final product should be 45.

The essential components required for preparation of jelly are pectin 0.5–1.0% (normally obtained from fruits), sugar 60–65%, fruit acid 1.0% (minimum 0.5%; optional 0.75%), and water 33–88%.

The protopectin is the water-insoluble substance present in the middle lamella of plant cells. The protopectin is hydrolyzed to water-soluble pectin when boiled in the presence of acids. Pectin is negatively charged particles and solution of pectin solution is more stable in neutral pH range. Increase in acidity or alkalinity decrease its stability. In jelly formation, sugar acts as a precipitating agent and the presence of acid help it. More the quantity of acid, the less sugar is required. Some salts also helps in precipitating pectin.[9,12]

Fruits can be divided into four groups according to their pectin and acid contents (Table 4.3). Further, the specifications for preparation of jam from different types of fruits base on their initial TSS content and acid content are presented in Table 4.4.

TABLE 4.3 Classification of Fruits Based on Their Pectin Content.

Rich in pectin and acid	Rich in pectin but low in acid	Low in pectin but rich in acid	Low in pectin and acid
Sour and crab apple, grape, sour guava, lemon, oranges (sour), plum (sour), and jamun	Apple of low acid varieties, unripe banana, sour cherry, fig (unripe), pear, ripe guava, peels of orange, and grapefruit	Apricot (sour), sweet cherry, sour peach, pineapple, and strawberry	Ripe apricot, peach (ripe), pomegranate, raspberry, strawberry, and any other over-ripe fruit

TABLE 4.4 Specifications for Preparation of Jelly from Various Fruits.

Fruit	Ingredients for 1 L extract	
	Sugar (kg)	Citric acid (g)
Gooseberry	0.80	—
Guava	0.75	3.0
Jamun	0.75	1.0
Karonda	0.75	—
Loquat	0.80	2.0
Papaya	0.75	3.0
Plum	0.75	2.5
Sour apple	0.75–1.00	2.0
Wood apple	1.00	—

4.5.1 PREPARATION OF JELLY (Fig. 4.2)

a. *Selection of fruits:* Fruits rich in pectin and acid such as guava, apples, lemons, and oranges should be utilized for jelly making. If the pectin or acid is deficient in fruit, it should be supplemented with the pectin or acid rich fruits. Both ripe and underripe fruit should be used because the earlier yields good flavor and the later good pectin. It is better to avoid overripe fruits because they may disintegrate during boiling and can affect the clarity of extract adversely.[11,12]

b. *Extraction of pectin:* Wash the fruits thoroughly and cut into pieces of about 1/8th to 1/4th of an inch in thickness for pectin extraction. About 1/2th to an equal volume of water is added to fruits in case of apples, whereas citrus fruits need 2–3 vol of water for each volume of sliced fruit due to prolong boiling required. Stainless steel utensils should be utilized for the preparation of jelly. However, fruits like grapes and berries do not require any addition of water. Fruits vary in their boiling time which is 20–25 min for apple, 5–10 min for grapes and other berries, 45–60 min for oranges, and 30–35 min for guava.

c. *Straining and clarification of extract:* Strain the boiled mixture through the coarse muslin cloth several times to get a clear pectin extract.

d. *Determination of pectin content of extract*

   i. *Alcohol test:* One spoon of cooled pectin extract is combined with three spoons of methylated spirit. Mix the solution well and kept undisturbed for 5 min. Observe and interpret the results based on the data in Table 4.5.

   ii. *Jelmeter test:* This method involves flow of pectin extract through a jelmeter on which graduations are marked as 1/2, 3/4, 1, and 1/4 exactly for 1 min. The level up to which the material has moved direct the ratio of extract and sugar to be added as 1:1/2, 1:3/4, 1:1, and 1:1/4.[11]

TABLE 4.5 Observations and Interpretation of the Alcohol Test for Determination of Pectin Content of Extract.

Type of clot formed	Pectin content	Recommended ratio of extract:sugar
Single transparent lump	Rich	1.00:1.00
Less firm and fragmented	Moderate	1.00:0.75
Numerous small granular clots	Poor	1.00:0.50

e. *Addition of sugar and acid:* Sugar is the essential component of jelly as it imparts sweetness and body to the product. Quantity of sugar to be added is calculated based on above described tests. Too high concentration of sugar can result in the formation of stiff jelly probably resulted from dehydration. Jelling of extract depends on the amount of acid and pectin present in the fruits. Tartaric acid found in fruits gives the best results compared to citric and malic acid. Final jelly should have 0.5–0.75% of total acid as higher acid will cause the syneresis problem. Strength of jelly increases with the increase in pH until optimum level is reached. The optimum pH for jelly containing 60%, 65%, and 70% sugar is approximately 3.0, 3.2, and 3.4, respectively.

f. *Cooking:* Boiling if continue for a long time can result in the formation of syrupy jelly and darken color. For reducing the boiling time, pectin extract should be brought to boil first followed by addition of sugar in it. During boiling, the scum formed is removed now and then when it rises to the top. Problem of foaming can be eliminated by addition of 1 table spoon of edible oil per 50 kg of boiling jelly. For proper inversion of sugar, the boiling should be completed within 20 min.[9]

g. *Judging of end point:* Prolonged boiling of jelly can result in a greater inversion of sugar and destruction of pectin. Therefore, it is necessary to judge the end point of jelly correctly by the methods described below:

- *Sheet or flake test:* Already described under jam preparation.

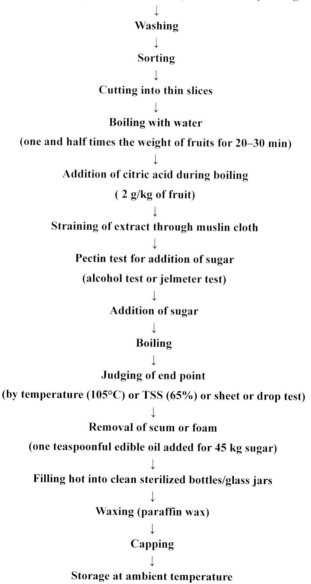

Firm not overripe fruit

(usually a combination of ripe and unripe fruits should be taken as ripe fruits impart good flavor and unripe fruits are rich in pectin necessary for a good set)
↓
Washing
↓
Sorting
↓
Cutting into thin slices
↓
Boiling with water

(one and half times the weight of fruits for 20–30 min)
↓
Addition of citric acid during boiling

(2 g/kg of fruit)
↓
Straining of extract through muslin cloth
↓
Pectin test for addition of sugar

(alcohol test or jelmeter test)
↓
Addition of sugar
↓
Boiling
↓
Judging of end point

(by temperature (105°C) or TSS (65%) or sheet or drop test)
↓
Removal of scum or foam

(one teaspoonful edible oil added for 45 kg sugar)
↓
Filling hot into clean sterilized bottles/glass jars
↓
Waxing (paraffin wax)
↓
Capping
↓
Storage at ambient temperature

FIGURE 4.2   Flowchart for the preparation of jelly. Modified significantly from Refs. [11, 12].

- *Drop test:* A drop of the concentrated mass is poured into a glass containing water. Settling down of the drop without disintegration denotes the end point.

- *Temperature test:* A solution containing 65% TSSs boils at 105°C. Heating of the jelly to this temperature would automatically bring the concentration of solids to 65%. This is the easiest way to determine the end point.[12]

h. *Packing:* Jelly is filled hot (85°C) into glass jars followed by its cooling to room temperature. Any appearance of foam on the top of jar is removed by spoon.[9]

4.5.2   QUALITIES OF A GOOD JELLY

Proper preparation of jelly can be judged by the following qualities[11]:

- Clear.
- Transparent.
- Sparkling.
- Attractive in color.
- It should not be gummy, sticky, or syrupy or have crystallized sugar.
- It should be free from dullness, with little or no syneresis.
- It should neither be tough nor rubbery.

4.5.3   OBSTACLES[11] AND CAUSES IN JELLY PREPARATION (SEE TABLE 4.6)

TABLE 4.6   Problems and Causes in Jelly Making.

Sr. no.	Problems	Possible causes
	Failure to set	1. Cooking below the end point 2. Lack of acid or pectin 3. Cooking beyond the end point 4. Slow cooking for a long time
	Cloudy (foggy) jellies	1. Use of immature fruit 2. Use of nonclarified juice 3. Overcooking 4. Overcooling 5. Nonremoval of scum

TABLE 4.6   *(Continued)*

Crystal formation	1. Use of excess sugar
Syneresis or weeping	1. Excess of acid 2. Too much concentration of sugar 3. Insufficient pectin 4. Premature gelatin
Fermentation	Caused by mold due to 1. Not covering the jelly properly 2. Breakdown of paraffin seal
Color changes	1. Storing in too warm place 2. Imperfect jar seal 3. Trapped air bubbles causing oxidative changes
Gummy jelly	1. More inversion of sugar due to overcooking
Stiff jelly	2. Overcooking 3. Excess of pectin

4.5.4 DIFFERENCES BETWEEN JAM AND JELLY (SEE TABLE 4.7)

TABLE 4.7   Differences Between Jam and Jelly.

Jam	Jelly
Prepared from fruit pulp	Prepared from clear fruit pectin extract
The final product obtained is not transparent	The final product obtained is transparent
Final TSS should not be less than 68% TSS as per FSSAI specifications	Final TSS should not be less than 65% TSS as per FSSAI specifications
Flavors and color are added to enhance the acceptability of the product	Original flavor of the fruit is preferred
Optimal pH of jam is 3.35–3.70	Optimal pH of jam is 3.30

4.6 MARMALADE

Marmalade is a semisolid or gel-like product similar to jelly prepared from citrus fruit together with other ingredients, namely, sweeteners, acids, pectin, and citrus fruit peel shreds. The ingredients are concentrated by cooking to such a point that the soluble solids content of the finished marmalade is not less than 65%. The peel shreds impart bitterness to the marmalade which gives the products distinct desirable characteristics.[7] It is to be noted that marmalades usually have slightly higher pectin and acid content compared to jelly. Generally, marmalade can be classified into three kinds: sweet marmalade, bitter marmalade, and sweet and bitter marmalade based on the varieties of oranges used for preparation of the marmalade.[13]

- *Sweet marmalade*: This is prepared using sweet varieties of oranges such as Navel and Valencia or other commercial dessert varieties other than tangerine. Thirty parts by weight of orange fruit ingredient to 70 parts by weight of sweetening ingredient are used for the preparation of sweet marmalade.
- *Bitter marmalade:* This is prepared using Seville or sour type of oranges other than tangerines. This type of marmalade is prepared using not less than 25 parts by weight of fruit ingredient to 75 parts by weight of sweetening ingredient.
- *Sweet and bitter marmalade:* This is prepared by blending 50% by weight of sweet oranges and bitter oranges other than tangerines. This type of marmalade is prepared using not less than 30 parts by weight of fruit ingredient to 70 parts by weight of sweetening ingredient.

Again, marmalades are further classified as jelly marmalades and jam marmalades based on their physical appearance. However, these two types of marmalades are prepared following the method of preparation of jellies.

4.6.1 PREPARATION OF JELLY MARMALADE

a. *Selection of fruits:* One should only use sun-ripened fruits. Fruits selected should be free of blemishes and other physical damage irrespective of size.[7]
b. *Preparation of the fruits:* The yellow portion of the peel is peeled off thinly with care so that very little of the white portion is removed. Meanwhile, the peeled fruits along with the white portion are cut into slices of 0.3–0.45 cm thickness or crushed into a pulp to facilitate the extraction of pectin.[7,9]
c. *Extraction of pectin:* The sliced or crushed fruits are boiled with 2–3 times their weight of water till sufficient pectin has been extracted, which usually takes about 45–60 min.[6,7] The process may be repeated a second or third time with relatively lesser amount of water is desired. If fruits are extracted more than one time, the different extracts obtained should be pooled out together and mixed to get uniform quality. The extract should have at least 1% acidity as citric acid.[9]

    d.   Clarification of pectin extract: In small-scale units, the pectin extract is placed in an aluminum or stainless steel vessel and let to stand overnight allowing the sediment to settle down therefore leaving a clear supernatant juice (Fig. 4.3).

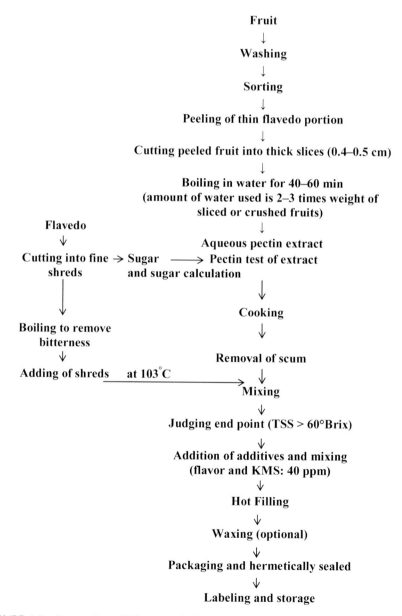

FIGURE 4.3   Preparation of jelly marmalade.

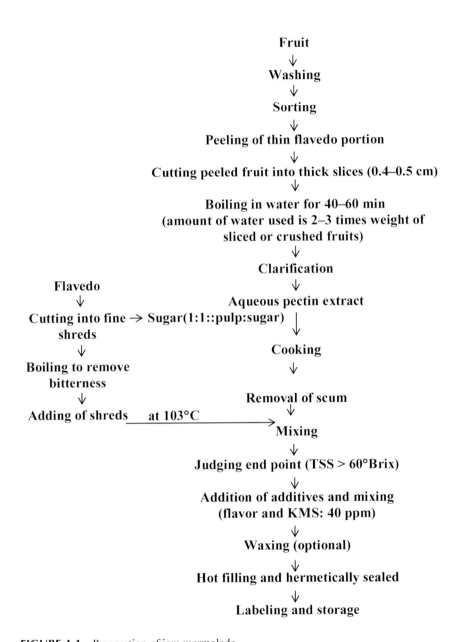

FIGURE 4.4   Preparation of jam marmalade.

e.   The clear juice is decanted carefully without disturbing the sediment. For large-scale production, the aqueous pectin extract is *separated and filtered* using filter aids and passed through jelly bags, or pumped through a plate-and-frame filter press to get a clear liquid.[6,7,9]

f. *Preparation of peel shreds:* Meanwhile, the thin yellow peel is cut into fine shreds of 1.9–2.5 cm long and 0.8–1.2 cm thick.[7] These shreds are boiled and water drained completely to remove bitterness before adding them to the marmalade.[6] The shreds may be kept in heavy syrup for sometimes to increase their bulk density to avoid floating on the surface of the marmalade.[9]

g. *Cooking:* The extract is brought to boiling in a steam-jacketed pan and sugar was added simultaneously at the desired requisite quantity. The requisite quantity of sugar to be added can be determined by testing the pectin extract with alcohol test or using jelly meter as performed during preparation of jelly.[7] When the mixture reaches the 103°C, then the prepared shreds are added to it. The amount of shreds suggested to be added to the pectin extract varied among different authors. Lal et al.[7] suggested addition of shreds at the rate of about 62 g to every kilogram of the original extract, whereas Sharma[9] indicated addition of prepared shreds of about 5–7% to the original weight of the pectin extract. Boiling is continued until the jellying point is reached, which can be determined in same manner as during preparation of jelly. The boiling process should not take more than 20 min.

h. *Cooling*: When the marmalade is ready, it is cooled in a shallow pan or in a water-cooled pan with continuous stirring to avoid settling down of shreds and maintain uniform distribution of the shreds in the marmalade. Marmalade is allowed to cool till temperature reaches 85°C and a thin slim is forms on the surface of the product which then prevents the shreds from coming to the surface.[6,7,9]

i. *Addition of additives*: During processing or cooking, the natural fruit flavors from the marmalade are lost due to their volatile nature. Therefore, a small quantity of orange oil or flavor may be added to the marmalade before filling into the containers to maintain the acceptability of the marmalade. Generally, potassium metabisulfite (KMS) is also added to avoid darkening and mold growth during storage, provided the product is not packed in tin containers (9 g KMS is added to 100 kg of marmalade resulting in 40 ppm of SO_2 in the product as per specifications).[7]

j. *Packaging:* After cooling to 85°C, the marmalade is filled into glasses or jars, which can be closed airtight, or is packed in cans which are hermetically sealed.[7]

4.6.2   PREPARATION OF JAM MARMALADE

The process for preparation of jam marmalade is exactly similar to that for making jelly marmalade except that clarification of the pectin extract of the fruit is not required in this method (Fig. 4.4). The peel is thinly removed leaving the white portion intact to the fruits. The fruits are prepared as described for preparing jelly marmalade. Here, the whole of the pulp mass is used. Extraction of pectin is done by boiling the prepared fruits with little portion of water until soft and sufficient amount of pectin has been extracted. Seeds and coarse particles are separated by passing the boiled fruit through a coarse sieve or pulper. Meanwhile the shreds are prepared as in jelly marmalade and then added to the sieved pulp. The mixture is boiled together with the requisite amount of added sugar to produce jam marmalade.

Another important consideration is that the pectin test is not required to be performed here as the test will not help to determine the amount of sugar needed. Moreover, since the whole pulp is used, marmalade with thicker consistency will be obtained. Therefore, in the case of jam marmalade, sugar is added on the basis of the weight of the fruit taken which is generally in the proportion of 1:1. The pulp sugar mixture is cooked until the marmalade contains 65% sugar. After cooking, a small quantity of orange oil is added to enhance the flavor of the marmalade. When ready, the marmalade is filled hot into cans which are then sealed hermetically. The cans are inverted for 5–10 min to sterilize the lids as well. Hence, in this case, no separate sterilization process is required.[6,7]

4.7   SUMMARY

Daily consumption of fruits and vegetables is highly recommended for significantly reducing the risk of many chronic diseases. Although preferred mostly in their fresh form, yet fruits and vegetables are susceptible to physical and pathological damage leading to reduced storage life. Appropriate processing techniques can help to preserve the quality and extend the storage life of fruits and vegetables. Different kinds of processed products are available in the market. Jams, jellies, and marmalades are processed products obtained from fruits and are collectively known as preserves. The processing of jams, jellies, and marmalades using fruits, sugar, pectin, and edible acids is one of the oldest food preserving processes known to mankind, where preservation is achieved using sugar as the preservative.

These products are very similar to one another but differ in the degree of gel attained, manner in which they prepared, and ingredient composition. Today these products form an important class of food products and are being manufactured extensively and prepared even at household level throughout the world. With the growing income of families and increasing trend toward using processed food products, the demand for jam, jellies, marmalade is in positive trade.

KEYWORDS

- **clarification**
- **flavedo**
- **gel**
- **jelmeter**
- **marmalade**
- **pectin**
- **total soluble solids**

REFERENCES

1. ASSOCHAM. *Horticulture Sector in India-state Level Experience*; The Associated Chambers of Commerce and Industry of New Delhi: India, 2013; p 50.
2. Cruess, W. V. *Commercial Fruit and Vegetable Products*; Agrobios: Jodhpur, India, 2009; pp 377–427.
3. FAO Statistical Yearbook 2013. *World Food and Agriculture*; Food and Agriculture Organization of the United Nations: Rome, 2013; pp 165–168.
4. Hedh, J. *The Jam and Marmalade Bible: A Complete Guide to Preserving*; Skyhorse Publishing Inc.: New York, 2013; pp 13–20.
5. Indian Horticulture Board, Ministry of Agriculture, Government of India. *Indian Horticulture Database*; Indian Horticulture Board, Ministry of Agriculture, Government of India, 2013.
6. Kale, P. N.; Adsule, P. G. Citrus (Chapter 3). In *Handbook of Fruit Science and Technology: Production, Composition, Storage, and Processing*; Salukhe, D. K., Eds.; Marcel Dekker Inc.: New York, 1995; pp 39–66.

7. Lal, G.; Siddappa, G. S.; Tandon, G. L. Jams, Jellies and Marmalades (Chapter 11). In *Preservation of Fruits and Vegetables*; ICAR: New Delhi, India, 1986; pp 156–197.

8. Sharma, H. P.; Sharma, S.; Vaishali; Prasad, K. Application of Non-Thermal Clarification in Fruit Juice Processing: A Review. *South Asian J. Food Technol. Environ.* **2015,** *1* (1), 15–21.

9. Sharma, S. K. Jam, Jelly, Marmalade and Preserve (Chapter 9). In *Postharvest Management and Processing of Fruits and Vegetables*; New India Pub. Agency: New Delhi, 2010; pp 132–149.

10. Sharma, S. K.; Nautiyal, M. C. *Postharvest Technology of Horticultural Crops*; New India Pub. Agency: New Delhi, 2009; pp 49–60.

11. Singh, I. S.; Singh, V. Jam, Jelly, Marmalade and Butter (Chapter 15). In *Post-Harvest Handling and Processing of Fruits and Vegetables*; Westville Pub. House: New Delhi, 2009; pp 137–142.

12. Srivastava, R. P.; Kumar, S. *Fruits and Vegetables Preservation: Principles and Practices*; International Book Distributing Co.: Lucknow, India, 2002; pp 215–225.

13. US Department of Agriculture. *United States Standards for Grades of Orange Marmalade*; US Department of Agriculture: Washington, DC, 1974; p 53.

PART II

Novel Processing Technologies: Methods and Applications

CHAPTER 5

ULTRAVIOLET LIGHT TREATMENT OF FRESH FRUITS AND VEGETABLES

BAZILLA GAYAS, BEENA MUNAZA, and
GAGANDEEP KAUR SIDHU

ABSTRACT

Fresh-cut fruits and vegetables exposed to atmospheric condition lead to rapid degradation. On this ground, promising technologies for sustaining produce quality and to inhibit the harmful microbial growth are being investigated to obtain longer shelf life. Nowadays, treatment with ultraviolet (UV) light is commonly used in food processing industry as it is lethal to most types of microorganisms and leaves no residue in the treated food. Postharvest fruit senescence has been delayed in many fruits when UV-C light was used as a treatment. In minimally processed fruits, use of UV-C light has increased the levels of polyphenols and flavonoids. Exposure of UV-C light activates the synthesis of health-promoting compounds in fruits and vegetables such as anthocyanin and stilbenoids.

5.1 INTRODUCTION

Fruits and vegetables have important role in human diet because of their health and nutritional benefits throughout world. According to WHO/FAO Report [2004], approximately 400 g of fruits and vegetables daily are recommended as preventive measure of many diseases such as obesity, diabetes, heart disease, and cancer.[76] However, fresh fruits and vegetables are more susceptible to spoilage by common foodborne pathogens such as *Salmonella*, *Shigella*.[47] Fresh-cut fruits and vegetables exposed to atmospheric condition lead to rapid degradation. On this ground, promising technologies for sustaining produce quality and to inhibit the harmful microbial growth are being investigated to obtain longer shelf life. Thermal technologies are the one which is being widely used for this purpose but these adversely affect the vitamin content as well as quality characteristics. Nonthermal technologies are alternate to this, which has no undesirable effect on sensory or nutritive properties of food produce.

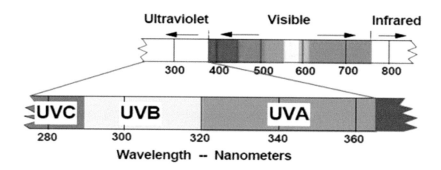

FIGURE 5.1 (See color insert.) Ultraviolet spectrum: Courtesy of Colorado State University, UV-B monitoring & Research Program funded by USDA (US Government). *Source*: http://www.cannabisgreen.com/evolutionary-effects-ultraviolet-radiation-b-power-psychoactive-cannabis/.

Nowadays, treatment with UV (ultraviolet) light is used in food processing industry as it is lethal to most types of microorganisms and no residues are left in treated food.[70] This treatment has been approved by the code of Food and Drug Administration (FDA) in the United States as it brings some benefits to fresh-fruit-cut industry.[56] UV light, part of electromagnetic spectrum,[66] occupies a wide band of wavelength between X-ray (200 nm) and visible light (400 nm). The spectrum (Fig. 5.1) has electromagnetic waves having frequencies higher than those that human eye recognize as violet color. UV light is usually classified into three types: short wave 200–280 nm UV light C (this range has germicidal effect and inactivates bacteria as well as viral microorganisms); medium wave 280–320 nm UV light B (this range can cause skin burning); and long wave 320–400 nm UV light A (causing change in human skin tanning).[72] Each band has different biological effects in crops.[64]

Majority of information in respect of biological effect of UV radiation is obtained from artificial UV at 254 nm radiations.[70] At 254 nm, UV irradiation is at its maximum and exhibits germicidal properties. It causes the formation of pyridine dimmers, which change and prevent microbial cell replication.[37]

The production of PAL is stimulated by UV treatment, which has important role in the synthesis of phytoalexins, phenolic compounds thus improving the resistance of fruits and vegetables to microorganism.[29] Also, the effect of UV radiation on microorganism may differ from species to species.[16] UV light has proven advantageous to delay postharvest fruit senescence and control decay of fruit and vegetable.[11]

On exposing fruits to low UV dose, various changes including formation of antifungal agent are induced in fruits. UV treatment can produce changes in chemistry of plants and in some cases enhances the nutraceutical properties of plant foods. Short UV-C is entirely absorbed in air within a few hundred meters. On collision of UV-C photons with oxygen, ozone is formed as result of exchange of energy. UV-C is absorbed so quickly, it is never observed in nature. UV light emitted within a UV reactor reacts with reactor components (such as reactor walls, lamp, and lamp sleeve) and the substance which is to be treated.

This chapter is focused on the application of UV light treatment to improve the quality and shelf life of fresh fruits and vegetables.

5.2 SOURCES OF ULTRAVIOLET LIGHT

UV light sources consist of a UV reactor either with open or closed vessel that contains UV lamps. UV intensity sensors, flow meters, and sometimes UV transmittance monitors are used to monitor dose delivery given to produce by the reactor. The characteristics of UV lamps, which are most important during treatment, include temperature during operation, electrical input (W/cm), germicidal UV output (W/cm), arc length, and rated lifetime (h).

Previously low-pressure mercury UV lamps were used having a range of 254 nm for treatment of food liquid and food-surface disinfection.[69] These lamps are monochromatic. Mercury-emission lamps contain a small quantity of elemental mercury (Hg) and an inert gas (e.g., argon). The low-pressure mercury lamps resemble fluorescent lamp but are lacking phosphor coating. Mercury lamps of medium wave are designed with pressure to produce maximum radiation in the UV-B region. They use glass bulbs in which this energy is freely transmitted. Medium-pressure mercury lamps are polychromatic. They have higher electrical potential and also operate at higher temperature. Mercury vapor lamps are efficient and have extended life. They are of potable size (Fig. 5.2). Apart from mercury lamps, other artificial sources of UV are also available that yield more intense emission than former.

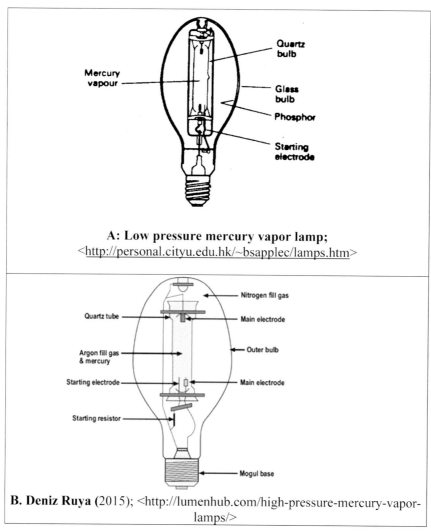

A: Low pressure mercury vapor lamp;
<http://personal.cityu.edu.hk/~bsapplec/lamps.htm>

B. Deniz Ruya (2015); <http://lumenhub.com/high-pressure-mercury-vapor-lamps/>

FIGURE 5.2   (See color insert.) Mercury lamps.

5.2.1   EXCIMER LAMPS

Excimer lamps can operate at low surface temperature and thus prevent fouling of liquid foods. These lamps emit quasi-monochromatic radiation. They have extremely low output and have application in sterilizing the packaging carton surfaces.[74] Excimer has abbreviation of excited dimer. Excimer lamps are formed of noble gas and noble gases/halogen mixture. Excimer lamps has electrodes that are separated by a dielectric barrier mainly quartz glass. Most often coaxial quartz tube arrangement is used as it is mechanically long lasting and can be easily manufactured. The excimer state is very short lived. According to choice of gas, numerous narrow band UV spectra can be made in single spectral line. Excimer combinations producing UV radiation varies in range between 120 and 320 nm.

5.2.2 PULSED LAMPS

Nowadays, pulsed lamps are applied to foods.[69] They are actually alternating current stored in a capacitor and high-speed switch is used for the discharge of energy forming a pulse of intense emission of light in 100 µs. In recent studies, pulsed lamps are mentioned for inactivation of *Aspergillus niger* in corn meal.[65] It has also seen to reduce psychrotrophic and coliform bacteria levels in lettuce without affecting sensory its quality.[6] These lamps are trustworthy because of robust packaging, high intensity, and with no mercury in lamps. They are said to penetrate opaque liquids in comparison with mercury lamps and thus treatment rate is enhanced in pulsed lamps.

5.3 ULTRAVIOLET DOSE MEASUREMENT

5.3.1 RADIOMETERS

Radiometers are used for the measurement of UV irradiance. Radiometers comprise (1) device that separates part of spectrum for measurement and (2) photosensitive detector.[54] For accurate dose estimation, the sensor should be precisely positioned to know coordinates within the UV field, but it is difficult to achieve practically. Therefore, two possible alternatives to radiometry include: chemical actinometery and biodosimetery.

5.3.2 CHEMICAL ACTINOMETERS

Actinometers are used for measuring a specific chemical change or quantity of photons in beam of light per unit time. With chemical actinometery, photochemical conversion is directly related to the amount of UV light absorbed by the treated product.[62] Mostly used chemical actinometers for UV processing of water include potassium ferrioxalate and potassium iodide.

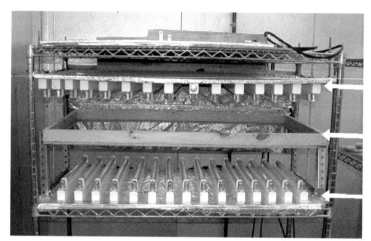

FIGURE 5.3 **(See color insert.)** UV-C radiation device.
Source: Reprinted with permission from Ref. [5]. © 2006 Elsevier.

5.3.3 BIODOSIMETRY

Biodosimetry is considered to be the most consistent technique for monitoring irradiance.[60] As this technique involves passing of inoculated liquid product through reactor, it is known to be a biological approach for dose measurement.[5] If compared with arithmetic mean of dose distribution, biodosimetry gives low volume average decimal reduction in case of microorganism.

Figure 5.3 shows the UV-C radiation device in which the upper and lower groups produce lights and the produce tray in the middle, which consists of polystyrene net and a wooden frame.

5.4 IMPACT OF UV LIGHT ON FRESH FRUITS AND VEGETABLES

UV treatments are of two types conventional UV treatment and hormetic UV treatment. In conventional UV treatment, the UV is directed at microorganism present on object's surface, whereas in hormetic UV treatment, the sample itself is the focused target of the incident UV. The goal is to elicit beneficial response in fresh fruits, vegetables, and other biological commodities. Both of the UV treatments are employed with same wavelength, but in case of hormetic treatment, only low doses are required. According to Shama,[63] "Hormesis involves the use of small doses of potentially harmful agents directed against a living organism or living tissue in order to elicit a beneficial or protective response." UV-C light is a nonthermal method, which improves the quality and shelf life of fresh produce by exploitation of UV light. Since the last two decades, UV-C light is in practice in horticultural sector as a tool against postharvest diseases.[59] UV-C light accomplishes two main important objectives:

- It reduces the initial count of microorganism on product surface.
- Host becomes resistant to microorganisms.[67]

Fresh fruits and vegetables are intended to be stored under fresh conditions for consumption depending upon their morphogenetic origin. However, they differ in their behavior and ability to respond to different postharvest treatments. As UV light also differ in its sensitivity to different horticultural crops, hence it is obvious that plant organs will react differently to different doses of UV-C light for prolonged life. For maximum hormesis effect, UV dose ranging from 0.125 to 9 kJ/m^2 is used to control growth of plant pathogens such as *Botrytis cynerea* in grapes and *Penicilium digtatum* in oranges.[19,51] UV radiation as abiotic stressors has been used in fresh produce, which stimulates the biosynthesis of defensive secondary metabolites with antimicrobial and antioxidant activity. These compounds prolong the shelf life of fresh produce and also help in maintaining the nutritional value and quality of UV-treated products. It has been seen to improve the level of phenolic compounds and carotenoids at low doses (<1 J/cm^2) thus increasing antioxidant capacity of harvested commodities.[58]

UV treatment in the range of 0.4–4.3 kJ/m^2 encourages the activity of numerous enzymes which has important role in plant antimicrobial defense actions. They are

- enzyme of peroxidases and reductases that are vital for oxidative burst and formation of lignin polymers which in turn generate structural barriers against invading pathogens,
- activity of glucanases and chitinases that exhibits lytic activities toward major fungal cell wall components, and
- 1-phenyl-alanine ammonia lyase (PAL) which is required in biosynthesis of phenolics (characterized by antioxidants and antimicrobial activities).[21,55]

Antioxidant capacity in fruits/vegetables is vital parameter relating to quality and by using UV-C light changes can be seen in antioxidant status of horticultural produce during their postharvest storage. In response to the damaging effect of free radicals both nonenzymatic antioxidants and enzymatic scavengers are activated by plant metabolism such a superoxide dismutase.[45,75] UV-C irradiation has found to have the capability to induce antioxidants which neutralize the effect of DNA and free radical damage associated with aging and senescence. Researches have reported the reduction in fruit softening when strawberries were exposed to UV-C light.[10] Postharvest fruit senescence has been delayed in many fruits when UV-C light was used as a treatment. In minimally processed fruit, use of UV-C light has increased the levels of polyphenols and flavonoids. Thus, by applying UV radiation during cutting process, shelf life of the minimally processed product can be improved.[7]

UV light has been reported to delay color changes in tomato, bell pepper, and broccoli by making chloroplast as a target sites for UV action. UV light also enhances anthocyanin pigments in strawberry and apples. In fresh-cut fruits, UV light is used for sanitization and thus reduction of microbial count without having any adverse effect on overall quality of sample. It was mentioned for surface disinfection of apples, kiwifruits, lemons, orange, and grapes.[38] If applied in moderate amounts, UV light has been found not to have any adverse effect on the food components.[36]

Very little attention has been paid on other wavelengths associated with UV-B which has wavelength range between 280 and 320 nm and UV-A having wavelength 320–400 nm for postharvest operations. UV-A has been applied for quality control within the food industry, particularly for the detection of aflatoxins from *Aspergillus flavus* and *Aspergillus parasiticus* on various grains and nuts like maize and peanuts during storage.[1]

Research studies were carried out to supplement UV-A and UV-B in green leaf lettuce which increased carotenoid level in the produce but the same treatment reduced the amount of carotenoid in red leaf lettuce.[14] UV-B treatment was also suggested in red clover, which increased levels of isoflavones, caffeic acid, and flavonols.[68] UV-B have seen to upregulate gene encoding of flavonoid biosynthesis pathways, like chalcone synthase and PALs, which are key enzyme in anthocyanin formation.[53] Table 5.1 shows the influence of UV light on fruit/vegetable's shelf life.

TABLE 5.1 Effects of UV Treatment on Fruits and Vegetables.

Fruit type	Hormetic effect	Dosage	References
Apples	Enhanced anthocyanin content	UV-B lamp (320 nm)	[71]
Blueberry fruit	Increased antioxidant capacity	0–4 kJ/m^2	[18]
Fresh-cut melon	Total soluble solids	3 min	[34]
Golden delicious apple	Reduced browning of minimal processed products	2.43×10^{-3} W/m^2	[9]
Grapefruit	Production of phytoalexins	0.50 kJ/m^2	[18]
Limon	Increased total phenolic capacity	UV-B lamp (280–400 nm)	[32]
Lollo rosso lettuce	Increased tissue brightness and reduced browning	2.44, 4.07, and 8.14 kJ/m^2	[4]
Mango	Increased shelf life and lower decay percentage	2.46 and 4.93 J/m^2	[28]
Peaches	Enhanced content of polyamine compounds	8.22 W/m^2	[27]
Persimmon fruit	Reduced postharvest disease	1.5 and 3 kJ/m^2	[35]
Red-leaf lettuce	Extended shelf life	1.18–7.11	[5]
Tomatoes	Retarded ripening	4 kJ/m^2	[57]
	Maintained better firmness and sensory attributes		
Tomatoes	Improved antioxidant properties	2 and 4 J/cm^2	[23]

5.4.1 CONTROL OF SENESCENCE

The process of aging in plant is termed as senescence. The major factor of senescence is damage by free radical. These free radical formed targets the cell membrane, enzymes, nucleic acids, and cell walls which results in increase in senescence.[13] UV irradiation can safeguard against the symptoms of senescence because of activating defensive mechanism of plant by hormetic UV doses. The UV doses are reported to control senescence and ripening in climacteric fruits.[46] But with prolonged exposure to UV, ripening and senescence will accelerate as a result of free radical generation.[42] Therefore, low levels of UV-C can be used in addition to refrigeration to preserve the horticultural crops.

The efficacy of UV-C treatment varies with different plant species, time of maturation, duration, and irradiation dose. UV treatment was given to tomato fruit which showed less production of CO_2 as compared to the control sample. Ethylene synthesis followed the similar pattern. The delayed climacteric rise was elucidated as indication of delayed senescence due to UV treatment.

5.4.2   CHANGES IN COLOR AND FLAVOR

Color can be a decisive factor for the acceptance or rejection of produce for consumers. The natural pigments imparting color quality are fat-soluble pigments (chlorophyll and carotenoids) and water-soluble pigments (anthocyanins, flavonoids, and betalins). Synthesis of carotenoids and loss of chlorophyll during ripening result in the color change of fruits.[25] To extend the shelf life of produce, it is very important that these changes take place at slower rate. These changes can be delayed by using UV-C doses for fruits and vegetables.

UV-C irradiation delayed color development and chlorophyll loss in ripe tomatoes and helped in lycopene development. Research studies were held for studying effect of different wavelength using florescent lamps to increase the flavonoid content in onions and strawberries after harvest. The shaded sides of strawberries and onions were irradiated and stored in dark due to which quercetin in onions and anthocyanins in strawberries were induced. The quercetin content in onion was doubled and the color difference between the sunlight and shaded portion of strawberry disappeared.[30]

Several researchers reported that UV-C exposure to maintain maximum chlorophyll level and thus preserving green color of broccoli.[17,40] UV-C treatment was given to shitake mushrooms, which reflects enhanced flavonoids and ascorbic acid levels.[33] According to D'hallewin,[18] UV-C was not effective method when compared with heat treatment or thiabendazole because UV-C-treated fruits showed poor visual appearance even though the decay control was significant. The reason for the loss of quality may be due to overexposure of UV-C irradiation. Similar results were observed in other crops, where the doses of irradiation were higher than the hormetic doses.[20,59] UV-A light treatment at 342 nm and fluence level of 4.5 and 9.0

kJ/m[2] was given to broccoli to delay the yellowing process, but the attempts were not found to be effective.[3]

The flavor in fruits is mainly contributed by the phenolic compounds. Phenols are the widely distributed category of phytochemical in the plant kingdom. These phenols prevail in higher plants in different forms, which include hydroxyl benzoic derivatives, flavonoids, lignins, and stilbenes. They affect the quality attributes of plants like flavor, appearance, and other health-benefiting properties.[61] Phenols also act as a substrate for browning reactions. UV-treated tomato fruit was found to be better in taste when compared with control or store-bought fruits as it resulted in increase in phenolic compounds.[46]

Beauliey[12] and Lamikanra[39] documented that when fruits treated with UV light, aroma was preserved to similar extent as in control samples. Fresh-cut tropical fruits (slices of banana and pineapple) were irradiated by UV-C light. Banana and guava showed an increase in total flavonoids and phenols, whereas for pineapple, significant increase was found in flavonoids but had no effect on the phenol content.[7]

5.5   USE OF COMBINED PRESERVATION TECHNIQUE

Nonionizing UV-C radiation has been used broadly for antimicrobial application such as disinfection of water, food and surface disinfection of vegetable commodities.[48,73] Exposure of UV-C light activates the synthesis of health-promoting compounds in fruits and vegetables such as anthocyanin and stilbenoids (C6–C2–C6).[15] There has been an trifling issue on to what should be the correct dose of UV-C light delivered to the produce, as exceeding the optimum UV dose will damage the produce and under dosing will lead to a failure in receiving maximum benefits and will result in reduced shelf life or lose in quality. To overcome this problem, combination of two or more treatments [such as chilling, modified atmosphere packaging (MAP)] is suggested for maintaining quality of fruit and vegetable produce.[48] The synergistic effect of UV-C light in combination with MAP has been tested on vegetable produce. Some studies have used more than two treatments, like fresh-cut fruits and vegetables were first disinfected using chlorinated water and then UV-C treatment was given which were further stored in MAP.[4,43] These studies resulted in microbial reduction and there was no harmful effect on organoleptic properties of the product.

UV-C light can also be combined with mild thermal treatments. The horticulture produce can be immersed in water at high temperature to improve the keeping quality and reduce the incidence of storage diseases. UV-C treatment at a dose of 10 kJ/m^2 was combined with thermal treatment at a temperature of 45°C for 15 min to achieve fungal inactivation in case of strawberries.

Light pulse is a new technique, which has been successfully used in combination with UV-C light to inactivate bacteria and fungi on food surface where the major component of the emitted spectrum is UV light. Pulsed light has more efficiency in terms of inactivation of fungi, bacteria, and virus.[22] According to McDonald,[50] spores of *Bacillus subtilis* were inactivated by pulsed UV light successfully than continuous UV light. Pulsed light of 4 mJ/cm^2 was used to inactivate *B. subtilis* and the same was done by continuous UV light at 8 mJ/cm^2. Marquine[49] demonstrated that UV light when combined with light pulses, an increase in inactivation of conidia of *Botrytis cinerea* and *Monilia fructigena* can be achieved. Pulsed UV light can be used as a nonthermal process if only applied for short duration. To minimize treatment temperature during testing procedure, following recommendation should be considered[52]:

- Lamps should be switched off and on instantly during sterilization.
- Short-duration pulses should be applied.
- In between pulses, cooling time should be allowed.
- Low infrared spectrum content should be confirmed.

UV-C sole and in combination with MeJa and salicylic acid was used to investigate the effect on stilbene biosynthesis in *Vitis vinifera*. UV-C irradiation for 20 min or MeJa at 100 µM was found effective in promoting stilbene accumulation; however, salicylic acid was less efficient than UV-C and MeJa. When UV-C and MeJa were combined, a higher induced intracellular stilbene production was observed to a maximum of 2005.05 ± 63.03 µg/g DW and also showed a synergistic effect on extracellular transresveratrol accumulation to 3.96 ± 0.2 mg/L.[2] UV-C irradiation was combined with antibrowning pretreatment step, which helped in preserving the color of slice apples better during storage at 5°C for 7 days.[24]

The combined use of UV-C treatment and modified atmosphere packaging were seen to be more effective in lowering psychotropic bacteria, coliform, and yeasts growth in lettuce without hampering its sensory properties.[4]

5.6 LIMITATIONS

UV light tends to penetrate only 50–300 nm into tissue. The reason behind is its inability to pass through physical barriers. Thus to achieve complete destruction of bacteria, complete surface should be exposed to help in the stimulation of defense mechanism against particular organisms. It was seen that waxing shields the bacteria from the rays of UV and hence reduces performance of UV-C light.[77] If UV-C light is used within the range of 330–480 nm, photoreactivation of the cells will result in the increase of viable microorganism.[41] To avoid such limitations, the produce should be refrigerated or should be stored in dark places.[31]

Many studies showed that the higher dose resulted in damage of fruits. Strawberries were treated with UV-C light within the range of 0.25–1.0 kJ/m^2 and it was reported that the higher doses were damaging the fruit. Likewise Gonzalez[26] reported that in case of mango fruits at a UV-C dose of 4.93 kJ/m^2 was beneficial but twice dose reflect in fruit damage. In fresh-cut mango fruits, reduction in the total ascorbic acid was reported when there was an increment in the exposure time of UV rays as it resulted in oxidation of ascorbic acid.[28]

In a study on bunch of grapes, it was noticed that exterior of the grapes from bunch could be irradiated but the center received less or no UV radiation. The effort to irradiate the correct dose to the core resulted in overdosing of the exterior grapes. The only way to treat the bunch was to remove the grape berries from bunch. At higher doses, recontamination or regrowth of certain microorganisms was noticed as compared to untreated produce. Varied results were obtained when arils of pomegranate were treated with UV-C; in some cases, UV-C-treated arils reflects higher microbial count.[43]

5.7 SUMMARY

To fulfill the requirements of safe food/natural flavor and texture, UV radiation can be an appropriate method for antimicrobial treatments for surface and perhaps food itself. UV technology is nonthermal, ecofriendly alternative in food industry compared to other treatments like thermal methods that can cause adverse effects and involve high energetic cost. Short-wave UV-C method is harmful to many microorganisms. The effect of UV radiation for fresh produce has been established at the laboratory level. However, to scale up the process for commercialization, further research is needed to ensure microbiological effectiveness and optimization of critical process factors.

Finally, UV radiation should satisfy FDA requirement along with Hazard Analysis and Critical Control Points (HACCP) programs to ensure the supply of safe food products to consumers.

KEYWORDS

- **actinometer**
- **biodosimeter**
- **excimer lamp**
- **hormesis**
- **mercury lamp**
- **radiometer**

REFERENCES

1. Ahmad, M. Seeing the World in Red and Blue: Insight into Plant Vision and Photoreceptors. *Curr. Opin. Plant Biol.* **1999**, *2*, 230–235.
2. Ai, X.; Ji-Cheng, Z.; Wei-Dong, H. Effects of UV-C, Methyl Jasmonate and Salicylic Acid, Alone or in Combination, on Stilbene Biosynthesis in Cell Suspension Cultures of *Vitis vinifera* L. cv. Cabernet Sauvignon. *Plant Cell, Tissue Organ Cult.* **2015**, *122*, 197–211.
3. Aiamla-or, S.; Yamauchi, N.; Takino, S.; Shigyo, M. Effects of UV-A and UV-B Irradiation on Broccoli (*Brassica oleracea* L. *Italica*) Floret Yellowing during Storage. *Postharv. Biol. Technol.* **2009**, *54*, 177–179.
4. Allende, A.; Artes, F. UV-C Radiation as a Novel Technique for Keeping Quality of Fresh Processed "Lollo Rosso" Lettuce. *Food Res. Int.* **2003**, *36*, 739–746.
5. Allende, A.; McEnvoy, J. L.; Luo, Y.; Artes, F.; Wang, C. V. Effectiveness of Two Sided UV-C Treatments in Inhibiting Natural Microflora and Extending the Shelf-Life of Minimally Processed 'Red Oak Leaf' Lettuce. *Food Microbiol.* **2006**, *23*, 241–249.
6. Allendre, A.; Artes, F. Combined Ultraviolet-C and Modified Atmosphere Packaging Treatments for Reducing Microbial Growth of Fresh Processed Lettuce. *LWT Food Sci. Technol.* **2003**, *36*, 779–786.
7. Alothman, M.; Bhat, R.; Karim, A. A. UV Radiation-Induced Changes of Antioxidant Capacity of Fresh-cut Tropical Fruits. *Innovat. Food Sci. Emerg. Technol.* **2009**, *10*, 512–516.
8. Alothman, M.; Bhat, R.; Karim, A. A. Effects of Radiation Processing on Phytochemicals and Antioxidants in Plant Produce. *Trends Food Sci. Technol.* **2009**, *20*, 201–212.
9. Anna, L.; Federica, T.; Marino, N. UV-A Treatment for Controlling Enzymatic Browning of Fresh-Cut Fruits. *Innovat. Food Sci. Emerg. Technol.* **2015**, *12*, 029.
10. Baka, M.; Mercier, J.; Corcuff, F.; Castaigne, F.; Arul, J. Photochemical Treatment to Improve Storability of Fresh Strawberries. *J. Food Sci.* **1999**, *64*, 1068–1072.

11. Barka, E. A.; Kalantari, J.; Makhlouf, J.; Arul, J. Impact of UVC Illumination on the Cell Wall-Degrading Enzymes during Ripening of Tomato (*Lycopersicon esculentum* L.) fruit. *J. Agric. Food Chem.* **2000**, *48*, 667–671.

12. Beaulieu, J. C. Effect of UV Irradiation on Cut Cantaloupe: Terpenoids and Esters. *J. Food Sci.* **2007**, *724*, 272–281.

13. Brady, C. J. Fruit Ripening. *Annu. Rev. Plant Physiol.* **1987**, *38*, 155–178.

14. Caldwell, C. R.; Britz, S. J. Effect of Supplemental Ultraviolet Radiation on the Carotenoid and Chlorophyll Composition of Greenhouse Grown Leaf Lettuce (*Latica sativa* L.) Cultivars. *J. Food Compos. Anal.* **2006**, *19*, 637–644.

15. Cantos, E.; Garcia-Viguera, C.; DE Pascual-Teresa, S.; Tomas-Barbera, F. A. Effect of Postharvest Ultraviolet Irradiation on Resveratrol and Other Phenolics of Cv. Napoleon Table Grapes. *J. Agric. Food Chem.* **2000**, *48*, 4606–4612.

16. Chang, J. C. H.; Ossoff, S. F.; Lobe, D. C.; Dorfman, M. H.; Dumais, C. M.; Qualls, R. G.; Johnson, J. D. UV Inactivation of Pathogenic and Indicator Microorganisms. *Appl. Environ. Microbiol.* **1985**, *49*, 1361–1365.

17. Costa, L.; Vecente, A. R.; Civello, P. M.; Chaves, A. R.; Martinez, G. A. UV-C Treatment Delays Postharvest Senescence in Broccoli Florets. *Postharv. Biol. Technol.* **2006**, *39*, 204–210.

18. D'hallewin, G.; Schirra, M.; Pala, M.; Ben-Yehoshua, S. Ultraviolet C Irradiation at 0.5 kJ/m² Reduces Decay without Causing Damage or Affecting Postharvest Quality of Star Ruby Grapefruit (*C. paradisi* Macf.). *J. Agric. Food Chem.* **2000**, *48*, 4571–4575.

19. D'hallewin, G.; Schirra, M.; Manueddu, E.; Piga, A.; Ben-Yehoshua, S. Scoparone and Scopoletin Accumulation and UV-C Induced Resistance to Postharvest Decay in Oranges as Influenced by Harvest Date. *J. Am. Soc. Hortic. Sci.* **1999**, *124*, 702–707.

20. Droby, S.; Chalutz, E.; Horev, B.; Cohen, L.; Gaba, C. L.; Wilson, M. Factors Affecting UV-induced Resistance in Grapefruit against the Green Mold Decay Caused by *Penicillium digitatum. Plant Sci.* **1993**, *42*, 418–424.

21. Erkan, M.; Wang, S. Y.; Wang, C. Y. Effect of UV Treated on Antioxidant Capacity, Antioxidant Enzyme Activity and Decay in Strawberry Fruit. *Postharv. Biol. Technol.* **2008**, *48*, 115–119.

22. Feuilloley, M. G. J.; Bourdet, G.; Orange, N. Effect of White Pulsed Light on *pseudomonas aeruginosa* Culturability and Its Endotoxin When Present in Ampoules for Injectables. *Eur. J. Parenter. Pharm. Sci.* **2006**, *11*, 37–43.

23. Gianpiero, P.; Giorgio, D.; Giovanna, F. Post-Harvest UV-C and PL Irradiation of Fruits and Vegetables. *Chem. Eng. Trans.* **2015**, *44*, 2382–9216.

24. Gomez, P. L.; Alzamora, S. M.; Castro, M. A.; Salvatori, D. M. Effect of Ultraviolet-C Light Dose on Quality of Cut Apple: Microorganism, Color and Composition Behavior. *J. Food Eng.* **2010**, *98*, 60–70.

25. Gong, Y.; Mattheis, J. P. Effect of Ethylene and 1-methyl-Cyclopropane on Chlorophyll Catabolism of Broccoli Florets. *Plant Growth Regul.* **2003**, *40*, 33–38.

26. Gonzalez-Aguilar, G. A.; Wang, C. Y.; Buta, J. G.; Krizek, D. T. Use of UV-C Irradiation to Prevent Decay and Maintain Postharvest Quality of Ripe "Tommy Atkins" Mangoes. *Int. J. Food Sci.* **2001**, *36*, 767–773.

27. Gonzalez-Aguilar, G.; Wang, C. Y.; Buta, G. J. UV-C Irradiation Reduces Breakdown and Chilling Injury of Peaches during Cold Storage. *J. Sci. Food Agric.* **2004**, *84*, 415–422.

28. Gonzalez-Anuilar, G. A.; Zavaleta-Gatica, R.; Tiznado-Hernandez, M. E. Improving Postharvest Quality of Mango 'Haden' by UV-C Treatment. *Postharv. Biol. Technol.* **2007,** *45,* 108–422.

29. Guerroro, B.; Barbosa, C. Review: Advantages and Limitations on Processing Foods by UV Light. *Food Sci. Technol. Int.* **2004,** *10,* 137–148.

30. Higashio, H.; Hirokane, H.; Sato, F.; Uragami, A. Effect of UV Irradiation after the Harvest on the Content of Flavonoid in Vegetables. *Acta Hortic.* **2005,** *682,* 1007–1012.

31. Hoyer, O. Testing Performance and Monitoring of UV Systems for Drinking Water Disinfection. *Water Supply* **1998,** *16,* 424–429.

32. Interdonato, R.; Rosa, M.; Nieve, C. B.; Gonzalez, J. A.; Hilal, M.; Prado, F. E. Effects of Low UV-B Doses on the Accumulation of UV-B Absorbing Compounds and Total Phenolics and Carbohydrate Metabolism in the Peel of Harvested Lemons. *Environ. Exp. Bot.* **2011,** *70,* 204–211.

33. Jiang, T.; Jahangir, M. M.; Jiang, Z.; Lu, X. Influence of UV-C Treatment on Antioxidant Capacity, Antioxidant Enzyme and Texture on Postharvest Shitake (*Lentinus edodes*) Mushrooms during Storage. *Postharv. Biol. Technol.* **2010,** *56,* 209–215.

34. Kasim, R.; Mehmet, U. K. Biochemical and Color Changes of Fresh Cut Melon (*Cucumis melo* L. cv. *Galia*) Treated with UV-C. *Food Sci. Technol.* **2014,** *34,* 547–551.

35. Khademi, O.; Zamani, Z.; Poor Ahmadi, E.; Kalantari, S. Effect of UV-C Radiation on Postharvest Physiology of Persimmon Fruit (*Diospyros kaki* thumb.) cv. 'Karaj' during Storage at Cold Temperature. *Int. Food Res. J.* **2013,** *20,* 247–253.

36. Krishnamurthy, K. Decontamination of Milk and Water by Pulsed UV-Light and Infrared Heating; *PhD Thesis*; Pennsylvania State University, 2006; Chapter 2, pp 23–44.

37. Lado, B. H.; Yousef, A. E. Alternative Food Preservation Technologies: Efficacy and Mechanisms. *Microbes Infect.* **2002,** *4,* 433–440.

38. Lagunas, S.; Pina, M. C.; MacDonald, J.; Bolkan, L. Development of Pulsed UV-Light Processes for Surface Fungal Disinfection of Fresh Fruit. *J. Food Product.* **2006,** *69,* 376–384.

39. Lamikanra, O.; Kueneman, D.; Ukuku, D.; Bett-Garber, K. L. Effect of Processing under Ultraviolet Light on the Shelf-Life of Fresh Cut Cantaloupe Melon. *J. Food Sci.* **2005,** *70,* 534–538.

40. Lemoine, M. L.; Civello, P. M.; Martinez, G. A.; Chaves, A. R. Influence of Postharvest UV-C Treatment on Refrigerated Storage of Minimally Processed Broccoli (*Brassica aleracea* var. *Italica*). *J. Sci. Food Agric.* **2007,** *87,* 1132–1139.

41. Liltved, H.; Landfald, B. Effects of High Intensity Light on Ultraviolet-Irradiated and Non-Irradiated Fish Pathogenic Bacteria. *Water Res.* **2000,** *34,* 481–486.

42. Liu, J.; Stevens, C.; Khan, V. A.; Lu, J. Y.; Wilson, C. L.; Adeyeye, O.; Kabwe, M. K.; Pusey, P. L.; Chalutz, E.; Sultana, T.; Drobt, S. Application of Ultraviolet-C Light on Storage Rots and Ripening of Tomatoes. *J. Food Protect.* **1993,** *56,* 868–872.

43. Lopez-Rubira, V.; Conesa, A.; Allende, A.; Artes, F. Shelf-Life and Overall Quality of Minimally Processed Pomegranate Arils Modified Atmosphere Packaged and Treated with UV-C. *Postharv. Biol. Technol.* **2005,** *37,* 174–185.

44. Lopez-Rubira, V.; Conesa, A.; Allende, A.; Artes, F. Shelf-Life and Overall Quality of Minimally Processed Pomegranate Arils Modified Atmosphere Packaged and Treated with UV-C. *Postharv. Biol. Technol.* **2005,** *37,* 174–185.

45. Maharaj, R.; Arul, J.; Nadeau, P. Effect of Photochemical Treatment in the Preservation of Fresh Tomato (*Lycopersicon esculentum* cv. *Capello*) by Delaying Senescence. *Postharv. Biol. Technol.* **1999,** *15,* 13–23.

46. Maharaj, R.; Arul, J.; Nadeau, P. UV-C Irradiation of Tomato and Its Effect on Color and Pigments. *Adv. Environ. Biol.* **2010,** *4*, 308–315.

47. Maria, Turtoi; Ultraviolet Light Treatment of Fresh Fruits and Vegetables Surface: A Review. *J. Agroaliment. Process. Technol.* **2013,** *19*, 325–337.

48. Marquenie, D.; Michiels, C. W.; Geeraerd, A. H.; Schenk, A.; Soontjens, C.; Van Impe, J. F.; Nicolai, B. M. Using Survival Analysis to Investigate the Effect of UV-C and Heat Treatment on Storage Rot of Strawberry and Sweet Cherry. *Int. J. Food Microbiol.* **2002,** *73*, 187–196.

49. Marquenie, D.; Michiels, C. W.; Van Impe, J. F.; Schirevens, E.; Nicolai, B. M. Pulsed White Light in Combination with UV-C and Heat to Reduce Storage Rot of Strawberry. *Postharv. Biol. Technol.* **2003,** *28*, 455–461.

50. McDonald, K. F.; Curry, R. D.; Clevenger, T. E.; Unklesbay, K.; Eisenstrack, A.; Golden, J.; Morgan, R D. Comparison of Pulsed and Continuous Ultraviolet Light Sources on the Contamination of Surfaces. *IEEE Trans. Plasma Sci.* **2000,** *28*, 1581–1587.

51. Nigro, F. A.; Ippolito; Lima, G. Use of UV-C to Reduce Storage Rot of Table Grape. *Postharv. Biol. Technol.* **1998,** *13*, 171–181.

52. Panico, L. Instantaneous Sterilization with Pulsed UV Light. *Workshop: Emerging Food Processing Technologies*; USDA, CSREES, Washington State University, 2010; pp 26–27.

53. Perkins-Veazie, P.; Collins, J.; Howard, L. Blueberry Fruit Response to Postharvest Application of Ultraviolet Radiation. *Postharv. Biol. Technol.* **2008,** *47*, 280–285.

54. Phillips, R. *Sources and Applications of Ultraviolet Radiation*; Academic Press: London, 1983; p 626.

55. Pombo, M.; Dotto, M. C.; Martniez, G. A.; Civello, P. M. UV-C Irradiation Delays Strawberry Fruit Softening and Modifies the Expression of Genes Involved in Cell Wall Degradation. *Postharv. Biol. Technol.* **2009,** *51*, 141–148.

56. Rhim, J. W.; Gennadios, A.; Fu, D.; Weller, C. L; Hanna, M. A. Properties of Ultravoilet Irradiated Protein Films. *Lebensum-Wiss U-Technol* **1999,** *32*, 129–133.

57. Robles, P.; De Campos, A.; Artes-Hernandez, F.; Gomez, P.; Calderon, A.; Ferrer, M.; Artes, F. Combined Effect of UV-C Radiation and Controlled Atmosphere Storage to Preserve Tomato Quality. In *V Congreso Iberoamericano de Technologia Postcosecha y Agroexportaciones* (*Ibero-American Congress on Postharvest Technology and Agroexports*); Cartagens: Spain; 29 May to 1 June, 2007 (CD ROM).

58. Rock, C. D. Trans-Acting Small Interfering RNA4: Key to Nutraceutical Synthesis in Grape Development? *Trends Plant Sci.* **2013,** *18*, 601–610.

59. Rodov, V.; Ben-Yehoshua, S.; Kim, J. J.; Shapiro, B.; Ittah, Y. Ultraviolet Illumination Induces Scoparone Production in Kumquat and Orange Fruit and Improves Decay Resistance. *J. Am. Soc. Hortic. Sci.* **1992,** *117*, 788–792.

60. Sastry, S. K.; Datta, A. K.; Worobo, R. W. Ultraviolet Light. *J. Food Sci., Suppl.* **2000,** *65*, 90–92.

61. Shahidi, F.; Chandrasekara, A.; Zhong, Y. Bioactive Phytochemicals in Vegetables. In *Handbook of Vegetables and Vegetables Processing*; Sinha, N. K., Ed.; Blackwell Publishing Ltd.: Ames, IA, 2009; pp 125–158.

62. Shama, G. *Ultraviolet Light. Encyclopedia of Food Microbiology—3*; Academic Press: London, 1999; pp 2208–2214.

63. Shama, G.; Alderson, P. UV Hormesis in Fruits: A Concept Ripe for Commercialization. *Trends Food Sci. Technol.* **2005,** *16*, 128–136.

64. Sharma, G. A New Role for UV. Extension to the Shelf-Life of Plant Foods by UV-Induced Effects. Paper Presented at 2007 IOA-IUVA Joint World Congress, Los Angeles, August 27–29, 2007; p 15.

65. Sharma, R. R.; Demirci, A. Inactivation of *Escherichia coli* O157:H7 on Inoculated Alfalfa Seeds with Pulsed Ultraviolet Light and Response Surface Modelling. *J. Food Sci.* **2003,** *68,* 1448–1453.

66. Snowball, M. R.; Hornsey, I. S. Purification of Water Supplies Using Ultraviolet Light. In *Developments in Food Microbiology*—3; Robinson, R. K. Ed.; Elsevier Applied Science Publishers: London, 1988; pp 171–191.

67. Stevens, C.; Khan, V. A.; Lu, J. Y.; Wilson, C. L.; Pusey, P. L.; Igwegbe, E. C. K.; Kabwe, M.; Mafolo, Y.; Liu, J.; Chalutz, E; Droby, S. Integration of Ultraviolet (UVC) Light with Yeast Treatment for Control of Postharvest Storage Rots of Fruits and Vegetables. *Biol. Cntrl.* **1997,** *10,* 98–103.

68. Swinny, C.; Khan, V. A.; Wilson, C. L;, Chalutz, E.; Droby, S.; Kabwe, M. K.; Huand, Z.; Adeyeye, O.; Pusey, L. P.; Tang, A. Y. A. Induced Resistance of Sweet Potato to Fusarium Root Rot by UV-C Hormesis. *Crop Protect.* **1999,** *18,* 462–470.

69. Tatiana, N. K.; Forney, L. J.; Moraru, C. I. *UV Light in Food Technology, Principles and Applications*; *Contemporary Food Engineering Series*; CRC Press: London, 2009; p 432.

70. Thomas, B.; Evanthia, L. T.; Richard, K. R. Review Existing and Potential Application of Ultraviolet Light in the Food Industry—A Critical Review. *J. Sci. Food Agric.* **2000,** *80,* 637–645.

71. Ubi, B. E.; Honda, C.; Bessho, H.; Kondo, S.; Wada, M.; Kobayashi, S.; Moriguchi, T. Expression Analysis of Anthocyanin Biosynthesis Genes in Apple Skin: Effect of UV-B and Temperature. *Plant Sci.* **2006,** *170,* 571–578.

72. Vicente, M.; Gomez, L.; Tatiana, K.; Kari, L. Ultraviolet and Pulsed Light Processing of Fluid Foods (Chapter 8). In *Novel Thermal and Non-thermal Technologies for Fluid Foods*; Cullen, P. J., Tiwari, B. K., Valdramidi, V., Eds.; *Food Science & Technology International Series*; Academic Press: New York, 2012; pp 185–223 (online).

73. Wang, C. Y.; Chen, C. T.; Wang, S. Y. Changes of Flavonoid Content and Antioxidant Capacity in Blueberries after Illumination with UV-C. *Food Chem.* **2009,** *117,* 426–431.

74. Warriner, K.; Kolstad, J.; Rumsby, J.; Waites, W. Carton Sterilization by UV-C Excimer Laser Light: Recovery of *Bacillus subtilis* Spores on Vegetable Extracts and Food Simulation Matrices. *J. Appl. Microbiol.* **2002,** *92,* 1051–1057.

75. Wellmann, E.; Hrazdina, G.; Grisebach, H. Induction of Anthocyanin Formation and of Enzymes Related to Its Biosynthesis by UV Light in Cell Cultures of *Haplopappus gracilis*. *Photochemistry* **1976,** *15,* 913–915.

76. WHO/FAO. *Report 2004:* Fruit and Vegetables for Health. *Report of a Joint FAO/WHO Workshop*; Kobe, Japan, 1–3 Sept 2004; p 310.

77. Yaun, L.; Yanqun, Z.; Chen, J.; Chen, H. Intra-specific Responses in Crop Growth and Yield of 20 Soybean Cultivars to Enhanced Ultraviolet-B Radiation under Field Conditions. *Field Crops Res.* **2002,** *78,* 1–16.

CHAPTER 6

APPLICATION OF PULSED LIGHT TECHNOLOGY IN PROCESSING OF FRUITS AND VEGETABLES

UMESH C. LOHANI, KHAN CHAND, NAVIN CHANDRA SHAHI, and ANUPAMA SINGH

ABSTRACT

Application of nonthermal processing in food nutrition and safety has gained a considerable attention of researchers in last few years. Pulsed light as one of the promising nonthermal technologies has been extensively practiced for the decontamination of microbes, nutritional and bioactive enhancement, and enzyme inactivation in different food materials. Pulsed light is a technique to treat the food materials using UV-C light having an intense broad spectrum pulses. For solid and liquid foods, this technology can replace continuous UV light treatments. This chapter provides a general principle of pulsed light, factors affecting its efficacy for reduction in surface microbial load, stimulating bioactive compound production and inactivation of polyphenol oxidase in different foods.

6.1 INTRODUCTION

Consumer demand for horticultural produces with fresh appearance and health-encouraging qualities is incessantly increasing all over the world. Recent developments regarding the demand of minimally processed and of course fresh fruits and vegetables with a prolonged shelf-life have motivated to ensure an appropriate preservation and safety of foodstuffs. Nutritious and healthy fruits and vegetables with high freshness can be obtained from the popular fresh-cut processing.[39] Although the fresh-cut fruit and vegetable industries have occupied their position in food service and retail markets, yet they are still trying to develop fresh-like quality of cut fruits and vegetables to attract consumers' interest.

The respiration and biochemical reactions of fresh-cut fruits and vegetables continue even after peeling, slicing, and cutting. The processing steps involved in the production of fresh-cut fruit and vegetable lead to increased microbiological and physiological activities, which affect the safety and quality of produce during storage.[35] Thus, the retained quality and safety of fresh-cut fruits and vegetables are deemed essential to ensure consumers' acceptance and safety.

Fresh-cut fruits and vegetables are convoyed by a public health threat due to foodborne illnesses. In the last two decades, more than 30 new infectious microbes have been recognized, which are associated with foodborne illnesses. These illnesses can occur when the product is handled, processed, and stored inappropriately inducing pathogenic and spoilage microorganisms' contamination. In particular, surface of the fresh-cut produce can easily be contaminated with microorganisms. Due to inadequate sanitation processes, their increased infective doses can cause food-poisoning outbreaks.[9,11]

Decontamination and microbial load reduction of fresh-cut produce generally cannot be achieved by conventional heating treatments, washing, and sanitizing methods.[43] Thermal processes are prejudicial to their sensory characteristics and antioxidants. Therefore, various nonthermal innovative techniques have been researched for surface decontamination of fresh-cut produce. Several innovative technologies have emerged for the surfaces decontamination to inactivate microorganisms such as the application of pulsed electric fields, gas plasma, and high pressure, the pulsed-light technology. Pulsed light (PL) particularly explores its great potential in extending shelf life of fresh-cut produces with their retained nutritional values.[1,7,16,33,36,37]

PL as a nonthermal technology works with intense pulses of wide spectrum and short duration to warrant microbial inactivation on the food surface for extending their shelf life. Typically, PL is produced using equipment consisted of one or more adjustable xenon lamps, and a high-voltage connection to generate an electrical pulse with high current (Fig. 6.1). The light is emitted with short and intense bursts when the current passes across the gas chamber of the xenon lamp. The lamp produces the light having the wavelength distribution ranges, that is, ultraviolet (UV) (100–400 nm), visible light (400–700 nm), and infrared (700–1100 nm). Short-duration (1 µs–0.1 s) light pulses typically emit flashes at the food surface with an energy density varied from 0.01 to 50 J/cm during processing applications.[4]

Electricity is stored in a capacitor for a relatively longer time (fractions of a second) followed by its releasing from the capacitor in a very short time (10th of fractions of second) and thus magnify the power of source.[15]

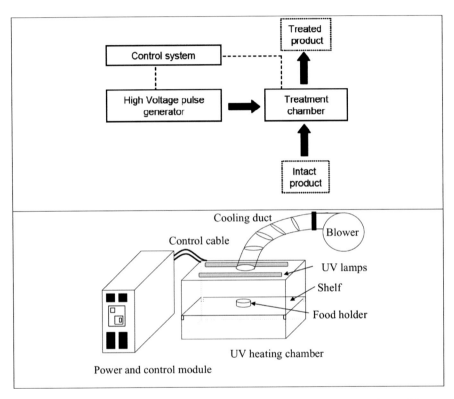

FIGURE 6.1   Pulsed UV light for processing of fruits and vegetables: Top—layout [*Source*: Maged Mohamed, E. A.; AmerEissa, A. H. Pulsed Electric Fields for Food Processing Technology (Chapter 11). In *Structure and Function of Food Engineering*; AmerEissa, A., Ed.; Intech, 2012 (online); Open access https://www.intechopen.com/books/structure-and-function-of-food-engineering/pulsed-electric-fields-for-food-processing-technology]. Bottom—typical example [*Source*: Murugesan, R.; Orsat, V.; Lefsrud, M. Effect of Pulsed Ultraviolet Light on the Total Phenol Content of Elderberry (*Sambucus nigra*) Fruit. *Food Nutr. Sci.* **2012**, *3* (6), 774–783 (Open Access). DOI: 10.4236/fns.2012.36104].

FDA[10] has approved its use for the decontamination of food and its rough and smooth surfaces. The cost-effectiveness, prompt substantial microbial reduction, low environmental affect, and its greater flexibility are some of the prime benefits claimed for this technology.[33] Beside these advantages, no residual compounds are required to get removed from foodstuff after

PL treatment.[15] The potential of commercial scale PL depends on fluence, fluence rate, exposure time, pulse width, pulse repetition rate, and penetrability of light, which may impart desirable effects on fresh-cut produce quality without spoiling the fresh like properties of treated produce.

This chapter explores the potential and adaptability of PL treatment to enhance the shelf life, safety, nutritive values, and bioactive components of fruits and vegetables.

6.2 APPLICATION OF PULSED LIGHT IN FRESH-CUT PRODUCE

6.2.1 MICROBIAL DECONTAMINATION

Although Hazard Analysis and Critical Control Points (HACCP) system is being broadly used in food processing industries, yet documented cases of foodborne illness are currently increasing. Bactericidal efficacy of PL is mainly ascribed to the photochemical effect of UV light and its ability to form thymine dimers in microbial DNA that inhibit cell replication.[13,15]

The lethal activity of PL can be on account of individual or combined effect of photothermal and photochemical mechanisms. The overall impact of individual or combined mechanisms would depend on the target microorganism and PL fluence. Microorganisms were inactivated by elevating the temperature less than 1°C concluding the importance of shorter UV wavelengths with higher photochemical action for PL lethality.[42] PL can expeditiously inactivate *Escherichia coli*, *Salmonella* spp., *Listeria innocua*, and native microflora and inoculated bacteria[6,17,37,38] without affecting product freshness.

Kramer and Muranyi[23] investigated that a PL of 1.08 J/cm^2 fluence inactivated *E. coli* and *L. innocua* inoculated in Tryptic soy agar by more than 6 log$_{10}$CFU. On the other hand, Aguirre et al.[3] found 1–1.5 log reduction in Tryptic soy broth inoculated *L. innocua* by using a PL with 15.75 J/cm^2 fluence, 90 pulses, and 250 µs pulse width. Moreover, Lasagabaster and Martínez de Marañón[24] also reported more than 7 log reduction in *L. innocua* inoculated in brain heart infusion broth by PL treatment of 0.4 J/cm^2 fluence.

Several in vitro studies have been conducted for the inactivation of high levels of human pathogens by PL in fresh fruits and vegetables surface. Hoornstra et al.[19] treated kale, carrots, paprika, and cabbage with PL of two pulses equivalent to a fluence of 0.30 J/cm^2. For carrots and paprika vegetables, the surface aerobic counts were reduced by 1.6 and 2.6 log CFU/

cm^2, respectively. When the vegetables were treated with three pulses, no significant microbial inactivation and no adverse impacts on sensory properties were observed when stored at 7 and 20°C for 7 days. Various minimally processed vegetables (spinach, carrot, cabbage, radicchio, green bell pepper, lettuce) were treated with more than 2000 pulses. After treatment, the mesophilic aerobic counts were reduced in the range of 0.56–2.04 log CFU/g.[18]

Botrytis cinerea and *Monilinia fructigena* fungi, responsible for major nutritional and economical losses of strawberries and sweet cherries during postharvest storage and transportation, were reported for a maximum inactivation of 3 and 4 log units[30] with the application of PL. For the shelf-life enhancement of fresh-cut produces, PL treatment stands as an alternative to other nonthermal and thermal process. Moreira reported an almost 2 log CFU/g reduction of microbial counts in pectin-coated and PL-treated apple slices at 4°C during 14 days of storage period in dark. The impact of PL of higher fluence (17.5, 52.5, 105.0, and 157.5 kJ/m^2) on mesophilic bacteria in apple slices was studied.[20] Inoculated bacteria and viable counts in apple slices were decocted by 1 and 3 logs, respectively, regardless of fluence.

Koh et al.[21] introduced repetitive PL treatment (0.9 J/cm^2) to minimize the negative effect of single PL treatment (11.7 J/cm^2) on the structure of tissues of fresh-cut cantaloupe. For the horticultural produces with different and irregular surfaces (cauliflowers, sweet peppers, tomatoes, strawberries, plums), high power PL was used for microbial control.[26] PL treatment inactivated the naturally distributed mesophilic bacteria by 1.0–1.3 log, while inoculated *Bacillus cereus* was inactivated by 1.3–2.0 log depending on the surface type because it was more susceptible to PL. Table 6.1 indicates the PL treatments effects on inactivation of microorganisms in various horticultural produces.

TABLE 6.1 Pulsed Light Treatments Effects on Inactivation of Microorganisms in Fruits and Vegetables.

Fruit/ vegetable	Microorganism	Pulse energy (J/cm^2)	Pulse number	Pulse width (µs)	Log reduction	Reference
Strawberry	*Botrytis cinerea*	7	3750	30	3	[30]
	Escherichia coli	64.8	180	–	3.3	[6]
	Salmonella spp.	64.8	180	–	4.3	
	Bacillus cereus	5.4	1000	112	1.5	[26]
Shredded spinach	Mesophilic aerobic	7	2700	30	0.9	

TABLE 6.1   *(Continued)*

Grated celeriac	Mesophilic aerobic	7	675	30	0.21	
Chopped green paprika	Mesophilic aerobic	7	2700	30	0.56	
Shredded radicchio	Mesophilic aerobic	7	2700	30	0.56	[14]
Grated carrot	Mesophilic aerobic	7	675	30	1.67	
Blueberry	*Escherichia coli*	32.4	180	–	4.9	[5]
	Salmonella enterica	32.4	180	–	3.8	
Raspberry	*Escherichia coli*	72	180	–	3.9	[6]
	Salmonella spp.	59.4	180	–	3.4	
Fresh-cut mushroom	Aerobic, mesophilic, psychrophilic microorganisms, yeast, mold	28	–	300	0.6–2.2	[33]
Plum	*Bacillus cereus*	5.4	1000	112	1.4	
Tomato	*Bacillus cereus*	5.4	1000	112	1.5	[26]
Cauliflower	*Bacillus cereus*	5.4	1000	112	1.3	
Strawberry	Mesophilic bacteria, *Listeria monocytogenes*, and *Bacillus cereus*	1.95, 3.9, and 3.9, respectively	500, 1000, and 1000, respectively	112	2.2, 1.5, and 1.1, respectively	[27]
	Listeria monocytogenes and *Lactobacillus brevis*	1.75–15.75	–	50	1–3	[20]
Apple slice	Mesophilic aerobic, psychrophilic, yeast, mold	12	30	300	0.3–1.0	[32]
	Mesophilic aerobic	12	30	–	3.0	[31]
	Mesophilic and psychrophilic aerobic	12	30	300	2.0	
Avacado cubes	Mesophilic microorganism	14	35	300	1.2	[2]
Apple juice, orange juice, strawberry juice	*Escherichia coli, Listeria innocua, Salmonella enteritidis, Saccharomyces cerevisiae*	71.6	3	360	0.3–2.6	[12]
Lamb's lettuce water	*Salmonella enterica, Listeria monocytogenes, Escherichia coli*	1.75	–	50	6.0	[28]
Green onion	*Salmonella*	1.27	–	–	1.0	[44]
Fresh-cut cantaloupe	Mesophilic aerobic	11.7	–	–	2.0–4.0	[21]

6.2.2   ENHANCEMENT OF NUTRITIONAL AND BIOACTIVE COMPOUNDS

PL might be useful in extracting the bioactive compounds from the plant tissues to minimize the damages caused by high-power radiations under stress conditions. In addition, PL technology attributed with a very short exposure time required to achieve the desired effects could significantly be promoted for its utilization on the industrial scale.[41] Very few studies have been carried out to investigate the potential applications of PL in postharvest processing of horticultural produces to improve their nutritional qualities.

PL application did not have an impact on the initial phenolic content of fresh-cut mushrooms.[33] However, phenolic content of samples was substantially decreased during storage because of accelerated cell decompartmentation, which allows phenolic compounds to come into contact with tyrosine. Phenolic content of fresh-cut mushrooms can be best maintained for 1 week by using PL at 4.8 and 12 J/cm^2 fluence. They also reported that lower levels of browning were correlated fairly well with phenolic content as a main factor of the discoloration process. Also, the PL having fluence of 0.2–10 J/cm^2 increased the vitamin D_2 effectively in mushrooms[22] and phenolic content and total anthocyanin in figs by 350% and 133%, respectively.[41] Pataro et al.[34] reported an increment in total carotenoid (156%), lycopene (161%), total phenolic content (26%), and antioxidant activity (38%) without affecting physicochemical and organoleptic properties of green tomatoes by using PL treatment of 1–8 J/cm^2 fluence.

Particular verities (Tommy Atkins) of mangoes were subjected to PL fluence of 0.6 J/cm^2 (2 pulses) for their improvement in hydrogen peroxide, total carotenoid content, and total antioxidant activity by 20%, 451%, and 130%, respectively.[25] Increase in total phenolic content was attributed with increase in phenylalanineammonia–lyase (PAL) activity during PL treatment.

Rock et al.[40] reported that PL (three pulses) exposure enhanced the antioxidant activity (oxygen radical absorbance capacity) and total phenolic content of fresh blueberries by 50% and 48%, respectively, relative to the control. Increased total anthocyanin contents due to the upregulation of PAL enzymes were also observed after PL treatment. The fresh-cut avocados exhibited minimal peroxide formation along with 1.3-fold increment of chlorophyll *a* and *b* when its lipidic fraction was subjected to PL treatments (6 J/cm^2).[2]

6.2.3   ENZYME INACTIVATION

Fresh-cut fruits and vegetables undergo browning due to presence of poly-phenol oxidase (PPO), one of the major enzymes involved in food decay. It also leads to unattractive appearance of fruits and vegetables and loss in their nutritional qualities. The prospect to exploit PL to inactivate PPO in horticultural produces has been largely investigated. The impact of PL sturdily depends on the PPO concentration. PPO would be easily inactivated up to 10 U concentrations as proteins could experience both intramolecular modification and photoreaction with surrounding molecules.[29]

Dunn et al.[8] demonstrated a PPO activity inhibition in treated potatoes using two to five flashes of PL at a fluence of 3 J/cm^2 that could be linked to a browning reduction of samples. The PPO extract was less active when it was recovered from the treated slices as compared to that from the untreated slices. On the other hand, PL with a fluence of 8 J/cm^2 increased PPO activi-ties and maintained PAL activity of fresh-cut mangoes after 3 days of storage at 6°C. The treatment retained the carotenoid content, firmness, and the color of fresh-cut mangoes.[7]

Few studies have focused on the influence of PL on physiological and quality characteristics of fruits and vegetables. PL was also reported for increase in respiration rate of vegetables. The respiration rate of minimally processed lettuce increased after PL (pulse duration of 30 and a pulse inten-sity of 7 J), whereas it was not affected in minimally processed cabbage.[18] It indicated different susceptibility of vegetables to PL.

6.4   SUMMARY

As a substitute to chemical and thermal methods, PL is an evolving, eco-friendly and nonthermal technology. It has potential to inactivate spoilage microorganisms and in vitro pathogenic on food surfaces rapidly and effec-tively but its ability on true foods is still under exploration. Advance studies are imperative to evaluate the PL effect on properties of food beyond its safety and spoilage. Also, study on optimization of critical process factors is required further to achieve the target inactivation level for specific food applications.

Though short treatment times, limited energy cost, deficiency of residual compounds, and greater flexibility are some of the major benefits of PL technique at laboratory scale, yet the equipment having better penetration and short-treatment times should be designed from commercial point of view. Furthermore, the commercial or industrial application of PL treatments needs to be compared with industrial nonthermal or conventional thermal processes. Control of food heating and homogeneous treatment of food is the most important technological problems of PL treatment that need to be overcome. Understanding the involvement of physiological process and response of plant tissues and their specific secondary metabolic pathways to different abiotic stresses during PL treatment is necessary to improve the food quality. Furthermore, more research is required on the nutritional changes, toxic by-products formation, and photosensitization applicability in foods.

KEYWORDS

- **bioactive compounds**
- **fluence**
- **microbial decontamination**
- **pulsed light**
- **pulse number**
- **pulse width**
- **UV light**

REFERENCES

1. Aguiló-Aguayo, I.; Charles, F.; Renard, C. M. G. C.; Page, D.; Carlin, F. Pulsed Light Effects on Surface Decontamination, Physical Qualities and Nutritional Composition of Tomato Fruit. *Postharv. Biol. Technol.* **2013,** *86*, 29–36.
2. Aguiló-Aguayo, I.; Oms-Oliu, G.; Martín-Belloso, O.; Soliva-Fortuny, R. Impact of Pulsed Light Treatments on Quality Characteristics and Oxidative Stability of Fresh-Cut Avocado. *LWT—Food Sci. Technol.* **2014,** *59* (1), 320–326.
3. Aguirre, J. S.; Hierro, E.; Fernández, M.; García de Fernando, G. D. Modelling the Effect of Light Penetration and Matrix Colour on the Inactivation of *Listeria innocua* by Pulsed Light. *Innov. Food Sci. Emerg. Technol.* **2014,** *26*, 505–510.
4. Barbosa-Canovas, G. V.; Pothakamury, U.; Palou, E.; Swanson, B. G. *Nonthermal Preservation of Foods*; Marcel Dekker: New York, 1998; pp 139–161.

5. Bialka, K. L.; Demirci, A. Decontamination of *Escherichia coli* O157:H7 and *Salmonella enterica* on Blueberries Using Ozone and Pulsed UV-Light. *J. Food Sci.* **2007,** *72* (9), M391–M396.

6. Bialka, K. L.; Demirci, A. Efficacy of Pulsed UV-Light for the Decontamination of *Escherichia coli* O157:H7 and *Salmonella* spp. on Raspberries and Strawberries. *J. Food Sci.* **2008,** *73* (5), M201–M207.

7. Charles, F.; Vidal, V.; Olive, F.; Filgueiras, H.; Sallanon, H. Pulsed Light Treatment as New Method to Maintain Physical and Nutritional Quality of Fresh-Cut Mangoes. *Innov. Food Sci. Emerg. Technol.* **2013,** *18*, 190–195.

8. Dunn, J. E.; Clark, R. W.; Asmus, J. F.; Pearlman, J. S.; Boyer, K.; Painchaud, F.; Hofmann, G. A. *Methods for Preservation of Foodstuffs*, 1989, US Patent number 4871559.

9. EFSA. Shiga Toxin-Producing *E. coli* (STEC) O104:H4 2011 Outbreaks in Europe: Taking Stock. *EFSA J.* **2011,** *9* (10), 2390.

10. FDA. *CFR—code of federal regulations: 21CFR179.41*, 1996. http://www.accessdata. fda.gov/scripts/cdrh/cfdocs/cfcfr/CFRSearch.cfm?fr=179.41/ (accessed Aug 1, 2017).

11. FDA. *Environmental Assessment: Factors Potentially Contributing to the Contamination of Fresh Whole Cantaloupe Implicated in a Multi-State Outbreak of Salmonellosis*, 2013. http://www.fda.gov/Food/RecallsOutbreaksEmergencies/Outbreaks/ucm341476. htm (accessed Aug 1, 2017).

12. Ferrario, M.; Alzamora, S. M.; Guerrero, S. Study of Pulsed Light Inactivation and Growth Dynamics during storage of *Escherichia coli* ATCC 35218, *Listeria innocua* ATCC 33090, *Salmonella enteritidis* MA44 and *Saccharomyces cerevisiae* KE162 and Native Flora in Apple, Orange and Strawberry Juices. *Int. J. Food Sci. Technol.* **2015,** *50* (11), 2498–2507.

13. Giese, N.; Darby, J. Sensitivity of Microorganisms to Different Wavelengths of UV Light: Implications on Modeling of Medium Pressure UV Systems. *Water Res.* **2000,** *34* (16), 4007–4013.

14. Gomez-Lopez, V. M.; Devlieghere, F.; Bonduelle, V.; Debevere, J. Factors Affecting the Inactivation of Micro-Organisms by Intense Light Pulses. *J. Appl. Microbiol.* **2005,** *99* (3), 460–470.

15. Gómez-López, V. M.; Ragaert, P.; Debevere, J.; Devlieghere, F. Pulsed Light for Food Decontamination: A Review. *Trends Food Sci. Technol.* **2007,** *18* (9), 464–473.

16. Gómez, P. L.; García-Loredo, A.; Nieto, A.; Salvatori, D. M.; Guerrero, S.; Alzamora, S. M. Effect of Pulsed Light Combined with an Antibrowning Pretreatment on Quality of Fresh Cut Apple. *Innov. Food Sci. Emerg. Technol.* **2012,** *16*, 102–112.

17. Gómez, P. L.; Salvatori, D. M.; García-Loredo, A.; Alzamora, S. M. Pulsed Light Treatment of Cut Apple: Dose Effect on Color, Structure, and Microbiological Stability. *Food Bioprocess Technol.* **2012,** *5* (6), 2311–2322.

18. Gomezlopez, V.; Devlieghere, F.; Bonduelle, V.; Debevere, J. Intense Light Pulses Decontamination of Minimally Processed Vegetables and their Shelf-Life. *Int. J. Food Microbiol.* **2005,** *103* (1), 79–89.

19. Hoornstra, E.; de Jong, G.; Notermans, S. Preservation of Vegetables by Light. *Frontiers in Microbial Fermentation and Preservation*; Wageningen, the Netherlands, 2002; pp 75–77.

20. Ignat, A.; Manzocco, L.; Maifreni, M.; Bartolomeoli, I.; Nicoli, M. C. Surface Decontamination of Fresh-Cut Apple by Pulsed Light: Effects on Structure, Colour and Sensory Properties. *Postharv. Biol. Technol.* **2014,** *91,* 122–127.

21. Koh, P. C.; Noranizan, M. A.; Karim, R.; Nur Hanani, Z. A. Repetitive Pulsed Light Treatment at Certain Interval on Fresh-Cut Cantaloupe (*Cucumis melo* L. *reticulatus* cv. Glamour). *Innov. Food Sci. Emerg. Technol.* **2016,** *36,* 92–103.

22. Koyyalamudi, S. R.; Jeong, S.-C.; Pang, G.; Teal, A.; Biggs, T. Concentration of Vitamin D2 in White Button Mushrooms (*Agaricus bisporus*) Exposed to Pulsed UV Light. *J. Food Compos. Anal.* **2011,** *24* (7), 976–979.

23. Kramer, B.; Muranyi, P. Effect of Pulsed Light on Structural and Physiological Properties of *Listeria innocua* and *Escherichia coli. J. Appl. Microbiol.* **2014,** *116* (3), 596–611.

24. Lasagabaster, A.; Martínez de Marañón, I. Impact of Process Parameters on *Listeria innocua* Inactivation Kinetics by Pulsed Light Technology. *Food Bioprocess Technol.* **2012,** *6* (7), 1828–1836.

25. Lopes, M. M. A.; Silva, E. O.; Canuto, K. M.; Silva, L. M. A.; Gallão, M. I.; Urban, L.; Ayala-Zavala, J. F.; Miranda, M. R. A. Low Fluence Pulsed Light Enhanced Phytochemical Content and Antioxidant Potential of 'Tommy Atkins' Mango Peel and Pulp. *Innov. Food Sci. Emerg. Technol.* **2016,** *33,* 216–224.

26. Luksiene, Z.; Buchovec, I.; Kairyte, K.; Paskeviciute, E.; Viskelis, P. High-Power Pulsed Light for Microbial Decontamination of Some Fruits and Vegetables with Different Surfaces. *J. Food, Agric. Environ.* **2012,** *10* (3 and 4), 162–167.

27. Luksiene, Z.; Buchovec, I.; Viskelis, P. Impact of High-Power Pulsed Light on Microbial Contamination, Health Promoting Components and Shelf-Life of Strawberries. *Food Technol. Biotechnol.* **2013,** *51* (2), 284–292.

28. Manzocco, L.; Ignat, A.; Bartolomeoli, I.; Maifreni, M.; Nicoli, M. C. Water Saving in Fresh-Cut Salad Washing by Pulsed Light. *Innov. Food Sci. Emerg. Technol.* **2015,** *28,* 47–51.

29. Manzocco, L.; Panozzo, A.; Nicoli, M. C. Inactivation of Polyphenoloxidase by Pulsed Light. *J. Food Sci.* **2013,** *78* (8), E1183–E1187.

30. Marquenie, D.; Michiels, C. W.; Van Impe, J. F.; Schrevens, E.; Nicolaï, B. N. Pulsed White Light in Combination with UV-C and Heat to Reduce Storage Rot of Strawberry. *Postharv. Biol. Technol.* **2003,** *28* (3), 455–461.

31. Moreira, M. R.; Álvarez, M. V.; Martín-Belloso, O.; Soliva-Fortuny, R. Effects of Pulsed Light Treatments and Pectin Edible Coatings on the Quality of Fresh-Cut Apples: A Hurdle Technology Approach. *J. Sci. Food Agric.* **2016.** DOI:10.1002/jsfa.7723.

32. Moreira, M. R.; Tomadoni, B.; Martín-Belloso, O.; Soliva-Fortuny, R. Preservation of Fresh-Cut Apple Quality Attributes by Pulsed Light in Combination with Gellan Gum-Based Prebiotic Edible Coatings. *LWT—Food Sci. Technol.* **2015,** *64* (2), 1130–1137.

33. Oms-Oliu, G.; Aguiló-Aguayo, I.; Martín-Belloso, O.; Soliva-Fortuny, R. Effects of Pulsed Light Treatments on Quality and Antioxidant Properties of Fresh-Cut Mushrooms (*Agaricus bisporus*). *Postharv. Biol. Technol.* **2010,** *56* (3), 216–222.

34. Pataro, G.; Sinik, M.; Capitoli, M. M.; Donsì, G.; Ferrari, G. The Influence of Post-Harvest UV-C and Pulsed Light Treatments on Quality and Antioxidant Properties of Tomato Fruits during Storage. *Innov. Food Sci. Emerg. Technol.* **2015,** *30,* 103–111.

35. Ragaert, P.; Devlieghere, F.; Debevere, J. Role of Microbiological and Physiological Spoilage Mechanisms during Storage of Minimally Processed Vegetables. *Postharv. Biol. Technol.* **2007,** *44* (3), 185–194.

36. Ramos-Villarroel, A.; Aron-Maftei, N.; Martín-Belloso, O.; Soliva-Fortuny, R. Bacterial Inactivation and Quality Changes of Fresh-Cut Avocados as Affected by Intense Light Pulses of Specific Spectra. *Int. J. Food Sci. Technol.* **2014,** *49* (1), 128–136.

37. Ramos-Villarroel, A. Y.; Aron-Maftei, N.; Martín-Belloso, O.; Soliva-Fortuny, R. Influence of Spectral Distribution on Bacterial Inactivation and Quality Changes of Fresh-Cut Watermelon Treated with Intense Light Pulses. *Postharv. Biol. Technol.* **2012,** *69,* 32–39.

38. Ramos-Villarroel, A. Y.; Aron-Maftei, N.; Martín-Belloso, O.; Soliva-Fortuny, R. The Role of Pulsed Light Spectral Distribution in the Inactivation of *Escherichia coli* and *Listeria innocua* on Fresh-Cut Mushrooms. *Food Cntrl.* **2012,** *24* (1–2), 206–213.

39. Ramos-Villarroel, A. Y.; Martín-Belloso, O.; Soliva-Fortuny, R. Bacterial Inactivation and Quality Changes in Fresh-Cut Avocado Treated with Intense Light Pulses. *Eur. Food Res. Technol.* **2011,** *233* (3), 395–402.

40. Rock, C.; Guner, S.; Yang, W.; Gu, L.; Percival, S.; Salcido, E. Enhanced Antioxidant Capacity of Fresh Blueberries by Pulsed Light Treatment. *J. Food Res.* **2015,** *4* (5), 89.

41. Rodov, V.; Vinokur, Y.; Horev, B. Brief Postharvest Exposure to Pulsed Light Stimulates Coloration and Anthocyanin Accumulation in Fig Fruit (*Ficus carica* L.). *Postharv. Biol. Technol.* **2012,** *68,* 43–46.

42. Rowan, N. J.; MacGregor, S. J.; Anderson, J. G.; Fouracre, R. A.; McIlvaney, L.; Farish, O. Pulsed-Light Inactivation of Food-Related Microorganisms. *Appl. Environ. Microbiol.* **1999,** *65* (3), 1312–1315.

43. Sapers, G. M. Efficacy of Washing and Sanitizing Methods for Disinfection of Fresh Fruit and Vegetable Products. *Food Technol. Biotechnol.* **2001,** *39,* 305–311.

44. Xu, W.; Chen, H.; Wu, C. Application of Pulsed Light (PL)-Surfactant Combination on Inactivation of *Salmonella* and Apparent Quality of Green Onions. *LWT—Food Sci. Technol.* **2015,** *61* (2), 596–601.

CHAPTER 7

USE OF HURDLE TECHNOLOGY IN PROCESSING OF FRUITS AND VEGETABLES

KIRAN DABAS and KHURSHEED ALAM KHAN

ABSTRACT

Considering the current prospect of growth and development, the demand for minimally processed food is increasing day-by-day. The demand from the consumers is not only limited to quantity of the food but they also focus on quality aspect of foods. Conventional preservation methods are based on single preservation parameter, which alters the quality aspect of food in term of sensory and chemical properties. Hence, hurdle techniques make minimum changes and preserve the sensory and chemical attributes of food products and have become a boon for the efficient preservation. It can be concluded that further advancement is required in designing generic prediction models (boundary and dynamic) and analytical microbial response toward the hurdles. Then only the classification of hurdles can be defined with the mode of attack.

7.1 INTRODUCTION

The concept of hurdle technology involves making food safer by introducing barriers/hurdles during various processing steps to avoid the pathogens growth in foodstuff. By introducing such hurdles, the shelf-life of food products will be extended and their consumption will be safe as microorganism growth is prevented. Generally, hurdle technology works by integrated approach of many preservation methods. Such set of preservative methods play a roll of hurdles in the surviving way of pathogens in food products.

The proper design of hurdles results in complete removal of pathogens in food products.[11]

With the increasing industrialization, income generation enhances for many people. This increase in income leads to consumption of processed foods. Also due to the concentrated employment opportunities in urban area, fresh fruits/vegetables availability in these areas is less as compared to rural area. Such lack of fresh fruits/vegetables also shifted consumers toward using processed foods. With the increase in processed food consumption, consumers also demand the food with minimum/proper processing so that nutritional content of the processed food will not be affected badly. Hurdle technology can show the promising results for such needs. With the concern to achieve the target of microbial stability, multiple hurdles (intelligent mix of hurdle) are combined to improve the nutritional and sensory qualities of food. Thus, the overall quality aspect of food can be improved with the incorporation of hurdle technology.[14] Therefore, it becomes imperative to understand the homeostasis of microorganism for implementation of hurdle technology.

The microorganism always works toward the creation of conducive environment for its overall development and that inherit tendency is called homeostasis. The variation of hurdles used in the form of temperature, refrigeration, and preservative disturb the internal environment of microorganism, thus preventing the multiplication and repairing mechanism, thereby microorganism either become inactive or die. The purpose of preservation can be best served through the disturbance of homeostasis of microorganism. For increasing the effectiveness of this mechanism simultaneously, disturbance of multiple homeostasis can be carried out. Another important method is auto-sterilization. In this process, a high thermal treatment or refrigerated storage condition is given to the ambient temperature-stable food, which accelerates the process of microbial growth. The growth is in such a way that there is not sufficient food for their growth and compels the repair mechanism to overcome the hurdles and loses metabolic energy and die. Thus due to auto-sterilization, food will be safe during storage even at room temperature.[17]

This technology has been explored both in developed and developing countries for effective preservation of food. Few decades back, this novice technology was not in much trend due to insufficient knowledge and practical implementation. However, this novel technique has now formed strong base in the market for serving the purpose of preservation. The behavior of microorganism and its homeostasis has provided a vast era of knowledge, which clears the concept of metabolic exhaustion, stress reaction that plays a crucial role in the delaying or killing of microorganism.[23]

This chapter discusses about the concept of hurdle technology, types of hurdles, recent advancement in hurdle technology, and their effectiveness in food preservation. In this chapter, potential hurdles for preservation of food are discussed with their hurdle effects. Also homeostasis, metabolic exhaustion, and stress reactions of various microorganisms in relation to hurdle technology are discussed. Applications of hurdle technology in dairy, meat, and fruit/vegetable products are discussed.

7.2   HURDLE TECHNOLOGY

Hurdle technology provides guidelines to generate a way of combining the present and novel preservation techniques for achieving the combination of hurdles, in which the spoilage-causing microorganism cannot sustain.

Hurdle technology comprises a selective combination of different barriers (hurdles), which eventually will ensure the chemical and biological safety of food for human consumption. This method makes sure that the growth of microorganism will either be controlled or delayed. Here, hurdle is the resistant which will prevent the attack of pathogens on the food commodity and prolong the keeping quality with maintenance of organoleptic attributes of food. Organoleptic attributes define the sensory characteristics of the food in term of taste, smell, texture, and appearance. Thereby, it becomes imperative to arrange hurdles in such a way so that the food remains stable for an extended period of time. The hurdle must not be overcome by the pathogens so that a harmless final product can be stored and distributed.[14]

Different types of hurdles can be used in food processing like high temperature during the thermal treatment; low temperature during storage; regulation of acidity, water activity maintenance or redox potential, or preservatives (Table 7.1). Also the selection of hurdles depends on the types of microorganisms and their severity. According to the requirements, hurdles will be adjusted within the system to meet the safety standards without compromising the consumer health in an economical way.[14]

7.3   HURDLES

The designed hurdle will either kill the microbes/pathogens or inactivate their growth. Different hurdles can be used like common salt or organic acid to fulfill the target of microbial control in foods. Other antimicrobials like bacteriocins, nisin, and natamycin can also be used for this purpose.

TABLE 7.1   Principal Hurdles Used for Food Preservation.

Parameter	Symbol	Application
Biopreservatives	–	Competitive flora such as microbial fermentation
High temperature	F	Application of heat energy
Increased acidity	pH	Addition of acid
Low temperature	T	Chilling and freezing effect
Other preservatives	–	Sorbates, sulfites, nitrites
Reduced redox potential	Eh	Oxygen amount is reduced
Reduced water activity	a_w	Drying and curing

Although many possible hurdles have been identified for food preservation, yet some of the important hurdles are summarized in Table 7.1. From Table 7.1, it can be observed that the important hurdles used in food preservation are either high or low temperature (sterilization, blanching, frying, baking, pasteurization, freezing, chilling), water activity (a_w), acidity (pH), and redox potential.[10]

7.4   HURDLE EFFECTS

- No. 1 illustrates the concept of hurdle effect in the combination of six hurdles incorporated for the purpose of preservation. The first hurdle is heat (F value), which has been used during processing; further followed with low temperature (t value), water activity (a_w), pH, redox potential (Eh), and preservatives (pres.).[17] The microorganism cannot overcome this hurdle which makes the food stable. This concept is only a theoretical concept (Fig. 7.1).
- In No. 2 concept, water activity and preservatives are considered main hurdles and other redox potential, storage temperature, and pH are less important. If the numbers of microorganism are few, the product will be stable.
- The aseptic packaging is based on No. 3 concept.

- No. 4 concept explains the unhygienic conditions and initially large number of microorganisms are present and all the hurdles cannot prevent the spoilage of food products.
- No. 5 concept explains the condition in which the food product is nutrient rich and helps in the faster growth of microorganism; and this growth is called as booster or Trampoline effect. Therefore, there is need of increase in the intensity of the hurdles.

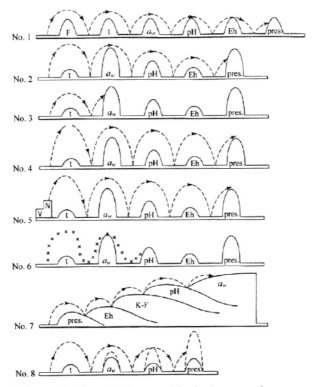

FIGURE 7.1   Effects of hurdle technology used for food preservation.
Source: https://www.slideshare.net/ammarbabar18/hurdle-technology-27567857
Legend:
a_w: water activity
Eh: redox potential
F: heating
K–F: competitive flora
N: nutrients
pH: acidification
pres.: preservative
t: chilling
V: vitamin

- No. 6 concept explains about the intermediate stage between the spoilage and microbes free product. If some microbes are killed sublethally by heat or cold, then only fewer hurdles will stop the growth of rest of the microorganism.
- No. 7 concept shows the level of hurdle used in fermented sausage, fermented vegetables, and ripened cheese.
- No. 8 concept shows the synergistic effect of hurdles; and this target the homeostasis of the microorganism.[14]

7.5   ASPECTS OF HURDLE TECHNOLOGY

During food preservation, microorganisms in a food have to face such environment, which inhibits or retards their growth or may result in death. The response of microorganism to this environment determines whether these microorganisms will grow or retard. Hurdle technique targets the physiological aspects of microorganism. This process involves four major mechanisms to delay or retard killing of the microorganism,[27] as below:

7.5.1   HOMEOSTASIS

Homeostasis is the tendency of microorganisms to maintain the stability and balance (uniformity) of their internal environment. For example, though the pH values of different foods are different, the microorganisms living in them keep their internal pH within a narrow limit. It is the natural maintenance process of microorganism through which internal environment is made conducive for the living cells. It can be illustrated with the homeostasis of the body, which is maintained with regulation of temperature and equilibrium between acidity and alkalinity.

These factors are basis of the development of microorganism. The understanding of homeostasis for higher organism is well known compared to microorganism. If this knowledge is incorporated in microorganism then spoilage or deterioration can be prevented through disturbance of homeostasis.[14] Once the homeostasis causing microorganism disturbance through the incorporation of hurdles, then the microbial spoilage will either be delayed or microorganism will be killed.

7.5.2   METABOLIC EXHAUSTION OR AUTO-STERILIZATION

Metabolic exhaustion of microorganisms is another crucial phenomenon, which could cause "auto-sterilization" of foodstuff. This concept was first observed when liver sausages adjusted to different water activities and then were inoculated with *Clostridium sporogenes* and kept at ambient temperature of 37°C.[18] Thereafter, the behavior of bacterial spore was taken into account during storage period. The outcome was quite surprising; instead of increasing the spore counts, it was decreased. In hurdle-treated products, microbes used energy in maintaining their homoeostasis. In the end, microbes will be metabolically exhausted and eventually die due to starvation. This process is called auto-sterilization. Therefore, the product is stable at ambient temperature.

7.5.3   STRESS REACTION

The stress reaction has been observed in microbes when they are under stress due to food shortage. Microbes become more resistant under the stress as they release stress shock proteins. The rate of formation of stress shock proteins depends on the pH, water activity, ethanol, and heat. The reactions of microbes under stress affect the process of food preservation. On other hand, the process of activation of genes for synthesis of stress shock proteins become more difficult if multiple hurdles are introduced at same time. This will lead to metabolic exhaustion of microbes. For the weakening of microbes, multiple hurdles should be given simultaneously to activate the energy utilization process. Therefore, the microbes become weak due to consumption of stress shock protein.

7.5.4   MULTITARGET PRESERVATION OF FOOD

It is an efficient way of food preservation. Hurdles used in food preservation act synergistically. Combination of different hurdles (pH, a_w, Eh) have synergistic effect on the cell of microorganism. This disturbs the homeostasis of microbes and restricts to synthesize stress shock proteins, which could maintain their homeostasis. Therefore with incorporation of different hurdles, optimal microbial stability and effective food preservation can be achieved.

7.6   APPLICATIONS OF HURDLE TECHNOLOGY IN DIFFERENT PRODUCTS

Examples of hurdles in a food system are high temperature during processing, low temperature during storage, increasing the acidity, lowering the water activity or redox potential, and the presence of preservatives. According to the type of pathogens and how risky they are, the intensity of the hurdles can be adjusted individually to meet consumer preferences in an economical way, without compromising the safety of the product. It is a novel concept with number of applications in the preservation of food products. Some of the sectors utilizing the potential of hurdle technology (Fig. 7.2) are discussed in this section.

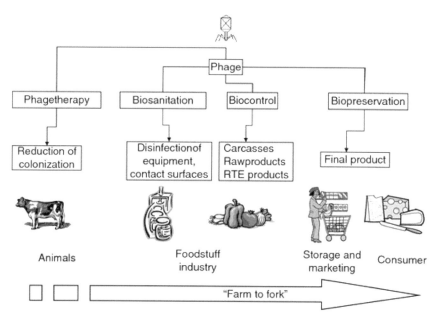

FIGURE 7.2   Examples of application of bacteriophage in food industry.
Source: García, P.; Martínez, B.; Obeso, J. M.; Rodríguez, A. Bacteriophages and Their Application in Food Safety, 2009 (Online). https://www.ncbi.nlm.nih.gov/pubmed/19120914. DOI:10.1111/j.1472-765X.2008.02458.x.

7.6.1   DAIRY PRODUCTS

The concept of hurdle technology can be used for improvement of the shelf life of dairy products. For example, paneer can be preserved by incorporation of different hurdles like pH, a_w, preservatives, and modified atmosphere packaging (MAP). Hurdle-treated paneer is stable for 1–12 days at room temperature and 6–20 days at refrigeration temperature. The sensory and chemical characteristics of paneer will remain unaffected during the storage period. In a similar way, the shelf life of paneer curry can be improved with control hurdles like water activity, pH, and preservatives. The product obtained after the application of hurdle technology was found to be of better quality compared to heat-sterilized product.

Another product brown-peda, milk-base product, was preserved using hurdle technology.[25] In this experiment, different packaging technology effects were studied (modified packaging and vacuum packaging) on the chemical sensory, textual, and microbiological characteristics of brown-peda during the storage of 40 days at 30°C. In the end, it was a stable product due to low moisture, high sugar, and strong heat treatment given during preparation. Therefore, it could be concluded that the hurdle-treated brown-peda could be preserved at ambient temperature approximately 40 days without affecting the quality.[2,32]

7.6.2   MEAT AND MEAT PRODUCTS

Hurdle technology is also used for meat and meat products preservation (Fig. 7.3). Number of experiments was conducted to study the effect of different hurdles in pork and some of hurdles that were taken namely pH, a_w, and vacuum packaging at refrigeration temperature.[16] Pork sausages, after the treatment with hurdles, reported the inhibition of growth of yeast and molds for approximately 12 days. Potassium sorbate dipping solution further increased the inhibition rate up to 30 days. Therefore, sausages were stable for 30 days without any deterioration. This technology has increased the shelf life of sausages up to 30 days. Similarly, ready-to-eat pickle-type buffalo meat products were also preserved through the combination of different hurdles pH, a_w, free fatty acid, thiobarbituric acid values, nitrite contents. Also, this technology has made possible to preserve the food used by space scientists and mountaineers.[21]

FIGURE 7.3 **(See color insert.)** Hurdle technology in meat processing.
Source: Irshad, A. *Laboratory Livestock Technology*; Veterinary College and Research Institute: Namakkal, TN, India, April 2015. https://pt.slideshare.net/irshad2k6/application-of-hurdle-technology-in-poultry-meat-processing-preservation/4

7.6.3 FRUITS AND VEGETABLES

Hurdle technology preservation technique can be considered for number of fruits and vegetables like carrot, pineapple, coconut, and papaya to improve the shelf-life. Grated carrots have been developed with application of hurdle technology.[36] Different hurdles like antimicrobials, partial dehydration, and polymeric bags were used to develop carrot (grated) that can remain fresh and safe for 6 months at ambient temperature. This technique can be applied on intermediate moisture pineapple to increase the shelf-life. For reducing the microbial load in pineapple slices, osmotic dehydration, infrared drying, and gamma radiation were used to increase shelf life up to 40 days. Another example is scrapped coconut, which was preserved using additives like humectants, acidulants, and preservatives. The coconut shelf life was increased to 3 months at 2–5°C and 1 month at the environmental temperature. High-moisture-grated papaya can also be prepared using different hurdles like mild heat treatment, a_w, pH reduction, and addition of preservatives. The obtained product is stable and safe microbiologically 5 months at ambient temperature.

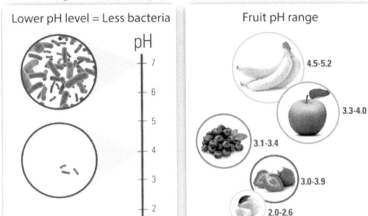

FIGURE 7.4   (See color insert.) Use of hurdle technology in fruits.
Source: http://foodingredients.treetop.com/resource-library/white-papers/
clearing-hurdles-with-real-fruit-ingredients/

7.6.4   FRUIT-DERIVED PRODUCTS

The target of the preservation of food products is effectively handled
through the combination of conventional hurdles and the novels one (Fig.
7.4). Different hurdles are UV light, pulse light, ultrasound, and high
hydroelectric pressure. Sugarcane juice is preserved with the application of
hurdles like heat treatment, preservatives, irradiation, and packaging material. The employed hurdles depict immense potential for improving the level
of product safety, thereby it can be applied to all kinds of food materials.
In fact, high-moisture fruit products such as pineapple, papaya, mango,
and banana can be effectively preserved using hurdle technology. Hurdle
combination used will be heat treatment, a_w, and antimicrobials for food
preservation.[1]

7.7   CONCEPT TO ACHIEVE THE TARGET OF FOOD PRESERVATION THROUGH HURDLE TECHNOLOGY

Food products are preserved to improve the shelf life and keeping quality.
There is a criterion on which preservation method is based. While designing

the method, few parameters are considered—microbiological, chemical, biochemical, and physical. Based on the level of changes in these parameters, time frame will be estimated for overall stability. Microbial and chemical safety aspects are more important compared to sensory properties. Other safety concern is microbial stability of both food used in the olden days and now processed foods, which use combination of different hurdles. Hurdles are arranged in such a manner that microorganism will not be able to overcome them.[15] List of hurdles used in industries includes heat treatment, water activity, pH, storage temperature, F-value concept, moisture content, property of glass transition phase, and these are some of the important factors in estimating food stability during storage. These hurdles are used according to the product requirement. For moisture-based food like canned products, F-value concept hurdle is used. Glass transition theory and water activity concepts hurdles are designed for very low moisture content products and products preserved at freezing temperature. F-value concept defines the sterility of the product, water activity is the state of water either free or bound and glass–rubber transition is the structural mobility. There are more than 60 hurdles used in food preservation.[14]

Considering only one hurdle in the safety of food product, means exploring high severity during processing, which can affect the nutritional and sensory attributes of food. Therefore, the concept of multiple hurdles came into picture for maintaining wholesome food product. The concept of hurdle has improved the preservation process as it controls microbial spoilage or other unwanted changes during storage.

Nowadays, the concept of hurdle technology is quite common among food manufacturers. The design of hurdle concept is estimated on the basis of empirical experiments. It is a challenge for food scientists to have a uniform method for analyzing the food shelflife with application of different hurdles. The stability MAP was given on the basis of the amount of water content and amount of solid content at different phase with the temperature.[19] For the target of safety, macro–micro areas concept helps in combining different hurdles and guides new path for deciding the use of hurdle in each of 13 micro-areas. Therefore, these developments provide guidelines in achieving the target of food preservation.[29]

7.7.1 USAGE OF HURDLE pH AND SALT

- The pH value less than 4.6 restricts the pathogens to multiply and is considered safe. This pH value is used as benchmark to segregate the food based on low- and high-risk foods. In fact, the Institute of Food Technologists (IFT) panel has summarized that the pH of 4.6 is the control value for spore-forming pathogens and pH of 4.2 is designated as control value for vegetative pathogens.[4]

- Salt (sodium chloride) is considered as most common preservative. In presence of high salt concentration, living of microorganism requires very high energy. According to the study,[11] salt concentration of 346 g/L (34.6%) was considered to stop all the microbial growths. But this hurdle cannot be used alone due to negative effect on health. In processed meat, the content of NaCl varies from product to product. About 2.8% for cooked sausages and 4.5% for cured meat product are considered as a high salt concentration. However, in general, salt shows good potential in combination with other hurdles.[24]

7.7.2 FACTORS USED FOR STABILITY OF FOOD PRODUCTS

7.7.2.1 THE F-VALUE CONCEPT

It helps to determine the time required for sterilization. In 1860s, the processing time was reduced to 30 min from 6 h due to number of research and developments.[35] The sterilization process depends on pH of food and type of bacteria (like heat resistant) to be inactivated. Food can categorized in three categories based on the pH values: pH values more than 4.6 as low-acid foods, pH value from 3.7 to 4.6 as medium-acid food, and pH values less than 3.7 as high-acid foods. In case of low-acid food, the microorganism *Clostridium botulinum* is taken as a target organism during thermal processing. For estimation of processing time, first-order kinetics equation is used as follows[3]:

$$t = \frac{1}{k} \ln\left(\frac{N_F}{N_0}\right) = \frac{D}{2.303} \ln\left(\frac{N_F}{N_0}\right) \tag{7.1}$$

where k is defined as the rate of destruction denoted as per second, N_0 is the initial number of microorganism, N_F is the number of microorganisms after thermal processing, and D is the time required to kill 90% of the microorganism at particular temperature in seconds.

Practically, it is not possible to kill the entire group of microorganisms and attain final concentration of the target organism as zero. If the final concentration approaches zero, then the time approaches to infinity.

However, it is not possible; therefore, the sterilization process is designed in finite frame of time to attain the safety level for the target microorganism. The economic and practical feasibility must be checked before confirming the process. The criteria should also be fulfilled: which states that "after processing the initial population of microorganism should be reduced up to 10^{12}." This term is called as 12D or "*botulinum* cook."[8] The process of sterilization is used for some specific purpose: (1) it defines the safety levels, (2) it fulfills the cooking need, and (3) it prevents the growth of spoilage causing microorganism like thermophilic.

According to the prediction or the probability distribution concept, there will be chances of finding one spore in the count of 10^{12} cans. Suppose the consumption of cans in a day is around 100 million cans. Therefore, for the consumption per day, it will be approx. 3.65×10^{12} cans per year; and according to the 12D concept, there are possibility of 3–4 outbreaks in every 100 years.[35] For the estimation of processing time to achieve 12D, the D value at the specified temperature must be known. The time required to achieve the 12D depends on the size of can used.

Considering the practicability of thermal processing treatment, a range of temperature is selected instead of a specific temperature during heat treatment and cooling treatment; hence, the time required to kill bacteria at specific temperature is denoted as F_r that is calculated during processing as follows[12]:

$$F_r = \int_0^t 10^{(T-T_r)/Z}\, dt \qquad (7.2)$$

where F_r denoted as the time required to kill bacteria at specific temperature in seconds, T_r as specific temperature in °C, and z is the thermal constant.

Suppose specific temperature is 121.1°C, then F_r can be written as F_0:

$$F_r = \int_0^t 10^{(T-121.1)/Z}\, dt \qquad (7.3)$$

Nowadays, sterilization treatment for *C. botulinum* at commercial level is used in range of 6–8 min. However, it is not same everywhere as some companies use temperature up to 10 min. It can be illustrated with an example: A can with 5 L capacity requires 6 min at 121.1°C to sterilize with 1.335×10^{-25} spores per package. According to probability, there is a chance of one outbreak in 100,000 billion years. In the last 50 years, there has been no outbreak directly due to sterilization.[35] Also, little research is done to find out the value of F_0 if other hurdles are used.

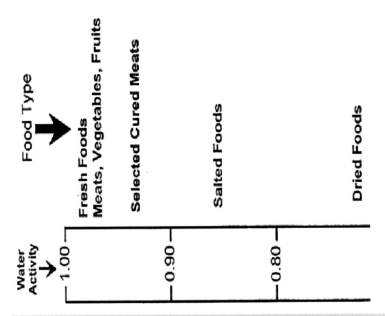

Water Activity Limits: Microbial Growth

Water Activity	Microorganism
<0.60	No microbial proliferation
0.61	Xeromyces bisporus (xerophilic fungi)
0.62	Zygosachharomyces rouxii (osmophilic yeast)
0.78-0.75	Aspergillus niger/flavus, Halobacterium halobium
0.84-0.81	Penicillium chrysogenum/glabrum, Paecilomyces variotti, Aspergillus fumigatus
0.86	Staphylococcus aureus
0.90	Bacillus subtilis, Saccharomyces cerevisiae
0.92	Mucor plumbeus, Rhodotorula mucilaginosa
0.93	Micrococcus lysodekticus, Rhyzopus nigricans
0.94	Enterobacter aerogenes
0.95	Bacillus cereus, Clostridium botulinum/ perfringens, Escherichia coli, Lactobacillus viridescens, Salmonella spp.
0.97	Pseudomonas aeruginosa

FIGURE 7.5 **(See color insert.)** Tolerance level of a_w (water activity) for microorganism multiplication.
Source: http://www.pharmaquality.com; http://www.meatsandsausages.com/sausage-types/fermented-sausage

7.7.2.2 WATER ACTIVITY

Water activity has important role for determining the shelf life of the food. The scientist W. J. Scott asserted that the active water has greater impact than the whole quantity of water in the food. The active water supports the food stability. This concept helped other scientists to determine regulations for the food stability,[32,34] such as:

- The food products follow the "BET—monolayer water activity" and are stable at this layer only.[9] Any deviation above or below will create unstability in the food products.[32]
- The specific limit of water activity is important for various biochemical reactions and defines the growth of microorganism.[13] Figure 7.5 indicates tolerance level of a_w (water activity) for microorganism multiplication. It can be illustrated with an example as follows:

Beyond a certain critical water activity, the growth of microorganism is almost nil. Harmful microorganism like pathogens cannot grow at water activity less than 0.85. However, yeast and molds require further reduction of water activity to around 0.6 for no growth. A MAP for the stability of food on the relation of growth of microorganism and water activity necessary for carrying out different biochemical reactions are used. The food stability MAP divides the growth trends in three zones: *Zone 1*: mono-layer Brunauer–Emmett–Teller (BET): BET-monolayer can only be achieved for dried foods; *Zone 2*: multilayer adsorption; and *Zone 3*: aqueous layer. In fact, food industries are implementing food stability MAP concept for designing food preservation for low-moisture-containing foods. Water activity is defined as $a_w = [RH/100] = p/p_0 = f/f_0$, where: RH is the relative humidity; p is the vapor pressure of water in the substance; p_0 is the vapor pressure of pure water; f is the fugacity of a substance; and f_0 is the fugacity of a pure material.

7.7.2.3 MOLECULAR MOVEMENT AND CRITICAL TEMPERATURE

The constant rate of reaction is plotted as function of the ratio of preservation temperature (T) and temperature at which glass transition take place (T_g) or the ratio of the moisture content of food material (X_w) and mono-layer BET (X_b). There will be deviation in the slope, above and below of critical ratio in the T/T_g graph at 1 only if glass transition theory exists and X_w/

X_b value will be 1 only if water activity theory exists.[11] Based upon this hypothesis, chemical reactions on many food products were completed and put forward that critical temperature is the ratio of T/T_g which varies with moisture content. The rate of reaction is not affected by moisture content and temperature below critical temperature, but above the critical temperature, it increases with increase in moisture content and temperature. The experimental value of T/T_g ranges from 0.78 to 1.5 and nearer to 1.0 indicates the glass transition.[33]

Molecular movement were studied with sophisticated instruments like saturation transfer electron paramagnetic resonance spectroscopy and proton nuclear magnetic resonance (1N-NMR) and results has small change is nearer to T_g and steep decline from solid phase to liquid phase called as critical temperature (T_c). Critical temperature (T_c) of sugar is 17–35°C higher than T_g. Compactness of hydrogen bonds and molecular structures alignment was analyzed with Fourier transform and infrared spectroscopy and was correlated higher value of T_c with steep decline. Sugar molecules got soft at 10–17°C above T_g with crystallization at temperature more than 30°C.[19]

In general, crystallization occurs above glass temperature but some scientists claim that it can be less than 30°C to glass temperature.[30] Alpha, beta, and gamma molecular relaxation below the glass transition were analyzed with dielectric and different spectroscopy. Nuclear magnetic resonance (NMR) is widely applicable in measuring glass transition at subatomic movement in the range of 1–2 nm over thermal and mechanical relaxation mobility of range of 20–300 nm.[23]

7.7.2.4 GLASS TRANSITION

Glass transition theory concept came after the water activity theory limitation was considered. Some of the basic concepts are as follows: (1) the stability of glass transition food is more at and below its glass transition (T_g or T_g'); and (2) more the $T - T_g$ or T/T_g, more will be the rate of decomposition.[19,37] The techniques which have been researched till now are not sufficient to explain all the experimental results; therefore, more research is needed in this direction. Although limitations are there in these techniques, yet it does not disqualify them completely, but remained a challenge to be used universally. As has been discussed before, the state of water presence is very important in the determination of stability of food. If the water is present in bound form with the food matrix or nonsolvent, then there will be fewer chances for the chemical reactions to take place and deterioration would be much less. In case of glass transition, the movements of molecules of reacting species are important at subatomic stage which slows down the diffusion of reactant through the system and assist in reaching the target of stability. A precise mix of water activity concept and glass transition theory can help for estimation of stability.[6,31]

7.7.2.5 STATE DIAGRAMS

State diagrams or phase diagram were developed for further advancement of glass transition technique. It is the modeling of stability MAP for various states and phases of food. The designing of MAP is based on the amount of water content concept and phase temperature. In many studies, the phase diagram was designed through the help of glass line, freezing curve, and their interaction as T_g'' by extrapolation. The major reason behind this phase diagram creation is to develop an understanding of complex changes. Those changes happen during the variation in the moisture content and temperature of foods.[19] Thereby, this concept will advocate the storage capability of food at particular moisture percentage level and temperature.[19]

7.7.2.6 DIFFERENT MICRO-AREAS IN PHASE DIAGRAM

Thirteen different micro-areas are shown on the phase diagram[31] based on the highest to lowest stability. Area 1 is considered nonreacting zone below the mono-layer BET and glass line and area 13 is highly reacting zone far from mono-layer BET and glass line. As the area number increases, the stability of food will decrease.[31] It was validated after studying thoroughly the chemical reaction of sucrose in the micro-areas 1 and 2.[28]

In this phase diagram [31], T_{ds}: temperature of solids decomposition; Tm_s: temperature at which solid melts, Tg_s: temperature of solids–glass transition; T_g^{iv}: temperature of end of solid plasticization; T_{gw}: temperature of glass transition of water; T_u: Temperature of solute crystallization during freeze concentration; T_m': temperature of end point of freezing; T_g''': temperature of glass transition of the solid contents in the frozen sample; T_g'': temperature of intersection of the freezing curve with glass line; T_g': temperature of glass transition at maximal-freeze concentration; and T_{bw}: temperature of boiling water.

There are following advantages of micro-areas concepts as below:

- Food stability standards can be made for each micro-area (food with low moisture content) with comparison to macro-area (food with high moisture content).
- The phase of the food material can be studied in each micro-area of the phase diagram. An intersection point of mono-layer BET and glass line will identify exact state or phase.[31]

7.8   ADVANCEMENT OF THE HURDLE TECHNOLOGY

The purpose of designing a range of hurdles during storage is to enhance the chemical and microbiological stability with the consideration of sensory and nutritional concept at economic cost. So, it is the intelligent mix of different hurdles. The major benefit of hurdle technology is its synergistic effect without the addition of additives. Previously, the knowledge regarding the hurdle technology was only superficial which could not able to bring out the governing principles. The physiological factors and behavior of microorganism are some of major aspect for applying the hurdle technology.[17] Three major behavior changes cause through changing the internal environment, changing the metabolism and providing shock. Microorganism has a natural tendency through their homeostasis to maintain uniformity and stability at the cost of spending energy for repairing their internal stability rather than in multiplication, thus needs more extra energy. This extra burden wouldn't allow microorganism to multiply and ultimately creates situation to die in course of attaining re-establishment by repair.

Data show that auto-sterilization occurred at unrefrigerated condition and further hurdles accelerate the metabolic exhaustion.[14] Another example is air-dried fermented sausage, which was prepared through traditional method in Germany. This product showed good record of safety because the temperature of fermentation was less than 15°C and prolong aging of the product was done.[20]

Preservation of food should be based on the concept of multitarget preservation technique. Till now, it has been known that there are 12 classes of biocides with different targets inside the microbial cell. Usually, the target is cell membrane and once the cell wall is damaged, the material inside the cell leaks out. Biocides also do the impairment of enzymes during synthesis of protein and deoxyribonucleic acid.[7] Indian paneer is indigenous milk product prepared with acid coagulation and heat treatment which usually spoils bacteriologically at room temperature within 1–2 days. It is the major disadvantage of this product in the path of industrial production.

Canned-sterilized paneer are less accepted by consumer due to sensory limitations in flavor, texture, and color. Application of hurdle technology increases the shelf life from 2 weeks to 3 months at different temperature with retaining sensory attributes.[27]

7.8.1 MODELING/PREDICTION OF MULTIPLE HURDLES

A mathematical model which describes the logarithm nature of thermal death was developed in 1930s and was unnoticed till 1980s. In 1980s, a new model came up, which could predict the behavioral changes of microorganism and can be opted as follows:

- Identify the upper boundary limitation of the growth when reaction takes place and similarly the state when no growth occurs and there is no reaction in stipulated time duration.
- Prediction of growth rate or reaction in the region where reaction take place.

A new channel opens up in the field of microbiology area, which was called predictive microbiology.[3] In this technique, different empirical growth models are used. It depends on the type of microorganisms, properties of food (physicochemical) and hurdle used. The thermal inactivation of *C. botulinum* showed that it depends linearly on the pH, and inversely on the temperature. The change in pH and water activity makes difficulties in the predicting effect of hurdles in combination. An experiment was also conducted on different sterile pork samples in isothermal environment and generated growth data of *Salmonella* and model was presented at the temperature for storage between 10°C and 40°C.

In 1970, a concept of probability model emerged. In this model, predictions were estimated for the formation of spore, spore growth, continuity, and toxin release in the given time period at particular storage temperature and the composition of food product.[5] Further progress also took place on the concept of growth (G)/no growth (NG) modeling. The modeling on G/NG was correlated with the boundary through an empirical equation, which was based on water activity and temperature limits for fungus species growth.

Probability and G/NG modeling are based on the logistic regression technique, which is widely accepted. Usually, probability models are based on the data and measured as positive and negative. It can be illustrated as follows: if toxin detection considered as variable, then two types of output in the form of response are as detectable or nondetectable or further positive response or negative response. It can be coded for convenience as 1 for positive response and 0 for negative response. The probability logistic regression denoted as "logit."[26]

$$\text{Logit } P = \log\left[\frac{P}{(1-P)}\right] \qquad (7.4)$$

The "growth/no growth model" is based on linear logistic regression and log it. It was correlated as the linear function of pH, water activity, amount of NaCl, and acetic acid added. The simpler form of growth/no growth model was derived from the gamma model. Variables or factors are related for determining the boundary logit. Scientists have also estimated growth/no growth boundary with the help of gamma function and found very little number of failure predictions. Therefore, the simpler G/NG model is recommended as the first estimated method in the unavailability of the experimental data. Along with this, practical aspect should be considered for the generation of boundary data. Considering the complexity of the food matrix, extrapolation in prediction is not advisable always.[22] Practical value is very important to find out the real changes that happen with time.

Currently, the stable food products usually follow the concept of one or more hurdles. Different hurdles are moisture, salt, heat treatment, storage temperature, and preservatives. Developing a unified concept for the food stability is not less than a challenge for the food scientists and engineers. Each product can be individually assessed. The stability of the food product can be achieved on the hurdles used through the logical tree. Due to limitation of availability of specific criteria for estimation of food stability, experimental data are more useful for multihurdles food. Previously discussed concepts on F-value, glass transition theory, and water activity concept help in determining criteria of food stability. It includes temperature treatment, moisture percentage, and combination of both in controlled way required to be maintained the targeted food products.

In the phase diagram, glass transition theory and water activity concept are used. The recent advancement in the phase diagram is the 13 macro–micro areas, which provide a way to find out various real challenges faced by scientist while designing different hurdles in the food preservation. To overcome these challenges, recent progress has been noted in the microbial reaction, prediction models, and molecular mobility can be used to combine intelligently to get required hurdles. Therefore, it can be concluded that further advancement is required in designing generic prediction models (boundary and dynamic) and analytical microbial response toward the hurdles. Then only, the classification of hurdles can be defined with the mode of attack.

7.9   SUMMARY

Considering the current prospect of growth and development, the demand for minimally processed food is increasing day-by-day. The demand from the consumers is not only limited to quantity of the food but they also focus on quality aspect of foods. Conventional preservation methods are based on single-preservation parameter, which alters the quality aspect of food in term of sensory and chemical properties. Hence, hurdle techniques make minimum changes and preserve the sensory and chemical attributes of food products and have become a boon for the efficient preservation.[27]

The new study in the area of water activity and glass transition theory along with the phase diagram can open the new channel for food stability. Water movement inside the food product can be estimated by NMR, spin probe electron spin resonance, and luminescence spectroscopy. This approach will further help to define the food stability parameters which in turn act as a tool to handle complex food matrices.[31]

KEYWORDS

- auto-sterilization
- exhaustion
- glass transition
- homeostasis
- hurdle
- microorganism
- molecular

REFERENCES

1. Alzamora, S. M.; Cerruti, P.; Guerrero, S.; López, A. M. Minimally Processed Fruits by Combined Methods. In *Food Preservation by Moisture Control: Fundamentals and*

Applications; Barbosa-Cánovas, G. V., Welt-Chanes, J., Eds.; Technomics Publishing: Lancaster, PA, 1995; pp 463–492.

2. Bhattacharya, D. C.; Mathur, O. N.; Srinivasan, M. R.; Samlik, O. Studies on Method of Production and Shelf-Life of Paneer. *J. Food Sci. Technol.* **1971,** *8* (5), 117–120.

3. Bigelow, W. D. The Logarithmic Nature of Thermal Death Time Curves. *J. Infect. Dis.* **1921,** *29,* 528–536.

4. Brown, M. H.; Booth, I. R. Acidulants and Low pH. In *Food Preservatives*; Russell, N. J., Gould, G. W., Eds.; Springer: Glasgow; 1991; pp 22–43.

5. Carrasco, E.; Del, R. S.; Racero, J. C.; Garcia-Gimena, R. M. A Review on Growth/No Growth *Salmonella* Models. *Food Res. Int.* **2012,** *47,* 90–99.

6. Champion, D.; Le, M. M.; Simatos, D. Towards an Improved Understanding of Glass Transition and Relaxation in Foods: Molecular Mobility in the Glass Transition Range. *Trends Food Sci. Technol.* **2000,** *11,* 41–45.

7. Denyer, S. P.; Hogo, W. B., Eds. *Mechanism of Actions of Chemical Biocides. Their Study and Exploitation*; Blackwell Scientific Publication: Oxford, 1991; p 343.

8. Esty, J. R.; Meyer, K. P. The Heat Resistance of Spores of Botulinus and Allied Anaerobes. *J. Infect. Dis.* **1922,** *31,* 650–663.

9. Fundo, J. F.; Qintas, M. A. C.; Silva, C. L. M. Molecular Dynamics and Structure in Physical Properties and Stability of Food System. *Food Eng. Rev.* **2015,** *7* (4), 384–392.

10. *Ganguly, S. Biologically Viable Methods for Food Preservation: A Review. Res. J. Chem. Environ. Sci.* **2013,** *1 (2), 1–2.*

11. Gray, T.; Killgore, J. ; Luo, J.; Jen, A. K. Y.; Overney, R. M. Molecular Mobility and Transitions in Complex Organic Systems Studied by Shear Force Microscopy. *Sci. J. Nanotechnol.* **2007,** *18* (4), 1–9.

12. Holdsworth, D.; Simpson, R., Eds. *Thermal Processing of Packaged Foods*; Springer Science: New York, 2008; pp 87–122.

13. Labuza, T. P.; McNally, L.; Gallagher, D.; Hawkes, J.; Hurtado, F. Stability of Intermediate Moisture Foods. *J. Food Sci.* **1972,** *37,* 154–159.

14. Leistner, L. Basic Aspects of Food Preservation by Hurdle Technology. *Int. J. Food Microbiol.* **2000,** *55,* 181–186.

15. Leistner, L. Hurdle Effect and Energy Saving. In *Food Quality and Nutrition*; Downey, W. K., Ed.; Applied Science Publishers: London, UK, 1978; pp 553–557.

16. Leistner, L. Hurdle Technology Applied to Meat Products of the Shelf Stable Products and Intermediate Moisture Food Types. In *Properties of Water in Foods Relation to Quality and Stability*; Simatos, D., Multon, J. L., Eds.; Martinus Nijhoff Publishers: Dordrecht, Netherlands, 1978; pp 309–329.

17. Leistner, L. Principles and Applications of Hurdle Technology. In *New Methods of Food Preservation*; Gould, G. W., Ed.; Blake Academic and Professional: London; 1995; pp 1–21.

18. Leistner, L.; Karan, S. D. Influence of the Stability of the Meat Preserves by Controlling Water Activity. *Meat Ind.* **1970,** *50,* 1547–1549.

19. Levine, H.; Slade, L. Water as Plasticizer: Physicochemical Aspects of Low-Moisture Polymeric System. In *Water Science Reviews*; Franks, F., Ed.; Cambridge University Press: Cambridge, 1998; Vol. 3; pp 79–186.

20. Lucke, F.; Vogeley, I. Traditional 'Air-Dried' Fermented Sausages from Central Germany. *Food Microbiol.* **2012**, *29*, 242–246.

21. Malik, A. H.; Sharma, B. D. Shelf-life Study of Hurdle Treated Ready-to-Eat Spiced Buffalo Meat Product Stored at 30°C ± 3°C for 7 Weeks under Vacuum and Aerobic Packaging. *J. Food Sci. Technol.* **2014**, *51* (5), 832–844.

22. Mertens, L.; Dang, T. D. T.; Geeraerd, A. H.; Vermeulan, A.; Van, D. E.; Cappuyns, A. M.; Debevere, J.; Devlieghere, F., Van, I. J. F. A Predictive Model for the Growth/No Growth Boundary of *Zygosaccharomyces bailii* at 7°C and Conditions Mimicking Acidified Sauces. *Food Bioprocess Technol.* **2012**, *5*, 2578–2585.

23. Mor, C. K.; Prajapati, J. P.; Pinto, S. V. Application of Hurdle Technology in Traditional Indian Dairy Products. In *National Seminar Paper at Indian Diary Industry—Opportunities and challenges XIth Alumni Convention*; Anand Agricultural University: Anand; Jan 8–9, 2015; pp 147–150.

24. Oren, A. Thermodynamic Limits to Microbial Life at High Salt Concentrations. *Environ. Microbiol.* **2011**, *13*, 1908–1923.

25. Panjagari, N. R.; Londhe, G. K.; Pal, D. Effect of Packaging Techniques on Shelf-Life of Brown Peda, A Milk Based Confection. *J. Food Sci. Technol.* **2007**, *47*, 117–125.

26. Pitt, R. E. A Descriptive Model of Mold Growth of Aflatoxin Formation as Affected by Environmental Conditions. *J. Food Protect.* **1992**, *56*, 139–146.

27. Pundhir, A.; Murtaza, N. Hurdle Technology—An Approach towards Food Preservation. *Int. J. Curr. Microbiol. Appl. Sci.* **2015**, *4* (7), 802–809.

28. Rahman, M. S. State Diagram of Foods: Its Potential Use in Food Processing and Product Stability. *Trends Food Sci. Technol.* **2006**, *17*, 129–141.

29. Rahman, M. S.; Siddiqui, M. W. Minimally Processed Foods. In *Food Engineering Series 2;* Springer International Publishing: Switzerland, 2015; pp 17–31.

30. Rahman, M. S. Plenary Lecture: Food Stability Determination: Challenges Beyond Water Activity and Glass Transition Concepts. In *6th Conference on Water in Food*, Reims, France; Mar 21–23, 2010; pp 8–34.

31. Rahman, M. S. Food Stability Determination by Macro–Micro Region Concept in the State Diagram and by Defining a Critical Temperature. *J. Food Eng.* **2010**, *99*, 402–416.

32. Rao, K. J.; Patil, G. R. Development of Ready to Eat Paneer Curry by Hurdle Technology. *J. Food Sci. Technol.* **1999**, *36*, 37–41.

33. Sacchetti, G.; Neri, L.; Bertolo, G.; Torreggiani, D.; Pittia, P. Influence of Water Activity and Molecular Mobility on Peroxidase Activity in Solution. In *Water Stress in Biological, Chemical, Pharmaceutical and Food Systems*; Gutiérrez-López, G., Alamilla-Beltrán, L., del Pilar Buera, M., Welti-Chanes, J., Parada-Arias, E., Barbosa-Cánovas, G., Eds; Food Engineering Series; Springer: New York, 2015; pp 289–298.

34. Scott, W. J. Water Relations of *Staphylococcus aureus* at 30°C. *Austr. J. Biol. Sci.* **1953**, *6*, 549–564.

35. Simpson, R.; Nunez, H.; Almonacid, S. Sterilization Process Design. In *Handbook of Food Process Design*, 1st ed.; Ahmed, J., Rahman, M. S., Eds.; Willey-Blackwell: Oxford, 2012; pp 362–379.

36. Vibhakar, H. S. J.; Gupta, D. K. D.; Jayaraman, K. S.; Mohan, K. S. Development of high-Moisture Shelf-Stable Grated Carrot Product Using Hurdle Technology. *J. Food Process. Preserv.* **2006,** *30*, 134–144.

37. White, G. W.; Cakebread, S. H. The Glassy State in Certain Sugar-Containing Food Products. *J. Food Technol.* **1996,** *1*, 73–82.

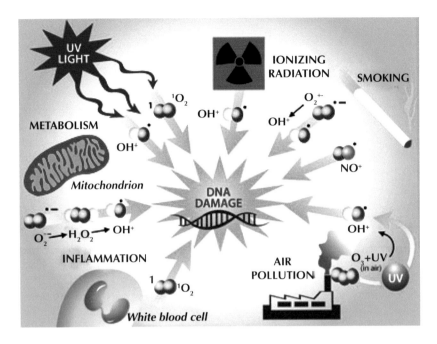

IMPLICATED DISEASE STATES

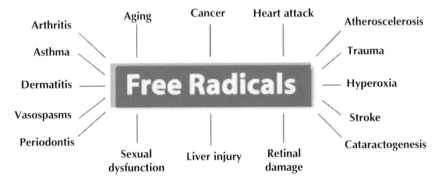

FIGURE 1.1 Factors responsible (top) for production of free radicals that cause various diseases (bottom).
Source: http://altered-states.net/barry/newsletter708/

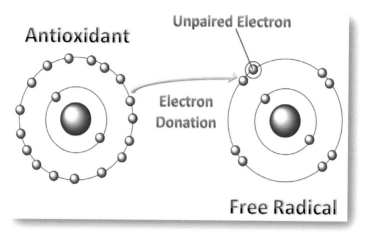

FIGURE 1.3   Mechanism of antioxidant activity by vitamin E.

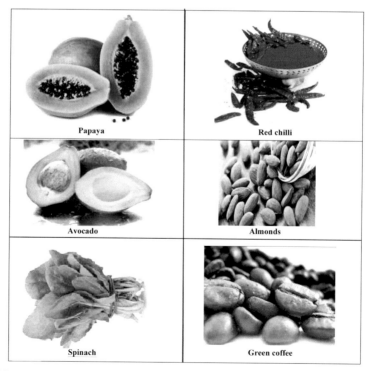

FIGURE 1.4   Structure of vitamin E (top) and selected examples of foods: sources of vitamin E.

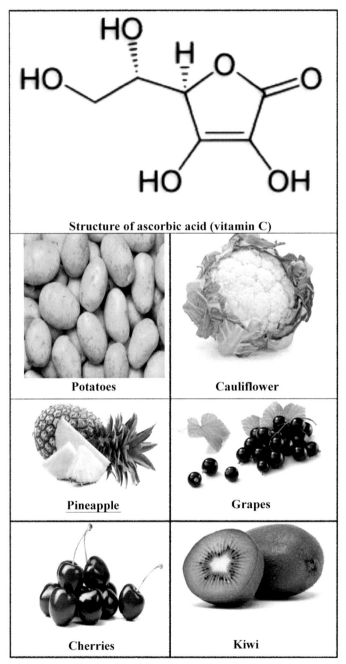

Structure of ascorbic acid (vitamin C)

Potatoes	Cauliflower
Pineapple	Grapes
Cherries	Kiwi

FIGURE 1.5 Structure of vitamin C (top); selected examples of foods: sources of vitamin

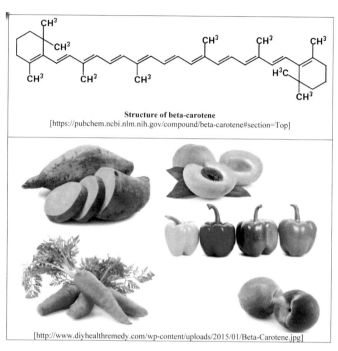

FIGURE 1.6 Structure of vitamin A (top); selected examples of foods: sources of vitamin A (β-carotene).

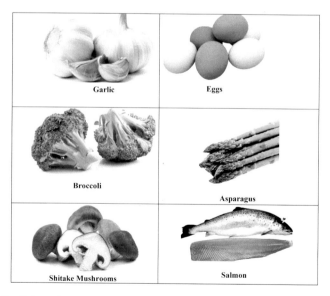

FIGURE 1.7 Selected examples of foods: sources of selenium.

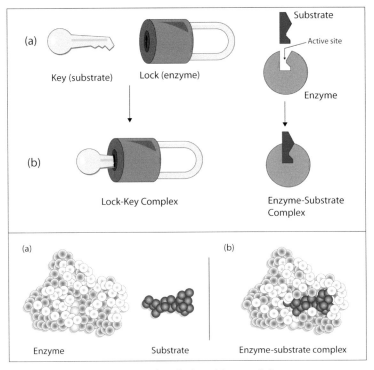

FIGURE 2.1 Enzyme–substrate reaction (lock-and-key model).
Source: https://saylordotorg.github.io/text_the-basics-of-general-organic-and-biological-chemistry/s21-06-enzyme-action.html

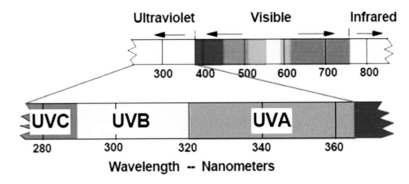

FIGURE 5.1 Ultraviolet spectrum: Courtesy of Colorado State University, UV-B monitoring & Research Program funded by USDA (US Government).
Source: http://www.cannabisgreen.com/evolutionary-effects-ultraviolet-radiation-b-power-psychoactive-cannabis/

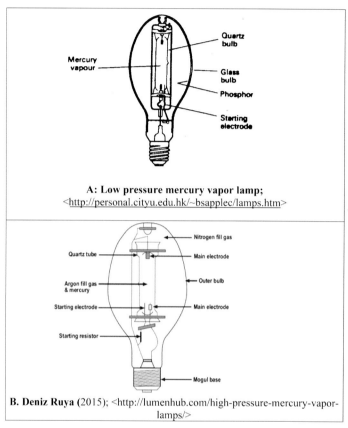

FIGURE 5.2   Mercury lamps.

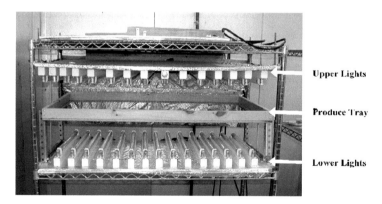

FIGURE 5.3   UV-C radiation device.
Source: Reprinted with permission from Ref. [5]. © 2006 Elsevier.

FIGURE 7.3 Hurdle technology in meat processing.
Source: Irshad, A. *Laboratory Livestock Technology*. Veterinary College and Research Institute: Namakkal, TN, India, April 2015. https://pt.slideshare.net/irshad2k6/application-of-hurdle-technology-in-poultry-meat-processing-preservation/4

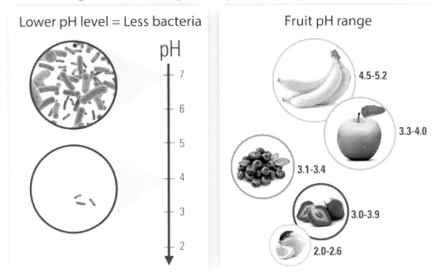

FIGURE 7.4 Use of hurdle technology in fruits.
Source: Reprinted with permission from Ref. [5]. © 2006 Elsevier.

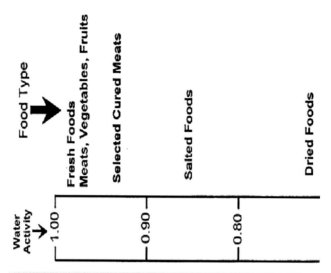

Water Activity Limits: Microbial Growth

Water Activity	Microorganism
<0.60	No microbial proliferation
0.61	*Xeromyces bisporus (xerophilic fungi)*
0.62	*Zygosachharomyces rouxii (osmophilic yeast)*
0.78-0.75	*Aspergillus niger/flavus, Halobacterium halobium*
0.84-0.81	*Penicillium chrysogenum/glabrum, Paecilomyces variotti, Aspergillus fumigatus*
0.86	*Staphylococcus aureus*
0.90	*Bacillus subtilis, Saccharomyces cerevisiae*
0.92	*Mucor plumbeus, Rhodotorula mucilaginosa*
0.93	*Micrococcus lysodekticus, Rhyzopus nigricans*
0.94	*Enterobacter aerogenes*
0.95	*Bacillus cereus, Clostridium botulinum/ perfringens, Escherichia coli, Lactobacillus viridescens, Salmonella spp.*
0.97	*Pseudomonas aeruginosa*

FIGURE 7.5   Tolerance level of a_w (water activity) for microorganism multiplication.
Source: http://www.pharmaquality.com; http://www.meatsandsausages.com/sausage-types/fermented-sausage

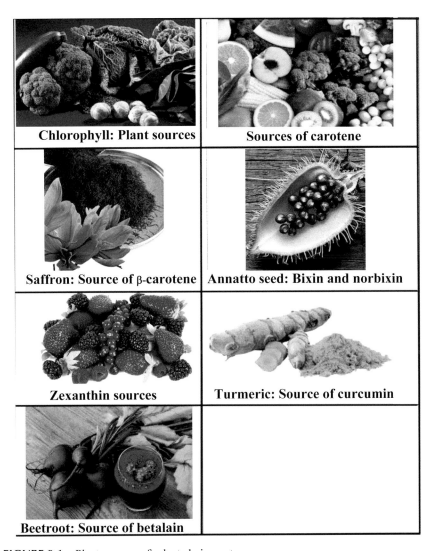

FIGURE 9.1   Plant sources of selected pigments.

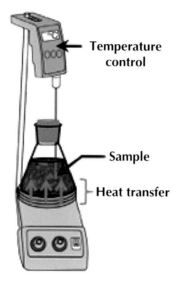

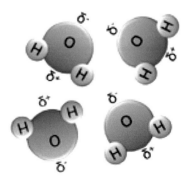

Water (dipole molecules)

FIGURE 9.2   Laboratory setup for extraction of pigments using conventional method.
Source: http://bestmicrowave.hubspace.org/4325/microwave-assisted-extraction.html

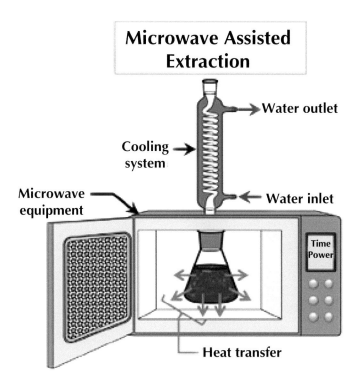

Microwave Assisted Extraction

Water outlet

Cooling system

Microwave equipment

Water inlet

Time
Power

Heat transfer

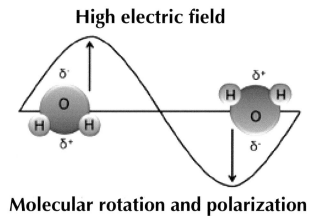

High electric field

δ^-

O

H   H

δ^+

δ^+

H   H

O

δ^-

Molecular rotation and polarization

FIGURE 9.3   Laboratory setup for extraction of pigments using microwave-assisted method.
Source: http://bestmicrowave.hubspace.org/4325/microwave-assisted-extraction.html

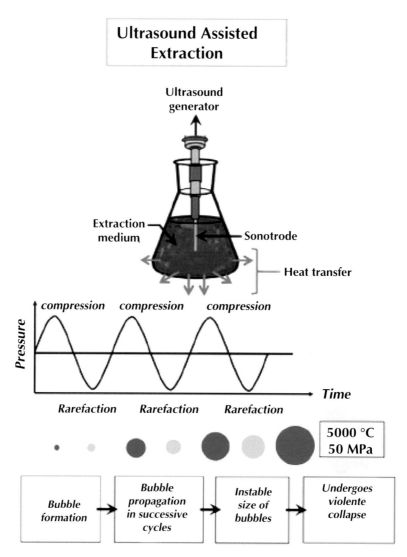

FIGURE 9.4   Laboratory setup for extraction of pigments using US-assisted method.
Source: http://bestmicrowave.hubspace.org/4325/microwave-assisted-extraction.html

FIGURE 10.2   Sorting system using *machine-vision system* are installed to sort fruit into different grades based on weight, dimensional size, color, shape, limited defects, and other parameters.
Source: http://tao.umd.edu/html/industry_exp_.html

FIGURE 10.3   Dual and triple angle gloss meter (Elcometer 407L).
Courtesy: http://www.elcometerusa.com/Laboratory/
Elcometer-407L-Dual-and-Triple-Angle-Glossmeter/

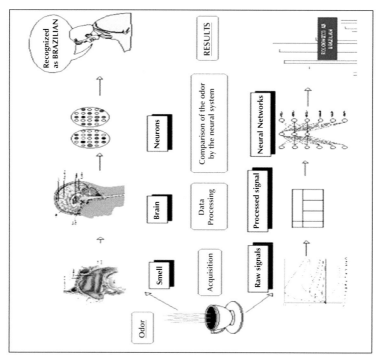

FIGURE 11.1   Differences between traditional nose and e-nose.
Source: http://www.enose.nl/rd/technology/

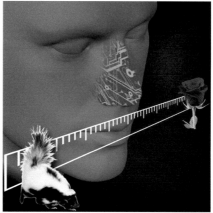

FIGURE 11.2   E-chip on a nose.
Source: Brodsky, G.; Sobel, N. Electronic Nose was Tuned to the Perceptual Axis of Odorant Pleasantness. *PLoS Comput. Biol. Issue Image* **2010**, *6* (4), ev06.i04 (Open Access): Creative Commons Attribution 2.5 Generic license. https://commons.wikimedia.org/wiki/File:An_Electronic_Nose_Estimates_Odor_Pleasantness.png

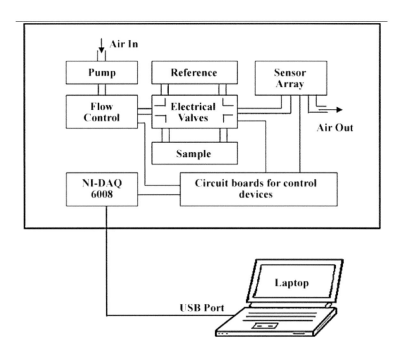

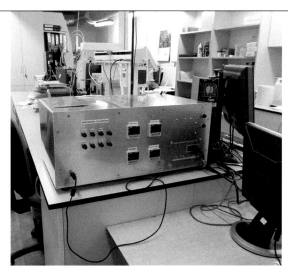

FIGURE 11.3   Principle of e-nose.

Source: https://www.elprocus.com/electronic-nose-work/; Katlewa, L. An Electronic Nose Prototype. This file is Licensed Under the Creative Commons Attribution-Share Alike 4.0 International License, 2013. https://commons.wikimedia.org/wiki/File:Enose_prototype_Analytical_Dept_Chemical_Faculty_GUT_Gdansk.jpg

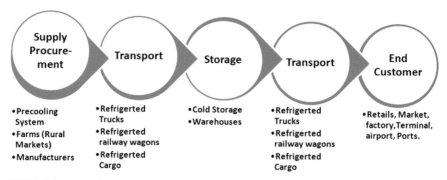

FIGURE 12.1   Infrastructures of the cold supply chain for fruits and vegetables.

FIGURE 14.3   Folate cycle conversion of tetrahydrofolate to methylene-THF.
Source: Online, open access. http://watcut.uwaterloo.ca/webnotes/Metabolism/FolateB12.
html

CHAPTER 8

USE OF VACUUM TECHNOLOGY IN PROCESSING OF FRUITS AND VEGETABLES

CHANDRAKALA RAVICHANDRAN and ASHUTOSH UPADHYAY

ABSTRACT

This chapter focuses on the concept of vacuum technology and role of vacuum techniques (such as vacuum cooling, vacuum drying, vacuum frying, etc.) in processing and preservation of fruits and vegetables. The chapter also includes discussion on principle and working of vacuum processing along with its application in cooling, drying, impregnation, frying, and packaging with special reference to fruits and vegetables. Vacuum technology is being adopted in food processing due to its high quality, extended shelf life, and retention of sensory quality of products. The vacuum technology can be future technology for preservation of fruits and vegetables. As the per capita consumption of fruits and vegetables increases every year, the need for preservation of fruits and vegetables for future will lie in hands of vacuum technology.

8.1 INTRODUCTION

The term "vacuum" is derived from a Latin word "Vacua" means "empty" or "void." An empty space without matter is called "vacuum;" however, practically such a complete empty space does not exist. Therefore, vacuum can be said as partially empty space where gases, molecules, and atoms are lesser comparatively than at normal atmospheric conditions. Greek philosopher Democritus (460–375 BC) in his work discussed about the existence of vacuum. Mankind has been studying vacuum since 17th century and it has drawn interest for several scientists in the past. However, today's modern world without vacuum technology in industries and research is unimaginable.

The absolute pressure of a system is a common way to measure the degree of vacuum. Based on the subatmospheric pressure range, vacuum has been classified into low, medium, high, very high, ultrahigh, and extremely high as shown in Table 8.1.[9] The vacuum technology can be defined as a process or physical measurements that are carried out under vacuum conditions. It is quite commonly used in various industries like semiconductors, chemical, electrical, food processing, metallurgy, etc. and plays important role in research in fields of biology, chemistry, and physics. However, the degree of vacuum employed will vary for each application, depending on their requirements. Generally in a system, the vacuum is created by vacuum pump, by means of suction the pressure decreases, the vacuum increases in it. There are number of national and international units of measurement of vacuum. The SI unit of pressure (i.e., Pascal (Pa)) is commonly used to define the degree of vacuum.

This chapter focuses on the concept of vacuum technology and role of vacuum techniques (such as vacuum cooling, vacuum drying (VD), vacuum frying (VF), etc.) in processing and preservation of fruits and vegetables (FAV). The chapter also includes discussion on principle and working of vacuum processing along with its application in cooling, drying, impregnation, frying, and packaging with special reference to FAV.

8.2   BASIC CONCEPTS IN VACUUM TECHNOLOGY

Air contains a mixture of gases, consisting of 71% of nitrogen, 21% of oxygen, and 1% of other gases like carbon dioxide, argon, etc. These particles are in continuous motion.

- *Pressure* (*P*) can be defined as force (*F*) exerted on surface (*A*) of wall when the particles are in motion.

$$P = \frac{F}{A} \tag{8.1}$$

To achieve vacuum, it is necessary to reduce the pressure in the vessel below the atmospheric pressure. Consequently, due to very low pressure, the vessel contains no or only small number of gas particles that exerts force as they collide against the walls.

TABLE 8.1   Pressure Ranges of Each Quality of Vacuum in Different Units.

Vacuum quality	Torr	Pa	Atmosphere (atm)
Atmospheric pressure	760	1.013×10^5	1
Low vacuum	760–25	1×10^5–3×10^3	9.87×10^{-1}–3×10^{-2}
Medium vacuum	25–1×10^{-3}	3×10^3–1×10^{-1}	3×10^{-2}–9.87×10^{-7}
High vacuum	1×10^{-3}–1×10^{-9}	1×10^{-1}–1×10^{-7}	9.87×10^{-7}–9.87×10^{-13}
Ultra high vacuum	1×10^{-9}–1×10^{-12}	1×10^{-7}–1×10^{-10}	9.87×10^{-13}–9.87×10^{-16}
Extremely high vacuum	$<1 \times 10^{-12}$	$<1 \times 10^{-10}$	$<9.87 \times 10^{-16}$
Outer space	1×10^{-6}–$<1 \times 10^{-17}$	1×10^{-4}–$<3 \times 10^{-15}$	9.87×10^{-10}–$<2.96 \times 10^{-20}$
Perfect vacuum	0	0	0

Source: https://en.wikipedia.org/wiki/Vacuum

- *Partial pressure* (P_p) of certain gas or vapor is defined as pressure exerted by each of these gases corresponding to its concentration if it alone were present in the vessel.
- *Total pressure* (P_T) in a vessel is a sum of partial pressure of all the gases and vapors within the vessel.
- *Standard pressure* (P_S) is equal to 1013.25 mbar = 101.33 Pa = 1 atm.
- *Gas pressure* (P)

$$P = n \cdot k \cdot T \tag{8.2}$$

The pressure (P, N/m²) exerted by the gas depends on particle number density (n, 1/m³) at thermodynamic temperature (T, K), where k is Boltzmann's constant ($k = 1.380 \times 10^{-23}$ J/K).

- *Gas density* (ρ, kg/m³) is defined as the product of particle number density (n) and particle mass (m_T).

$$\rho = n \cdot m_T \tag{8.3}$$

- *General gas equation* (ideal gas law)

$$P \cdot V = \frac{m}{M} \cdot R \cdot T \quad \text{or} \quad n \cdot k \cdot T \cdot V = \frac{m}{M} \cdot R \cdot T \quad \text{or}$$

and for $m/M = 1$, we have

$$P \cdot V = R \cdot T \hspace{4cm} (8.4)$$

where m is the mass (kg); M is the molar mass (kg/kmol); V is the volume (m^3); R is the general gas constant ($R = 8.314$ kJ/kmol K).

8.2.1   VACUUM GENERATION

The vacuum generation is a most complex phenomenon, which requires characteristics equipment called "vacuum pumps." There are many types of vacuum pumps commercially available in market. The vacuum generation could be done by two ways:

(1) By evacuation of gas from closed space into atmosphere.
(2) Gas can be combined with vacuum system by absorbing or condensing.

The vacuum generators are of three types:

- *Vacuum ejectors*: They operate by venturi nozzle principle, which means vacuum generation takes place by pneumatically driven nozzles without moving parts.
- *Gas transfer vacuum pump*
  - *Positive displacement pump*: They displace gas from closed space to atmosphere.
  - *Kinetic vacuum pump*: They displace gas by mechanical drive system by accelerating through pumping action.
- *Gas-binding vacuum pump*: They bind the gas in closed space to an active substrate or condense gas at suitable temperature.

8.3   IMPORTANCE OF VACUUM IN PRESERVATION OF FRUITS AND VEGETABLES

Research data and clinical trials suggest that regular consumption of FAV decreases the risk of cardiovascular diseases and cancer. Consuming FAV as a part of diet also prevents from noncommunicable diseases, obesity, etc. Insufficient consumption of FAV resulted in deaths around 14% due to gastrointestinal cancer and 11% due to heart diseases (WHO). Improper handling, poor storage associated with biochemical activity are main causes of FAV spoilage that can only be prevented by proper handling, processing, and storage. Precooling and packaging of fresh produce would decrease the metabolic reactions and enhance its shelf life. There are various technologies for cooling and storage of horticultural produce, but vacuum technology plays a vital role in providing excellent quality products. Vacuum cooling and vacuum packaging (VP) are best suited for many fresh FAV to preserve sensory and nutritional qualities.

Due to changing trends of consumer preference, ready-to-cook and ready-to-serve products are encouraged in market. There are a number of FAV products (like jam, jelly, marmalade, juice, nectar, squash, leather, pickles, chips, chutney, sauces, etc.) available in market. However, these products must be processed under moderate conditions so that they stay safe, healthy, nutritious, and of high quality. Consumers are also more concerned about the fresh foods in natural form without addition of chemical preservatives. To meet all these consumer demands, minimal processing by using relatively mild preservation technologies came into existence. This minimal processing could be achieved in combination with vacuum technology to produce safe and nutritious products. In recent years, the interest on vacuum technology in various aspects of food processing has increased due to lesser heat treatment involved, high-quality products, preservation of natural characteristics, enhanced shelf life, etc.

8.4 ROLE OF VACUUM TECHNIQUES IN PROCESSING OF FRUITS AND VEGETABLES

There are many traditional food preservation techniques like cooling, drying, pasteurization, frying, etc. However, due to extreme temperatures involved in these processing techniques, the sensory and nutritional qualities of produce get degraded. When these techniques are combined with vacuum, effective processing is achieved even at low temperatures under vacuum. At atmospheric pressure, the molecules are crowded and they collide with one another. This collision may result in number of chemical reactions. However at reduced pressure, the molecules can travel without collision resulting in fewer reactions. The vacuum technology is used in number of unit operations like cooling, drying, impregnation, frying, packaging, etc.

This chapter discusses in detail about the application of vacuum technology in FAV processing. The principle and working of vacuum processing along with its application in cooling, drying, impregnation, frying, packaging with special reference to FAV is also included.

8.4.1   VACUUM COOLING

The quality of FAV depends upon its efficient production practices, proper handling during harvest and postharvest. The production practices include type of soil used for growth, stresses in plants due to drought, climatic conditions, and mechanical damage to crop. During harvesting, several factors (like stage and time of harvesting, gentle handling, and appropriate packing in crates or bins with suitable bedding material) must be taken care. Majority of deteriorative reactions initiate during postharvest stage. Only sufficient postharvest handling practices can preserve the crop quality. Specifically, low temperature plays a vital role in maintaining the fresh nature of crops and prevents it from postharvest spoilage. Precooling can be carried out as an on-farm practice so as to preserve the natural quality attributes and to increase its shelf life. Here, the product temperature is reduced from 3 to 6°C for safe storage and transport.[2]

Precooling is carried out to remove the field heat generated in the product after harvest, which may cause undesirable changes in it. During precooling, the respiration rate is decreased, microorganism growth is inhibited, nutritional and sensory quality is maintained, ripening is delayed, etc. Researchers and farmers have developed number of precooling methods (like room cooling, forced air cooling, hydro-cooling, liquid icing, evaporative cooling, vacuum cooling, etc.) for cooling of fresh produce. Among them, evaporative cooling and vacuum cooling are the techniques that work by evaporation of surface water.

8.4.1.1   ROLE OF VACUUM IN COOLING

The basic principle, under which the vacuum cooling and VD takes place, is similar. It works on the principle that "as pressure decreases the boiling point of water also decreases." At normal atmospheric pressure (101.32 kPa), the water boils at 100°C; however, the water can even boil at 0°C at 0.61 kPa. Under vacuum, a difference in partial pressure of water vapor inside the product and partial pressure of water vapor in surrounding air exists. This is called water vapor pressure deficit (WVPD).[39] Due to WVPD, the normal transpiration process is enhanced under vacuum. Therefore, the water in the product gets evaporated by means of latent heat of vaporization from product, leading to cooling of produce. At higher pressure, lesser cooling takes place. In 1947, vacuum cooling was employed for cooling of iceberg type of lettuce at pilot plant in Salinas, California.[4] Later, its application was extended to horticultural and floricultural produce due to its rapid cooling rates and retention of quality. Later in 1972, these have been commercially used for cooling of mushrooms at United States.[24]

Vacuum cooling can be used only for the products with large surface to mass ratio and ability to lose some amount of moisture. The product should possess sufficient structural capabilities to withstand with the reduced pressure without causing shrinkage. It is also suitable for the porous products resulting in better quality.[56] Generally, highly perishable FAV produce (MC, 90%), which can withstand moisture loss, are suitable for vacuum cooling. With every 5–6°C decrease in temperature, around 1% weight loss due to loss in moisture has been noticed.[5] This may lead to quality loss in produce, but this can be prevented by spraying the water onto the product prior to vacuum treatment.

A typical vacuum cooler consists of vacuum chamber and vacuum pump. The cooling chamber can be horizontal or rectangular in construction, constructed with hardcore metal like steel that must be capable of withstanding reduced pressure without leaking. The vacuum pump should be of size depending on the size of vacuum chamber. Mostly, the refrigerant used for cooling under vacuum would be water and there are different types of vacuum refrigeration systems using water as a refrigerant.[40]

The vacuum pumps generally include mechanical rotary pump and steam jet or barometric condenser. The vacuum cooling consists of placing the product in the cooling chamber after which two main steps takes place: first, evacuation occurs by which the air present in cooling chamber is removed completely, until flash point is reached. This step ensures that no air is present in the chamber; however, cooling is not yet started. The time of evacuation purely depends on size of vacuum chamber and capacity of vacuum pump. Second, as the pressure is reduced further subsequent evaporation takes place to achieve cooling. The pressure is reduced until the desired temperature required for cooling is attained. The vapor generated during the process is removed by the condenser to increase the cooling efficiency.[23] There are also coolers called "hydrovac coolers," similar to vacuum cooler but they are equipped with water sprayer system designed to spray water at the end of the cooling process, to minimize the weight loss in produce.

8.4.1.2 STUDIES ON VACUUM COOLING OF FRUITS AND VEGETABLES

Among several precooling methods, air-blast cooling, hydro-cooling, and vacuum cooling are rapid cooling methods. However, vacuum cooling was found to be most efficient technique than hydro-coolers, water spray vacuum coolers, and forced air coolers for certain reasons.[54] Vacuum cooling is product specific; it has been primarily used for cooling of lettuce for decades in the United States. Because of large surface to mass ratio and high moisture content, lettuce was suited for vacuum cooling. It is widely employed for cooling of leafy vegetables, lettuce, mushrooms, spinach, green onion, carrot, cauliflower, Brussels sprout, Belgian endive, Chinese cabbage, celery, lima bean, snap bean, snow bean, sweet corn, etc.[45] In recent days, its application has increased to various other products like meat products, bakery products, fish products, viscous products, particulate food products, sauces, cooked meal, ready meal, etc.

Studies on effect of pressure on vacuum cooling of iceberg lettuce concluded that vacuum cooling at 0.7 kPa is 11.5 times faster than conventional cooling. The mass loss is high for low vacuum pressure and minimum mass loss was attained at vacuum pressure of 1.5 kPa up to 10°C was reached within 950 s. The mass loss at 6°C of conventional cooling is equal to mass loss at 10°C of vacuum cooling at 1.5 kPa.[36] Studies on pressure reduction rates on physical, chemical quality, and ultrastructure of vacuum cooled and stored iceberg lettuce revealed that moderate pressure favors in superior quality characteristics in case of iceberg lettuce.[19] Moreover, the cooling characteristics of vacuum cooled lettuce at different pressure reduction rates showed that changes in pressure reduction has not significantly changed the mass loss.[41] There are several factors affecting the vacuum cooling rate like capacity of vacuum pump, pumping speed, minimum pressure attained, regulation of condenser temperature, packaging of product, prewetting, etc.[28]

A study on effect of vacuum cooling on anatomy, mushroom quality, and weight loss has concluded that vacuum cooling has brought better color attributes than conventional one. However, the weight loss was higher at 5°C for vacuum cooled than conventional one.[7] The efficiency of vacuum cooling is also affected by type of packaging employed during cooling. Studies reveal that vacuum cooled mushroom can be best stored with MAP.[52] However, vacuum cooling has resulted in lower cooling rate for nonleafy vegetables and nonporous products, and higher cooling rate for leafy vegetables.

8.4.1.3 ADVANTAGES AND DISADVANTAGES OF VACUUM COOLING

Many researchers have carried out a detailed review on vacuum precooling and have projected some of its advantages,[6,15,18] such as:

- Improved product quality and safety.
- Proper temperature control is possible when compared to other precooling methods.
- Rapid cooling rate of 0.5°C/min, which is 60 times that of slow air cooling,[27] resulting in rapid cooling time.
- Uniform cooling of produce for the porous product.
- When compared to air cooling and hydro-cooling, it has lowest energy cost per kilogram cooled product and the operating cost of vacuum cooler was comparable to other conventional cooling.[10]

Some of the disadvantages include:

- Batch operation was seen in most of the units.
- Initial capital investment is high because it employs heavy duty cooling chamber to withstand reduced pressure.
- Suitable only for porous products and leafy crops and not suited for thick cuticle crop (e.g., papaya, tomato, apple) and bulky products like melons, pumpkins, etc.
- There is around 3–4% weight loss in FAV, which can be prevented by installing water sprayer in vacuum cooler.

8.4.2 VACUUM IMPREGNATION

The FAV processing industry focuses mainly on preservation of existing natural compounds or incorporation of some vitamins, minerals, etc. by minimal processing. Vacuum impregnation (VI) is a technique used for pretreatment during minimal processing, drying, freezing, etc. It is done on porous products to preserve color, flavor, and texture and also to strengthen them with functional compounds. VI is completely different from osmotic dehydration (OD) as it involves infusion of liquid under vacuum conditions based on pressure differences. VI has been previously used in combination with OD, which involves countercurrent flow of water from product into the osmotic solution and solute from osmotic solution to the product.

The OD is based on concentration difference to accelerate the water loss and solid gain in product. The movement of water and solute by osmotic process takes considerably longer time and there may be leaching of nutrients. However, VI when used with OD helps in faster impregnation of osmotic solution into the product. Under vacuum, the air or liquid associated with the porous structure of product is replaced by solutes of osmotic solution. It occurs due to positive potential difference when atmospheric conditions are restored.[46] However in recent years, there is an increasing interest among researchers to use this technique for active incorporation of functional ingredients into the products. Moreover, they also improve the nutritional value and increase the shelf life of product by preserving its physicochemical quality.

8.4.2.1 *ROLE OF VACUUM IMPREGNATION IN FRUITS AND VEGETABLES*

VI is the mass transfer phenomenon that works on the principle of hydrodynamic mechanism and deformation–relaxation phenomenon.[18,46] Basically, the impregnation includes two stages: one is under reduced pressure, another under atmospheric pressure:

- At first stage, the product is immersed in osmotic solution under vacuum. Due to the reduced pressure surrounding the product, deformation and expansion of capillaries take place until the equilibrium is reached. This means that the gases and liquids bounded within capillaries and pores of the product are released and removed. Later, hydrodynamic mechanism takes place whereby these capillaries are filled with surrounding liquid until the equilibrium is reached.
- At the second stage, the product is brought to atmospheric conditions, wherein there will be relaxation process resulting in shrinkage of capillaries. Moreover, excessive flow of solutes from outside to inside of capillaries occurs at this stage. The efficiency of the process depends upon the vacuum pump and degassing system employed in it.

VI technique is used as pretreatment to achieve dehydration without heat in FAV for energy saving and quality improvement.[58] VI also alters the physicochemical properties of the fruit and vegetable tissues specifically texture and a detailed review has also been presented.[15] VI is being used for reduction of pH in tissues, thereby preventing the growth of microorganisms and spores. When tissues are soaked in several organic acids, VI leads to increased diffusion of hydrogen ions thereby causing excessive pH reduction. By replacing the intercellular spaces with osmotic solution, it results in OD by reducing their water activity. VI results in modification of the tissue structure that enhances its thermal conductivity and quality of the product. Moreover, by impregnating with calcium ions, the tissues have enhanced its texture and rigidity. Not only the calcium ions, other compounds like polyamines also retain tissue firmness but also help in inhibition of ethylene synthesis. It also preserves the color, aroma, and taste characteristics of fruits by inhibiting the enzymatic browning. In recent years, VI is being used in production of novel functional foods by facilitating the incorporation of minerals and vitamins, probiotics, etc. Some of the recent works on VI for improvement of FAV quality is given in Table 8.2.

8.4.2.2   FACTORS TO BE CONSIDERED FOR VACUUM IMPREGNATION

(1) *Process conditions*: There are a number of processing parameters that interfere in efficient impregnation.[38] Among which, pressure is the foremost criteria of concern; OD operate at atmospheric pressure, whereas VO impregnation takes place under vacuum conditions at 100 mbar.

TABLE 8.2   Some of the Recent Findings on Vacuum Impregnation in Fruits and Vegetables.

Application	Solution composition	Treatment conditions	Quality improvement	References
Enrichment of iron in potatoes	0.4 g/100 g iron solution	1000 Pa for 10–120 min	• 1 h VI cooked unpeeled potatoes showed 6 times more iron content than raw potatoes	[16]
			• Iron content was found to be increasing with increase in vacuum time and restoration time	

Incorporation of folic acid in apple products	200 mg/L of folic acid and 2 mg/L of potassium sorbate and 1.13 g/L of calcium chloride	1. VI: 5 kPa for 5 min at 40°C 2. VI/OH: VI treated and OH at 13 V/cm for 105 min at 30, 40, and 50°C 3. Air dried after treatment 50, 60, and 70°C	• The higher folic acid content of 293.026 ± 4.026 μg/100 g was obtained during VI treatment at 40°C. The VI/OH at 50°C also resulted in 286.934 ± 16.986 μg/100 g • Increase in firmness of dried apples and minimum color change was seen at VI/OH at 50°C • BI is minimum for the apples dried at 60°C	[30]
Fortification of calcium in pineapple snack	1 g Ca^{2+}/100 mL	1. Atmospheric pressure 2. VI: Vacuum pressure (13.33 kPa) for 10 min and atmospheric pressure for 10 min 3. VI followed by vacuum pulse (6.70 kPa) 4. After impregnation, convective, and freeze drying was carried out	• VI samples show higher calcium content of 26,815.6 ± 3220.4 μg/g followed by vacuum pulse treatment • Vacuum pulse treatment after VI resulted in draining of residual impregnation liquid on pineapple surface without affecting the calcium content • VI resulted in 20% higher volume of final product than other treatment • VI followed by drying resulted in porous crunchy products with high glass transition temperature and preserved the flavor of the final product	[26]

| Sodium incorporation reduction in tomatoes | 1. Maltodextrin and 10% NaCl 2. 27.5% sucrose and 10% NaCl | 1. OD at atmospheric pressure 2. Pulsed vacuum OD followed by vacuum pulse at 100 mbar for 20 min was given | • Water loss was higher for solution of maltodextrin–NaCl; however, solid gain was higher in sucrose NaCl mixture • Maltodextrin–NaCl solution resulted in low sodium incorporation nearly 25% less than sucrose NaCl mixture • Vacuum pulse is essential for reducing NaCl incorporation in tomatoes | [11] |
| Preservation apple | Sorbitol, glucose, fructose, sucrose, trehalose, and maltose | VI: Vacuum at 857–50 mbar for 10–1000 s | • VI treatment of apples at 738 mbar for 10 s has showed minimal solute loss and preserved the physical properties of fruit but they show limited loss to mechanical properties and color • Firmness was higher for apples processed using sorbitol • Sorbitol, sucrose, and trehalose showed positive effect on preservation of color during storage • The hue values are higher in the order of trehalose > sorbitol > glucose > sucrose > maltose | [33] |

VI, vacuum impregnation; *OH*, Ohmic heating; *OD*, osmotic dehydration; *BI*, Browning index.

Stronger the vacuum more porous the product becomes and more efficient the penetration takes place. Therefore, the vacuum pressure is responsible for structure of product. Second most important factor is processing time, which is related to economy of the process, temperature that affects the mass transfer, and energy savings. The volume of solution that is impregnated depends on the impregnation time. Other conditions include sonication, ratio of product and solution, the pretreatment method, etc. When highly concentrated sugar solutions are used as an impregnating agent, fruit pieces will float on solution leading to improper impregnation; in such case, stirring or agitation must be employed, because agitation and stirring increases the mass transfer rate during the process.[1,37]

    (2) *Product characteristics*: The process of VI and factors that affect the final product quality also depend on the product characteristics like porosity, firmness, cell arrangement, intercellular spaces accessibility, surface-to-volume ratio, etc. The porous fraction of biological tissues varies from one food to another. The higher the porosity, greater will be the impregnation level into food tissues. Other microscopic property is tortuosity being high in certain fruits leading to hindrance of influx of impregnation solution. Some authors have also dealt that internal variables like number and diameter of pores in tissue, its spatial distribution of cells and type of liquid or gas on intercellular spaces also affect the VI process.[12]

    (3) *Impregnation solutions*: The selection of VI solutions is also a major factor that influences the final product characteristics. VI solution can be a solute or miscible solvent that should possess the properties of being edible, nontoxic, economic, and better sensory attributes. There are wide range of solutes or solvents being employed as dipping solution for fruit and vegetable preservation. These impregnation solutions can be classified as isotonic, hypotonic, and hypertonic solutions. The deformation and shrinkage of tissues will be based on type of solution employed. Chlorides are effective antibrowning agents capable of inhibiting the enzyme polyphenol oxidase (PPO) that causes enzymatic browning. Specifically, the sodium chloride can be best alternative to ascorbic acid as it is most cost effective and allowable food additive.[25]

        (1) *Dipping solution*: Similarly, honey, protease, benzoic acids, cinnamic acids, ascorbic acid, sodium and calcium ascorbates, etc. are also used as dipping solution for preserving the native color of FAV. Other solutions include lactose,

maltodextrin, sorbitol, fructose, glycerol, ethanol, trehalose, sucrose, glucose starch syrup, and combination of solutes that can be used for impregnation.[17,55] The texture firming of FAV during storage is achieved by using calcium chloride impregnation, which can improve the cell wall rigidity and firms the cellular tissues. Moreover, calcium plays a role in delaying senescence and controlling enzymatic reactions. The concentration of VI solutions also affects the process; increase in solute concentration leads to increase in water loss and solid gain rates.

(2) *Other factors*: Factors like osmotic pressure, molecular weight, pH, solubility, and ionic state of impregnation solutions also affect the final properties of product. The penetration into tissues is higher for the osmotic agent with low molecular weight. The solutes enriched with active compounds are also used as impregnation solution for production of functional foods. *Aloe vera* being rich in physiologically active compounds has been used in endive, cauliflower, broccoli, and carrot by VI.[44] The rheological characteristic of osmotic solution is major criteria that affects it impregnation of solutes into tissues. Low-viscosity solutions highly favor impregnation, whereas high-viscosity solutions lead to water loss.

8.4.2.3   APPLICATIONS OF VACUUM IMPREGNATION

- Candying of fruits and salting of food products like pickles, meat, etc.
- Development of functional food by enriching with active compounds from natural sources.
- Development of minimally processed FAV with extended shelf life.
- Introduction of chemical and biological compounds to microstructure of food to improve its quality.
- Pretreatment before freezing, canning, drying, pasteurization, etc. to increase the mass transfer.

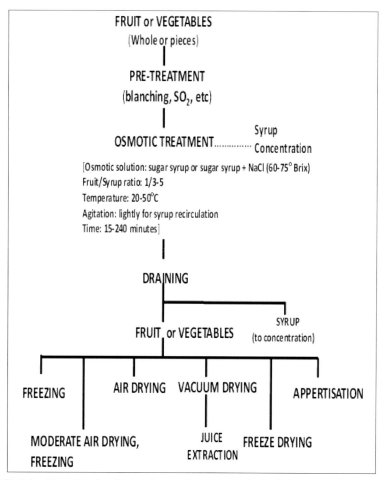

FIGURE 8.1   Flowchart for the continuous freeze dryer; by authors.

8.4.3   VACUUM DRYING

Drying of FAV is the oldest preservation technique. In ancient times, sun drying and salting was basically employed to dry the food products to increase its shelf life.

However, it is not possible to control the drying rate by sun drying and it purely depends on environmental conditions. The drying is a complex process involving heat and mass transfer phenomenon by which the moisture in the food is reduced to desired level thereby preventing undesirable chemical and biochemical reactions.

With subsequent development, different drying methods and drying equipments came into practice that controls the temperature, humidity, and drying time. India is a prominent producer and exporter of dried fruits and preserved vegetables and has exported nearly 63,701.78 MT during the year 2014–2015 to major destinations like the United States, United Kingdom, Europe, and Russia (Agricultural and Processed Foods Export Development Authority (India), APEDA). Due to increasing demand and acceptability, the market for dried FAV has been increasing day by day.

Drying not only increases the shelf life of the product but also reduces the transportation cost. The dried FAV have been used widely in bakery, confectionary, puddings, sauces, etc. The dried fruit juice powders have huge application in beverage industry and they are used as fortifiers in food recipes. But the dried powder quality depends upon the drying conditions like drying temperature, drying time, operating pressure, type of dryer, etc. With recent developments in the field of drying technology, numerous dryers are commercially available for drying of different food products (Fig. 8.1). Each of them has its own unique parameters, which may suit to a specific product characteristic. Depending on criteria of operating pressure employed for drying, they can be classified as vacuum and atmospheric dryers.[21]

8.4.3.1   ROLE OF VACUUM IN DRYING

FAV are an excellent source of vitamins, minerals, fibers, phytochemical like flavanoids, phenolic compounds, carotenoid, anthocyanin, and ascorbic acid. These compounds have the capability to neutralize the harmful effects of free radicals in our body and prevent aging. Fruits are main sources of polyphenols like quercitin, myricetin, kaempferol, catechins, tannins which contribute to antioxidant activity.[3] However, the processing conditions cause subsequent changes in phenolic content leading to loss of many bioactive compounds. With respect to drying, hot air convective drying causes loss of many phenolic compounds like anthocyanins, ascorbic acid, etc. To preserve all, the heat sensitive vitamins and bioactive compounds, drying under vacuum came into practice in early 1920s.

Atmospheric dryers operate under normal atmospheric pressure of 101.3 kPa. At this pressure, the water in food boils at 100°C and gets evaporated. However, the vacuum dryers operate at subatmospheric pressure (between 0.005 and 1 atm) wherein the water in food boils at lower temperature (<100°C). In VD, the evaporation takes place when the food has been placed in the vacuum chamber and its total pressure has decreased to reach boiling pressure at prevailing temperature. At a total pressure of 0.05–0.19 atm, water in food boils at approximately 30–60°C; therefore, the drying time is reduced with enhanced drying rates. The foods that are heat sensitive when dried at normal hot air dryer cause loss of color, flavor, and nutritional value. Such losses can be reduced in vacuum drier because of low temperature being employed. Coriander leaves dried under vacuum was found to be effective for moisture removal, vitamin C, and dietary fiber retention thereby maintaining its nutritional quality.[53] The VD is an expensive technique as compared to other conventional drying as it employs vacuum pump and maximum portion of total energy consumption is due to vacuum pump.[31] Moreover, construction of vacuum drier must be robust to withstand the pressure difference between outside and inside of drying chamber. The energy requirement is maximum in case of VD when compared to other conventional dryers. However, by increasing temperature of drying chamber corresponding drying time can be reduced for lowering energy consumption.[32]

The vacuum drier consists of a vacuum oven coupled with vacuum pump and condenser. The vacuum oven is generally constructed using cast iron, consisting of hollow plates with that provide necessary heat for conduction and radiation to occur. They are fitted with steam supply to heat the plates in oven; vacuum gauge, temperature control was also fitted with them. The food placed on the plates in vacuum oven gets dried under vacuum and the water vapor generated from food is removed via condenser as condensates (water) and maintains vacuum inside the drying chamber.

8.4.3.2   ADVANTAGES AND DISADVANTAGES OF VACUUM DRYING

The major advantage of VD is drying of heat-sensitive and oxygen-sensitive foods. Foods rich in unsaturated fatty acids are susceptible to oxidative rancidity when exposed to oxygen. Such foods could be dried under vacuum conditions to retain its final quality. They also have faster drying rates at low temperature so they assure to maintain the nutritional quality of product. They have some limitations which include high cost of installation and operation, high energy consumption, and low energy efficiency.

Vacuum dryers are used in combination with other driers to increase their performance and drying rates. The advantages of other VD techniques and its underlying principle are presented in Table 8.3, which also presents principle of operation of dryers.

The most important factor in VD is heat source. In VD, the best possible way to heat the food material is by conduction or radiation. However, conductive heating causes uneven heating on food surfaces and center leading to improper moisture removal from core of drying material. To overcome this scenario, heating of food material by the application of electromagnetic waves (far infrared, microwaves, etc.) are employed thereby to achieve uniform and faster heating of food material. These combination methods have resulted in satisfactory drying rates and reduced the drying time considerably.

TABLE 8.3   Basic Principles of Vacuum Drying Techniques.

Drying technique	Principle of working	Advantages
Far-infrared-vacuum drying	• Similar to microwaves, far-IR radiations are employed to heat up the material • IR rays will penetrate faster and uniformly throughout the material and drying takes place under vacuum at faster rate	• Less drying time than other vacuum drying methods • The drying rates are quite higher
FD	• Removal of water takes place by ice sublimation • Initially the product is frozen, the pressure in chamber is brought down below 0.006 atm, and low temperature is controlled in chamber so that ice from product directly sublimates to vapor. This is called primary drying • Second, temperature of chamber is increased so that bound unfrozen water absorbs the heat of desorption to become free water and finally gets evaporated to vapor. This is called secondary drying	• Long shelf life due to low moisture content • No additives are added during drying process • The final product will be of best quality • The product becomes porous after FD, so it easy to rehydrate

TABLE 8.3 *(Continued)*

Drying technique	Principle of working	Advantages
MW-VD	• Removal of water takes place by dielectric loss properties in microwave field • When the product is heated by microwaves, the water molecules present in food having a largest dielectric constant will absorb these microwaves and vibrate faster. Intense vibration produces heat at a faster and uniform rate throughout the product • Under vacuum, mass transfer occurs at low temperature, when it combines with rapid energy transfer by microwave heating it result in low temperature drying at faster rate	• Better quality products can be produced at intermediate cost • Easy rehydration • Greater retention of vitamin C, β-carotene, and soft texture • Helps to preserve natural pigments
LPSSD	• The steam becomes superheated at low temperature under vacuum. This superheated steam is used to heat the food product and evaporate the moisture from it • The evaporated moisture can be recycled to be used as drying medium because the steam is superheated (not saturated), so it does not condense even at temperature drop	• Color of products is well preserved • Effective in reducing microbial load • Effective in retaining aromatic compounds • Retention of ascorbic acid is higher
Vacuum osmotic dehydration	• Under vacuum, the osmotic dehydration occurs at a faster rate because vacuum accelerates capillary flow function thereby aids for solute exchange and higher water extraction rates	• Improves the color, texture, and rehydration properties • Prevents enzymatic browning and inhibits polyphenol oxidase
Vacuum spray drying	• Developed to dry the sticky fruit juices to produce fruit powders • In vacuum chamber, the product is sprayed with carrier agent is exposed to superheated steam as a heating medium. The low-temperature drying would be possible due to subatmospheric pressure inside the chamber	• Best color retention in fruit juice powders • Easy product recovery when juice is dried with some carrier agent like maltodextrin • Heat-sensitive functional ingredients are preserved • Shelf stable products are produced

IR, infrared; MW-VD, microwave vacuum drying; LPSSD, superheated steam vacuum drying.

Drying at subatmospheric pressure has been increasing interest among researchers these days. Numerous drying methods are used in combination with vacuum technology to achieve best results. Among these, superheated steam vacuum drying and far-infrared vacuum drying (FIR–VD) show best possible results in drying of FAV for retaining its bioactive compounds and providing best physicochemical properties. Many researchers have carried out comparative studies between different drying methods and their outcomes are presented in Table 8.4. Among all techniques, the conventional VD was found to be costly and gave products of inferior quality. Specially, VD in carrots has resulted in more nonuniform shrinkage, because the product surface became dry and rigid before its center had dried.[13] However, when used in combination with superheated steam drying, VD provided better quality products but the drying time was longer than VD alone.

The FIR–VD has been used to dry the products at faster rate in short duration of time, but eventually the quality of product was also not at appreciable levels due to the radiation intensity, higher temperature developed due to FIR.[34] However, FD was the most suitable technique to obtain high quality products but it is costly.

Therefore, the VD technology, in combination with other techniques like microwaves, infrared, superheated steam, and OD, was found to have numerous applications in drying of heat sensitive and oxygen sensitive food products. To preserve the bioactive components and valuable vitamins present in FAV, new generation drying technologies like above techniques can be employed. Further research work should be focused on designing of cost-effective vacuum driers, research studies to improve drying efficiency of VD, and pretreatment to improve the product characteristics of vacuum-dried products.

8.4.4   VACUUM FRYING

Frying has been one of the commonly practiced food preservation techniques for decades due its unique texture and flavor characteristics. Frying is a heat and mass transfer process by which the edible oil/fat at high temperature above the boiling point of water is used to remove water from the product as vapors, by getting absorbed into the product. Simultaneous drying and cooking takes place in the oil as a heat transfer medium. Frying can be carried out at different temperatures and can be called shallow frying and deep frying.

TABLE 8.4 Recent Research Findings on Vacuum Drying Applications for Fruits and Vegetables.

Type of product	Drying method	Response studied	Major findings	Reference
Figs	Vacuum osmotic dehydration	Effective diffusivity, activation energy, sensory analysis	Pretreatment by vacuum impregnation before dehydration resulted in decreased drying period than nonpretreated figs The shrinkage was higher in nonpretreated figs than pretreated one: The effective moisture diffusivity for pretreated figs varied from 2.75×10^{-10} to 10.25×10^{-10} m²/s; however, nonpretreated figs showed less diffusivity The nonpretreated figs showed lower activation energy than pretreated figs. Except flavor, there is no significant difference between traditionally dried and treated figs Traditionally dried figs has gained higher rank, so osmotic dehydration technique should be further modified by changing sugar type, sugar concentration, temperature, time, and vacuum pressure levels	[43]
Green peas	Combined multistage heat-pump fluidized bed AFD and MW-VD, hot air CD	Drying parameters, morphological characteristics, internal porosity and cotyledon structure changes, microstructural changes, color	The AFD takes considerably longer time than MW-VD and CD, their corresponding drying rates was also lower compared to other The effective moisture diffusivity was lower (6.94×10^{-11} and 8.78×10^{-11} m²/s) in case of AFD dried green peas. However, HACD and MW-VD has diffusivity of 1.43×10^{-9} and 1.92×10^{-9} m²/s. HACD peas were wrinkled due to high shrinkage but MW-VD peas have less shrinkage but with less deformation Bulk density was lower for MW-VD peas but porosity was higher for HACD peas The peas dried by MW-VD has more uniform structure because microwave energy causes rapid moisture transfer from center of material to surrounding vacuum thereby preventing structural collapse HACD has caused complete starch gelatinization in peas showing diluted starch protein matrix but MW-VD has resulted in incomplete gelatinization Green peas dried by combination of AFD and MW-VD exhibit less change in hue, chroma, and saturation with high chlorophyll in it	[59]

TABLE 8.4 *(Continued)*

Jujube	CD[1], MW-VD, combination method of CPD–MVFD and FD	Flavonols, vitamin C, antioxidant capacity, sensory attributes	The flavan-3-ols content was quite higher in FD and MW-VD samples; however, high temperature in CD has decreased it significantly	[57]
			The flavonols, quercitin, and kaempferol derivatives were well preserved at CD at 50°C and CPD-VMFD and MW-VD	
			Since jujube is an excellent source of vitamin C, FD jujube showed higher content of vitamin C than other drying methods	
			FD and MW-VD being a most suitable method to retain the antioxidant capacity of jujube fruits	
			FD and CD at 50°C was shown to provide essential sensory attributes to dried jujube; however, MW-VD has imparted some bitter taste which may be due to burnt bitter flavor developed at high temperatures	
			FD and MW-VD can be regarded as best drying technique for jujube due to its positive abilities	
			CD at 70°C can be regarded as poor method for drying jujube to retain pulp color, sweetness, and bioactive compounds	
Kiwi fruit	VD, hot air drying	Moisture content, Hardness, vitamin C, antioxidant activity, surface color	The hardness of fruit increased after 8 h of VD at 50°C and 10 h of hot air drying at 50°C. Therefore, with increase in temperature, gradual hardening and softening occurs due to denaturation of middle lamellae	[35]
			AA was retained higher in VD samples than hot air ones because decomposition of AA by nonenzymatic oxidation is inhibited under vacuum	
			The antioxidant activity was 4.3–5.2 for VD and 4.5–5.5 for hot-air-dried kiwi fruits	
			The total color change in dried kiwi fruit was lower in VD than hot air dried because Maillard reaction was inhibited at lower temperatures	
Mango (Tommy Atkins)	FD, VD, CD[2], infrared drying	Total phenolic, carotenoid, ascorbic acid (AA) contents, antioxidant activity, and physicochemical analysis	Highest phenolic content was observed in freeze-dried mango powder, whereas CD and VD exhibited lower phenolic content	[49]
			The AA content and carotenoid content was higher in freeze-dried samples and lower in case of IRD mango powder	
			With regard to moisture content, the VD reduced the moisture from 86.85 g/100 g bw to 5.86 g/100 g bw	

TABLE 8.4 *(Continued)*

Orange juice	Vacuum spray drying	Moisture content, hygroscopicity, water activity, particle size, particle morphology, color, rehydration, AA retention	The moisture content and water activity obtained by vacuum spray drying of orange juice is comparatively lower than conventional spray drying (MC: 2.29–3.49%, aw 0.15–0.25) because the high heat transfer causing excessive moisture removal Increase in maltodextrin addition to orange juice resulted in decreased moisture content, since the maltodextrin increases the total solid content and reduces water available for evaporation Hygroscopicity of VSD powders decreased with increasing maltodextrin Powder produced from orange juice; TSS and maltodextrin solids at 50:50 has retained maximum vitamin C of 71% which was comparatively higher than conventional spray drying Powder produced from orange juice; TSS and maltodextrin solids at 50:50 has bright yellow color and exhibited dented surface with deformation and wrinkles; however, powders of 40:60 exhibited smoother and smaller particles Stability and recovery was good in case of powders produced by 30:70 Sorption isotherm of VSD orange juice powder exhibit like Type III sigmoidal curve	[22]
Plums	FD, VD, CD², MW-VD, combination method of CPD–MVFD	Bulk density, porosity, color, phenolic compounds, anthocyanins, flavonols, HMF, antioxidant capacity	Bulk density is higher for MW-VD plum powder; however, it is low in case of FD. But in case of porosity and solubility, the FD plum powders are quite superior The main group of phenolic compounds in plum powders like 5-hydroxycinnamic acids and flavonols was higher in CPD at 70°C and MVFD at 1.2 W/g. The degradation was higher in MW-VD powders due release of phenolic compounds by cell wall disruption caused by excessive heat generated by MW The highest level of anthocyanin was seen in plum powders dried under vacuum conditions (FD, VD, MW-VD) and maximum degradation was in CD The antioxidant activity was quite higher in case of MW-VD powders and HMF is higher in CD rather than MW-VD powders	[29]

TABLE 8.4 (Continued)

Redcurrants	VD, lyophiliza-tion, and CD³	Moisture content, total color change, firmness, rehydration power, phenols, flavanoids, monomeric anthocyanin, AA, antioxidant activity	When compared to CD (78°C for 8 h) and lyophilized (−30°C, 0.01 mbar, 72 h), VD redcurrants showed lower moisture content of 8.09% and color change was also minimal in VD due to low temperature Rehydration power of VD samples and lyophilized samples are same; however, CD samples show unsatisfactory results Total phenolic compounds are retained higher in VD redcurrants than lyophilized and CD Prolonged VD has resulted in degradation of total flavanoids and content of anthocyanin was also similar to lyophilized samples; however, drying parameters affects their content Under low pressure and less drying time during VD resulted in significant increase in AA and its corresponding antioxidant activity	[51]

AA, ascorbic acid; AFD, atmospheric freeze drying; CD¹, convective drying; CD², cabinet drying; CD³, conventional drying; CPD–MVFD, convective predrying and microwave finish drying; FD, freeze drying; HACD, hot air convective drying; HMF, hydroxymethyl furfural; IRD, infrared drying; MC, moisture content; MW-VD, microwave vacuum drying; VD, vacuum drying; VSD, vacuum spray drying.

However, one of the major issues with conventionally fried foods is high oil content and acryl amide formation leading to development of coronary heart disease, diabetes, cancer, hypotension, etc.; and high oil content in fried product will also increase the cost of the unit operation. Since last few decades, the percentage of health conscious consumers has increased certainly. Due to consumer driven health demands, alternative technology to fabricate the low fat snacks similar to fried snacks in case of color, texture, flavor, mouth feel, etc. was developed. Such processes (extrusion, drying, baking) have not been successful in the market due to lack of characteristics of fried snacks. Therefore, researchers have developed some alternative frying technologies such as VF, pressure frying, microwave frying, etc. Among these techniques, VF is most popular technique to produce high-quality snacks possibly at low temperature with characteristic fried flavor, texture, and color with reduced oil uptake.

8.4.4.1 ROLE OF VACUUM IN FRYING

Similar to other vacuum techniques like drying and cooling, VF also works at pressure less than 6.65 kPa. In case of VD and cooling, the boiling point of water in food gets reduced at reduced pressure. Similarly in VF, the boiling point of frying oil and water in product gets reduced thereby resulting in frying at lower temperatures (~120–130°C). Due to low temperatures and reduced oxygen content during the process, they can retain natural color, texture, and flavor. Moreover, vacuum-fried products are snacks with low oil content compared to conventionally fried snacks. This technique is feasible for frying of FAV due to predominant enzymatic browning. Sufficient pretreatments like blanching, osmotic treatment with antibrowning agents, followed by VD have been practiced to enhance its color characteristics.

In the early 1970s, Florigo (H&H Industry systems BV, Netherlands) had developed the continuous VF concept for production of high quality French fries. Now the company has extended the production line for FAV snacks, delicate snacks, etc. by automatic continuous vacuum fryers. Vacuum fryers operate at both batch and continuous modes. Depending on type of product and scale of manufacture, their applications may vary. The basic components of vacuum fryer include (1) frying vessel/pan, (2) vacuum pump, (3) condenser, and (4) centrifuge.

The frying vessel is equipped with frying basket and centrifuge. Vacuum pump is employed to create vacuum in frying vessel and condenser are refrigeration coils placed next to vessel to collect vapors. Centrifuge is used for de-oiling the fried product by centrifugal force. In case of batch operation, the products are placed on frying basket and vessel is allowed to reach vacuum. After sufficient pressure (<6.65 kPa) has been reached, the basket is immersed into the frying oil. Frying takes place at 100–120°C under vacuum and taken to de-oiling unit to remove the excess oil from product. Later, the products are cooled, packed, and stored.

The continuous vacuum fryer operates in a vacuum tube and transport of material to fryer is through conveyor belts. They are also equipped with frying oil circulation and filtration systems. In case of FAV frying, the raw material should be subjected to preprocessing and postprocessing operations. Preprocessing operations include washing, peeling, slicing, impregnation/blanching, and postprocessing operations are de-oiling, packaging, and storage. The quality of final product also depends on the initial quality of raw material, sound mature fruit, and high quality of frying oil must be used for frying. Sufficient pretreatments like blanching and impregnation prevent the product from browning by preserving its natural color and make them porous are being suitable for further processing. In case of postprocessing, packaging plays a major role. As fried products are oxygen and moisture sensitive products, the packaging material should possess barrier properties against oxygen, moisture, and light. In practice, nitrogen is filled into the package for preserving from shock and vibration during transport or handling.

8.4.4.2   STUDIES ON VACUUM FRYING ON PRODUCTION OF CHIPS FROM FRUITS AND VEGETABLES

Vacuum frying is commonly used for frying of FAV like apple, grapes, papaya, banana, guava, jackfruit, mango, peaches, potato, sweet potato, pumpkin, etc. However, it is widely used in the production of potato chips and French fries. The main characteristic of VF is reduction in oil uptake by products and low temperature frying. A comparative study between atmospheric frying and vacuum fried carrot crisps revealed that carotenoid degradation pathways like isomerization and oxidation were found to be prevented by VF technique. Moreover, this technique has resulted in fried carrot crisps with oil content reduced to around 50% when compared to atmospheric frying. The natural color was preserved with approximately 90% retention of *trans*-α-carotene and 86% of *trans*-β-carotene.[14] The oil absorption in fried foods also depends upon several factors like oil quality, composition, product's shape, temperature, frying time, moisture content, porosity, pre- and posttreatment methods.

Postfrying treatment (like superheated steam drying and hot air drying) reduced the oil absorption after frying. Moreover, reducing the moisture content by microwave, hot air treatment and baking as pretreatments also resulted in reduction of oil absorption. Several studies indicate that pressurization plays a vital role in oil absorption during VF. It was also seen that oil absorption was higher at the surface of the product due to high heat and mass transfer rates occurring during the process.

Pre- and postprocessing also plays a major role in quality of vacuum fried snacks. The studies on effect of pretreatment on quality of vacuum fried carrot snacks indicated the importance of pretreatment prior to frying.[2] This study used high pressure processing and freezing as two pretreatment methods, wherein both the methods were able to improve the nutritional and organoleptic qualities of the VF carrot snacks. During storage, pretreatment has maintained the antioxidant capacity and phenolic content in the product. Recent days, researchers have come up with combination frying techniques. Vacuum frying is used in combination with other frying techniques to achieve higher quality products. A comparative study between VF and microwave vacuum frying (MW-VF) on oil uptake, moisture content, texture, color, and microstructure of potato chips has been investigated.[50]

It was found that oil uptake was lower in MW-VF chips (29.35 g oil/100 g dry solid) than VF chips (39.14 g oil/100 g dry solid). With regard to crispiness of potato chips, MW-VF chips were crispier in texture than VF chips. It is because at higher MW power, faster evaporation leading to formation of higher pore density thereby causing crispy texture. Microstructural studies using SEM also revealed that MW-VF chips maintained their cell structural integrity; however, cell structure was collapsed and damaged in case of VF chips. It was concluded that MW-VF would be best option for producing healthier chips of low oil content and crispier texture.

Vacuum frying is one of the promising techniques for the production of novel snacks from FAV by retaining their natural color, nutritional quality, with reduced oil content and acrylamide formation. They can be called new generation frying technology by fulfilling the needs of health conscious consumers.

8.4.4.3   ADVANTAGES AND DISADVANTAGES OF VACUUM FRYING

Advantages

- Due to less exposure to oxygen, it prevents the undesirable changes in products like oxidative rancidity, enzymatic browning, etc.

- Healthier snacks with less oil content.
- Higher evaporation rates resulting in faster dehydration of products.
- It can be employed for frying of heat sensitive foods (suitable for frying of FAV to preserve its natural color and flavor and preventing it from darkening).
- It retains the nutritional quality of the products with better texture, flavor, color, and mouth feel similar to conventional frying.
- Low-temperature frying prevents acrylamide formation.
- Oil deterioration is reduced.

Disadvantages

- High cost of the equipment.
- It requires de-oiling unit to remove excess oil.

8.4.5 VACUUM PACKAGING

FAV being a most perishable commodity with high moisture content are susceptible to spoilage by chemical and microbial reactions. Even after harvest these commodities continue their physiological and biochemical activities leading to wilting, shriveling, and weight loss. These postharvest losses can be prevented only by sufficient packaging and storage techniques. Consumer demands for fresh, safer, and convenient products could be delivered only through proper packaging. Packaging serves as medium for protection of product from physical and chemical damage, convenience for handling, maintain its quality, extend its shelf life, prevent from shock and vibration during transportation and handling, prevent from microbial spoilage, containment of product, convey information about the product, etc.

There are wide ranges of packaging materials used in packaging of foods; each material has its own characteristics, quality criteria, feasibility with product, etc. Selection of packaging material will depend upon the product characteristics, packaging method, product handling, and storage conditions. There are a number of novel packaging technologies such as modified atmospheric packaging (MAP), active packaging, intelligent packaging, etc. MAP could be a VP, which removes most of the air surrounding the product, before the product could be enclosed in barrier material, and flushed with different proportion of gas mixture and then sealed. Vacuum packaging has however been the best method of packaging for dried and frozen FAV.

8.4.5.1   ROLE OF VACUUM IN PACKAGING

Vacuum packaging is a technique for packaging food commodity under reduced pressure. In the early 1950s, the cryovac VP was brought up commercially for packaging of Turkeys. Later in the 1970s, the technique of introducing inert gas into vacuum packing has evolved resulting in extension of shelf life in perishable products.

During VP, the air is completely removed from the package and sealed by vacuum sealers. By doing this, the packed product will be devoid of oxygen thereby preventing oxidative reactions and growth of aerobic microbes causing food spoilage. Vacuum packaging is commonly used for frozen foods and type of packaging material employed for VP should have specific characteristics. The material should possess barrier against oxygen and light, which may cause enzymatic browning, loss of ascorbic acid, color, flavor change, etc. They should also have ability to withstand with shock, vibration, and must be economical. Vacuum packaging helps to enhance the shelf life of many vegetables like asparagus, celery, green peas, green onions, potatoes, etc. under frozen storage. The method of dehydration decides the choice of packaging material. Similarly, vacuum packing is the best method suitable for packing of dried FAV to preserve its flavor, texture, and color. The best suited packaging material would be glass container because during vacuum sealing the suction causes shrink skin package leading to puncture of packaging material by dried products. Banavac is a patented system for vacuum packing of green bananas using 0.04-mm thick polyethylene bags.

8.4.5.2   RESEARCH STUDIES ON VACUUM PACKAGING OF FRUITS AND VEGETABLES

Several research studies have investigated the effect of VP on fresh-cut vegetables for shelf life. The fresh-cut FAV are most susceptible to spoilage, browning, damage, etc. Selection of packaging material and method for these fresh cut is tedious because they require specific packaging require-ments. Those fresh-cut produce were found to be well preserved by vacuum packing under refrigerated storage. VP helps to preserve the natural characteristics of minimally processed potatoes. The activity of enzyme PPO is effectively inhibited by this technique and has increased the shelf life up to 1 week under refrigerated storage. It has been investigated that 100-μm PE or PA bags can be best for VP material, since PE possess water and vapor permeability and PA possess mechanical strength and impermeability.[42] Similarly, VP leads to increased shelf life of peeled potatoes by reducing PPO activity and retaining the natural color.[48] The studies on microbial and sensory quality of potato strips treated with different sanitizers and stored under MAP and VP resulted in preserving the natural color, texture, and aroma of potato strips.[6] The sensory qualities of potatoes can be stored best for 14 days at 4°C by VP than the MAP.

Moderate vacuum packaging (MVP) is most interesting method in MAP. This technique is most commonly employed for packing of fresh FAV. The product is packed using rigid container or pouch under reduced pressure (~40 kPa) and stored at refrigeration temperature. The product is now surrounded by normal air; due to lesser oxygen content in surrounding air, biochemical and metabolic reactions are hindered and inhibit the growth of spoilage microorganisms. Many researchers have studied the application of this MVP on FAV preservation. These minimally processed litchi arils were packed in MVP and stored at 40°C to study the effects on shelf life. It was found that minimal processing followed by MVP resulted in shelf life of 24 days, by preserving their color and inhibiting microbial growth and chemical changes.[47] Similarly, other study on iceberg lettuce also proved that MVP using 80-μm polyethylene PE) bags has prevented enzymatic browning and increased shelf life to 10 days.[20]

8.4.5.3 ADVANTAGES AND DISADVANTAGES OF VACUUM PACKAGING

Advantages

- Best suited for packaging of fresh meat and meat products to enhance its shelf life.
- Enhanced shelf life.
- It is also being employed for packing of poultry, dairy products, ready meals, cheeses, soup, etc.
- Optimal for refrigerated and frozen storage, etc.
- Superior product quality.

Drawbacks

- Encouragement of anaerobic bacteria.
- High cost of equipment.
- Product-specific application.
- Specific packaging material required.

8.5   ROLE OF VACUUM IN OTHER PROCESSING TECHNIQUES

The vacuum technique has been used in several food-processing applications, like vacuum pasteurization, vacuum concentration/evaporation, vacuum distillation, etc. Vacuum pasteurization is used for pasteurization using direct steam at reduced pressure inside a vacuum chamber. The equipment employed for vacuum pasteurization is known as "vacreator" and it is commercially employed for milk and cream pasteurization. This helps to achieve pasteurization effect at low temperatures without degradation of nutrients when compared to conventional pasteurization. In case of vacuum evaporation under reduced pressure, the evaporation takes place at lower temperature than the boiling temperature. Therefore, it is commonly employed for concentration of sucrose solution and milk evaporation to preserve heat-sensitive components. Vacuum distillation is used for concentration of essential oils for extraction of flavors. However, these vacuum techniques are not used commonly for processing of FAV.

8.6   SUMMARY

Vacuum environment can play a major role in preservation of FAV. For the preservation of fresh produce and fruit products like: chips, candies, dried fruits, etc. Due to processing at low temperature, possible chemical interactions and generation of undesirable compounds can be prevented. Moreover, the demands of health conscious consumers toward freshness, safety, nutritious and quality are increasing. Therefore, the vacuum technology is being adopted in food processing due to its high quality, extended shelf life, and retention of sensory quality of products. Their applications are getting wider day-by-day not only in food processing but also in other like semiconductors, electrical, metallurgy, etc. The vacuum technology can be future technology for preservation of FAV. As the per capita consumption of FAV increases every year, the need for preservation of FAV for future will lie in hands of vacuum technology.

KEYWORDS

- **dielectric loss**
- **infrared vacuum drying**
- **vacreation**
- **vacuum cooling**
- **vacuum drying**
- **vacuum evaporation**
- **vacuum frying**

REFERENCES

1. Ahmed, I.; Qazi, I. M.; Jamal, S. Developments in Osmotic Dehydration Technique for the Preservation of Fruits and Vegetables. *Innov. Food Sci. Emerg. Technol.* **2016,** *34,* 29–43.
2. Albertos, I.; Martin Diana, A. B.; Sanz, M. A.; Barat, J. M.; Diez, A. M.; Jaime, I.; Rico, D. Effect of High Pressure Processing or Freezing Technologies as Pretreatment in Vacuum Fried Carrot Snacks. *Innov. Food Sci. Emerg. Technol.* **2016,** *33,* 115–122.
3. Balasundram, N.; Sundram, N.; Samman, S. Phenolic Compounds in Plants and Agri-Industrial By-Products: Antioxidant Activity, Occurrence, and Potential Uses. *Food Chem.* **2006,** *99* (1), 191–203.
4. Barger, W. R. *Factors Affecting Temperature Reduction and Weight-Loss in Vacuum-Cooled Lettuce*; USDA: Washington, DC, 1961; pp 1–28.
5. Barger, W. R. *Vacuum Precooling: A Comparison of Cooling of Different Vegetables*; USDA: Washington, DC, 1963; pp 1–20.
6. Beltran, D.; Selma, M. V.; Tudela, J. A.; Gil, M. I. Effect of Different Sanitizers on Microbial and Sensory Quality of Fresh-Cut Potato Strips Stored under Modified Atmosphere or Vacuum Packaging. *Postharv. Biol. Technol.* **2005,** *37* (1), 37–46.
7. Burton, K. S.; Frost, C. E.; Atkey, P. T. Effect of Vacuum Cooling on Mushroom Browning. *Int. J. Food Sci. Technol.* **1987,** *22* (6), 599–606.
8. Cánovas, G. V. B.; Moulina, J. J. F.; Alzamora, S. M.; Tapia, M. S.; Lopez-Malo, A.; Chanes, J. W. *Handling and Preservation of Fruits and Vegetables by Combined Methods for Rural Areas*; Food and Agricuture Organization: Rome, 2003; pp 1–98.
9. Chambers, A.; Fitch, R.; Halliday, B. S. *Basic Vacuum Technology*; IOP Publishing Ltd.: Bristol, UK, 1998; pp 1–30.
10. Chen, Y. L. Vacuum Cooling and Its Energy Use Analysis. *J. Chin. Agric. Eng.* **1986,** *32,* 43–50.

11. Corrêa, J. L. G.; Ernesto, D. B.; Mendonça, K. S. Pulsed Vacuum Osmotic Dehydration of Tomatoes: Sodium Incorporation Reduction and Kinetics Modeling. *LWT—Food Sci. Technol.* **2016,** *71,* 17–24.

12. Derossi, A.; De Pilli, T.; Severini, C. The Application of Vacuum Impregnation Techniques in Food Industry (Chapter 2). In *Scientific, Health and Social Aspects of Food Industry;* Valdez, B., Eds.; IntechOpen: London, 2012; pp 26–56. https://www.intechopen.com/books/scientific-health-and-social-aspects-of-the-food-industry.

13. Devahastin, S.; Suvarnakutaa, S.; Soponronnarit, P.; Mujumdar, A. S. Comparative Study of Low-Pressure Superheated Steam and Vacuum Drying of a Heat-Sensitive Material. *Dry. Technol.—Int. J.* **2004,** *22* (8) (online).

14. Dueik, V.; Robert, P.; Bouchon, P. Vacuum Frying Reduces Oil Uptake and Improves the Quality Parameters of Carrot Crisps. *Food Chem.* **2010,** *119* (3), 1143–1149.

15. Elzbieta, R. K.; Roza, B. M.; Marcin, K. Applicability of Vacuum Impregnation to Modify Physicochemical, Sensory and Nutritive Characteristics of Plant Origin Products: Review. *Int. J. Mol. Sci.* **2014,** *15,* 16557–16610.

16. Erihemu, Hironaka, K.; Koaze, H.; Oda, Y.; Shimada, K. Iron Enrichment of Whole Potato Tuber by Vacuum Impregnation. *J. Food Sci. Technol.* **2014,** *52* (4), 2352–2358.

17. Ferrando, M.; Spiess, W. E. L. Cellular Response of Plant Tissue during the Osmotic Treatment with Sucrose, Maltose, and Trehalose Solutions. *J. Food Eng.* **2001,** *49* (2–3), 115–127.

18. Fito, P.; Andrés, A.; Chiralt, A.; Pardo, P. Coupling of Hydrodynamic Mechanism and Deformation–Relaxation Phenomena during Vacuum Treatments in Solid Porous Food-Liquid Systems. *J. Food Eng.* **1996,** *27* (3), 229–240.

19. He, Y. S.; Li, F. Y. Experimental Study and Process Parameters Analysis on the Vacuum Cooling of Iceberg Lettuce. *Energy Conserv. Manage.* **2008,** *49* (10), 2720–2726.

20. Heimdal, H.; Kuhn, B.; Leif, P.; Lone, M. L. Biochemical Changes and Sensory Quality of Shredded and Ma-Packaged Iceberg Lettuce. *J. Food Sci.* **2006,** *60* (6), 1265–1268.

21. Hu, G. Q.; Zhang, M.; Mujumdar, A. S.; Xiao, N. G.; Jincai, S. Drying of Edamames by Hot Air and Vacuum Microwave Combination. *J. Food Eng.* **2006,** *77* (4), 977–982.

22. Islam, M. Z.; Kitamura, Y.; Yamano, Y.; Kitamura, M. Effect of Vacuum Spray Drying on the Physicochemical Properties, Water Sorption and Glass Transition Phenomenon of Orange Juice Powder. *J. Food Eng.* **2016,** *169,* 131–140.

23. Kumar, S.; Kumar, M. Precooling of Horticultural Produce. *Postharvest Management and Value Addition;* Daya Publishing House, New Delhi, 2007; Chapter 9, pp 155–210.

24. Lane, W. C. Pre-Packaging and Marketing of Fresh Mushrooms. *Mushrooms Sci.* **1972,** *8,* 763–775.

25. Li, Y.; Wills, R. B. H.; Golding, J. B. Sodium Chloride, a Cost Effective Partial Replacement of Calcium Ascorbate and Ascorbic Acid to Inhibit Surface Browning on Fresh-Cut Apple Slices. *LWT—Food Sci. Technol.* **2015,** *64* (1), 503–507.

26. Lima, M. M. D.; Tribuzi, G.; de Souza, J. A. R.; de Souza, I. G.; Laurindo, J. B.; Carciofi, B. A. M. Vacuum Impregnation and Drying of Calcium-Fortified Pineapple Snacks. *LWT—Food Sci. Technol.* **2016,** *72,* 501–509.

27. Longmore, A. P. The Pros and Cons of Vacuum Cooling. *Food Ind. S. Afr.* **1973,** *26,* 6–11.

28. Mcdonald, K.; Sun, D. Vacuum Cooling Technology for the Food Processing Industry: A Review. *J. Food Eng.* **2000,** *45* (2), 55–65.

29. Michalska, A.; Wojdylo, A.; Lech, K.; Lysiak, P. G.; Figiel, A. Physicochemical Properties of Whole Fruit Plum Powders Obtained Using Different Drying Technologies. *Food Chem.* **2016,** *207,* 223–232.

30. Moreno, J.; Espinoza, C.; Simpson, R.; Petzold, G.; Nuñez, H.; Gianelli, M. P. Application of Ohmic Heating/Vacuum Impregnation Treatments and Air Drying to Develop an Apple Snack Enriched in Folic Acid. *Innov. Food Sci. Emerg. Technol.* **2016,** *33,* 381–386.

31. Motevali, A.; Minaei, S.; Banakar, A.; Ghobadian, B.; Khoshtaghaza, M. H. Comparison of Energy Parameters in Various Dryers. *Energy Convers. Manage.* **2014,** *87,* 711–725.

32. Motevali, A.; Minaei, S.; Khoshtagaza, M. H. Evaluation of Energy Consumption in Different Drying Methods. *Energy Convers. Manage.* **2011,** *52* (2), 1192–1199.

33. Neri, L.; Biase, L. D.; Sacchetti, G.; Mattia, C. D.; Santarelli, V.; Mastrocola, D.; Pittia, P. Use of Vacuum Impregnation for the Production of High Quality Fresh-Like Apple Products. *J. Food Eng.* **2016,** *179,* 98–108.

34. Nimmol, C. Vacuum Far-Infrared Drying of Foods and Agricultural Materials. *J. KMUTNB,* **2010,** *20* (1), 37–44.

35. Orikasa, T.; Koide, S.; Okamoto, S.; Imaizumi, T.; Muramatsu, Y.; Takeda, I. J.; Shiina, T.; Tagawa, A. Impacts of Hot Air and Vacuum Drying on the Quality Attributes of Kiwifruit Slices. *J. Food Eng.* **2014,** *125* (1), 51–58.

36. Ozturk, M. H.; Ozturk, K. H. Effect of Pressure on the Vacuum Cooling of Iceberg Lettuce. *Int. J. Refrig.* **2009,** *32* (3), 395–403.

37. Phisut, N. Factors Affecting Mass Transfer during Osmotic Dehydration of Fruits. *Int. Food Res. J.* **2012,** *19* (1), 7–18.

38. Pterson, B. Fortification and Impregnation Practices in Food Processing (Chapter 14). In *Conventional and Advanced Food Processing Technologies*; Bhattacharya, S., Ed.; Wiley Publishing Ltd.: Chichester, UK, 2015; p 744 (online). DOI:10.1002/9781118406281.ch14.

39. Ramaswamy, H. S. *Postharvest Technologies of Fruits & Vegetables*; DEStech Publications Inc.: Lancaster, PA, 2014; pp 1–317.

40. Rao, G. C. *Engineering for Storage of Fruits and Vegetables: Cold Storage, Controlled Atmosphere Storage, Modified Atmosphere Storage*; Academic Press: Cambridge, MS, 2015; pp 1–859.

41. Rennie, T. J.; Raghavan, G. S. V.; Vigneault, C.; Gariepy, Y. Vacuum Cooling of Lettuce with Various Rates of Pressure Reduction. *Trans. ASAE* **2001,** *44,* 89–93.

42. Rocha, A. M. C.; Coulon, E. C.; Morais, A. M. M. Effects of Vacuum Packaging on the Physical Quality of Minimally Processed Potatoes. *Food Serv.* **2003,** *1,* 81–88.

43. Sahin, U.; Ozturk, K. H. Effects of Pulsed Vacuum Osmotic Dehydration (PVOD) on Drying Kinetics of Figs (*Ficus carica* L). *Innov. Food Sci. Emerg. Technol.* **2016,** *36,* 94–111.

44. Sanzana, S.; Gras, M. L.; Vidal-Brotóns, D. Functional Foods Enriched in *Aloe vera*. Effects of Vacuum Impregnation and Temperature on the Respiration Rate and the Respiratory Quotient of Some Vegetables. *Proc. Food Sci.* **2011,** *1,* 1528–1533.

45. Sargent, S.; Ritenour, M.; Brecht, J.; Bartz, J. *Handling, Cooling and Sanitation Techniques for Maintaining Postharvest Quality: Bulletin HS-719*; Institute of Food & Agricultural Sciences, University of Florida: Gainsville, FL, 2007; p 17.

46. Saurel, R. The Use of Vacuum Technology to Improve Processed Fruit and Vegetables (Chapter 18). In *Fruits and Vegetable Processing Improving Quality*; Series in Food

Science, Technology and Nutrition; Woodhead Publishing Ltd.: Cambridge UK, 2002; pp 363–380.

47. Shah, S. N.; Nath, N. Changes in Qualities of Minimally Processed Litchis: Effect of Antibrowning Agents, Osmo-Vacuum Drying and Moderate Vacuum Packaging. *LWT—Food Sci. Technol.* **2008,** *41* (4), 660–668.

48. Snoeck, D.; Raposo, J. D. M.; Morais, A. M. M. Polyphenol Oxidase Activity and Color Changes of Peeled Potato (cv. Monalisa) in Vacuum. *Int. J. Postharv. Technol. Innov.* **2011,** *2* (3), 233–242.

49. Sogi, S. D.; Siddiq, M.; Dolan, K. D. Total Phenolics, Carotenoids and Antioxidant Properties of Tommy Atkin Mango Cubes as Affected by Drying Techniques. *LWT— Food Sci. Technol.* **2015,** *62* (1), 564–568.

50. Su, Y.; Zhang, M.; Zhang, W.; Adhikari, B.; Yang, Z. Application of Novel Microwave-Assisted Vacuum Frying to Reduce the Oil Uptake and Improve the Quality of Potato Chips. *LWT—Food Sci. Technol.* **2016,** *73*, 490–497.

51. Sumic, Z.; Vakula, A.; Tepic, A.; Cakarevic, J.; Vitas, J.; Pavlic, B. Modeling and Optimization of Red Currants Vacuum Drying Process by Response Surface Methodology (RSM). *Food Chem.* **2016,** *203*, 465–475.

52. Tao, F.; Zhang, M.; Hangqing, Y.; Jincai, S. Effects of Different Storage Conditions on Chemical and Physical Properties of White Mushrooms after Vacuum Cooling. *J. Food Eng.* **2006,** *77* (3), 545–549.

53. Thirugnanasambandham, K.; Sivakumar, V. Enhancement of Shelf-Life of *Coriandrum sativum* Leaves Using Vacuum Drying Process: Modeling and Optimization. *J. Saudi Soc. Agric. Sci.* **2014,** *1*, 1–7.

54. Thompson, J. F.; Chen, Y. L. Comparative Energy Use of Vacuum, Hydro and Forced Air Coolers for Fruit and Vegetables. *ASHRAE Trans.* 1988, *94*, 1427–1432.

55. Tortoe, C. A Review of Osmodehydration for Food Industry. *Afr. J. Food Sci.* 2010, *4* (6), 303–324.

56. Wang, L.; Sun, D. Rapid Cooling of Porous and Moisture Foods by Using Vacuum. *Trends Food Sci. Technol.* **2002,** *12*, 174–184.

57. Wojdylo, A.; Figiel, A.; Legua, P.; Lech, K.; Carbonell-Barrachina, A. A.; Hernandez, F. Chemical Composition, Antioxidant Capacity and Sensory Quality of Dried Jujube Fruits as Affected by Cultivar and Drying Method. *Food Chem.* **2016,** *207*, 170–179.

58. Zhao, Y.; Xie, J. Practical Applications of Vacuum Impregnation in Fruit and Vegetable Processing. *Trends Food Sci. Technol.* **2004,** *15* (9), 434–451.

59. Zielinska, M.; Zapotoczny, P.; Filho Alves, O.; Eikevik, T. M.; Blaszczak, W. A Multi-Stage Combined Heat Pump and Microwave Vacuum Drying of Green Peas. *J. Food Eng.* **2013,** *115* (3), 347–356.

CHAPTER 9

EXTRACTION TECHNIQUES OF COLOR PIGMENTS FROM FRUITS AND VEGETABLES

DEBABANDYA MOHAPATRA and ADINATH KATE

ABSTRACT

Colors are one of the most important attributes of sensory science. Natural colors have most significant roles in foods and therapeutic concentrates. There are lots of safety standards and regulations for the use of synthetic colors and pigments regarding its ill effects. Therefore, there is always crucial demand of the natural colors from the consumers. There are varieties of the pigments present in the nature having differential properties regarding its solubility and chemical reactions. The existing traditional extraction methods are inefficient for extraction of these pigments from their parent source. There are some novel extraction methods (like aqueous extraction, vegetable oil extraction, solvent extraction, microwave-assisted extraction, pulse electric field extraction, ultrasound-assisted extraction, supercritical fluid extraction, etc.) proven to be superior for efficient extraction of naturally bound colored pigments. This chapter deals with the various color pigments available in natural habitat and potential methods for their extraction in usable form.

9.1 INTRODUCTION

Colors have long fascinated the human and affected their social, cultural, food, and personal choices. Colors are the extensions of human nature and have very powerful effect on human psychology. Colors not only add to the appearance but also have therapeutic value; hence, it is considered one of the most important parts of sensory science. Plants produce more than 1000 of compounds, including the compounds which impart color to them.

These colored compounds are usually known as plant pigments, which can be grouped under four major categories such as[49] chlorophylls, carotenoids, anthocyanins, and betalains.

These pigments are found in flowers, fruit, and vegetable skins, seeds, barks, etc. These colored pigments not only enhance the attractiveness but also protect plants from UV light.[49] Curcumin, a yellow color pigment found in turmeric, has antiseptic and insecticidal properties.[8] Moreover, the colored fruits and flowers often attract bees, birds, and animals that help in propagating the pollens and seeds, thus help in completing the life cycle of the plant. Pigments like chlorophyll helps in photosynthesis. Most of the plant pigments are group of polyphenols, which have health beneficial properties as they are rich in antioxidants.[1]

Plant pigments were long been used as dyeing material in textile, food, and other, but with the discovery of synthetic colors, use of these coloring agents had seen a gradual decrease. As the awareness of consumer regarding the ill effects of synthetic colors have increased, the focus is now shifting back to organic colors with no toxic effect on our ecosystems. Some of these plant pigments are used for dyeing of textile and now coming back to organic products line.[45] Food industry is especially keen on using plant-based natural coloring agents in food, instead of synthetic food colors,[50] because they are natural and impose no health risks besides having therapeutic value.

This chapter deals with the various color pigments available in the natural habitat and potential methods for their extraction in usable form.

9.2   PLANT PIGMENTS

Based on their chemical structure, plant pigments can be grouped into four categories: (1) tetrapyrroles (e.g., chlorophyll); (2) carotenoids (e.g., β-carotene), (3) flavonoids (e.g., anthocyanins), and (4) N-heterocyclic compounds (e.g., betalains) such as bixin and curcumin. Chemical structure of each of these pigments can be found on Google Search or in any book on plant biocompounds. These compounds along with their health benefits are briefly discussed in this section.

9.2.1   CHLOROPHYLLS

Chlorophyll is the green-colored primary pigment of plants, which is responsible for converting solar energy into food through photosynthesis.

It is considered as the blood of plant. These two molecules are similar in structure except for the centrally bound Mg ion, which is the integral part of chlorophyll molecule as contrast to hemoglobin, where Fe ion is the central part (Fig. 9.1). They are also present in algae and cyanobacteria. The major chlorophylls found in higher plants are chlorophyll-*a*, which impart blue-green color, absorbs light, and converts it into energy, and chlorophyll-*b* which imparts yellow-green color to the plants and increases the absorption of spectrum. Both are structurally similar except for CHO⁻, which is present in chlorophyll-*b* (Fig. 9.1). Chlorophyll-*c* and chlorophyll-*d* are associated with photosynthesis in some of the prokaryotes or cyanobacteria.[26]

This chapter focuses only on the extraction of chlorophyll-*a* and *b* from the higher plants. Both chlorophyll-*a* and *b* are lipophilic in nature.[13] Although chlorophylls are not water soluble, yet their water-soluble derivatives are currently in demand because of their anticancer properties.[13] Water-soluble chlorophyll proteins are found in *Chenopodium*, *Atriplex*, *Polygonum*, *Amaranthus*, *Brassica*, *Raphanus*, and *Lepidium* species.[44]

9.2.2   CAROTENOIDS

Carotenoids are found in higher plants. They help in absorbing excess solar energy and prevent plants from UV light damage. These fat-soluble polyunsaturated hydrocarbons are generally concentrated in the outer peel or skin of the fruits and vegetables. They are responsible for red–orange–yellow colors in plants. This group includes pigments like: α-carotenes, β-carotenes, γ-carotenes, lycopene, β-cryptoxanthin, delta-carotene, zeta-carotene, rubyxanthin, zeaxanthin, lutein, neoxanthin, and violaxanthin.[11,41,52]

9.2.2.1   α-CAROTENE

It is responsible for the orange color. The anticancer properties have made it one of the desirable bioactive compounds derived from plants. It acts as a precursor to vitamin A. This conversion is very important for good vision, healthy skin, healthy bones, and proper immune system. It is a rich source of flavonoids, which act as antioxidants. However, alpha-carotene plays an effective role of antioxidant than beta-carotene. They are found in pumpkin, broccoli, carrots, sweet potatoes, kale, tomato, cantaloupe, winter squash spinach, mangos, kiwi, etc. (Fig. 9.1).

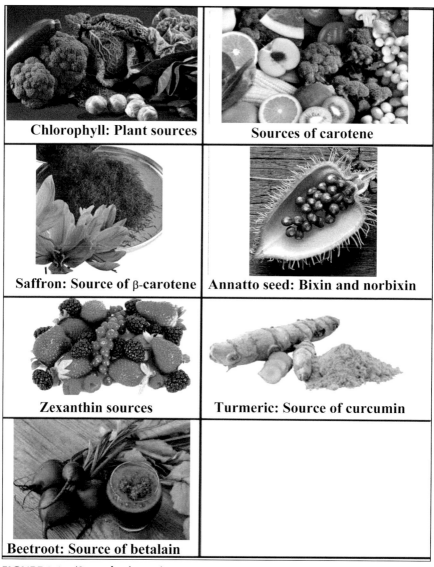

Chlorophyll: Plant sources	**Sources of carotene**
Saffron: Source of β-carotene	**Annatto seed: Bixin and norbixin**
Zexanthin sources	**Turmeric: Source of curcumin**
Beetroot: Source of betalain	

FIGURE 9.1 **(See color insert.)** Plant sources of selected pigments.

9.2.2.2 β-CAROTENE

Beta-carotene is one of the most abundantly naturally found pigments among the carotenoids family. It acts as precursor to vitamin A. Beta-carotene reduces the risk of diseases like lung cancer, heart diseases, skin health, and age-related eye disorders. It is responsible for the red–orange color of the fruits and vegetables. It is found in the plant sources such as apple, apricot, asparagus, cantaloupe, cilantro, jujube, mango, pumpkin, broccoli, Brussels sprout, carrot, tomato, West Indian cherry, sweet potato, etc.

9.2.2.3 γ-CAROTENE

It is the intermediate compound formed during the biosynthesis of other cyclized carotenoids in plants. It is also formed from cyclization of lycopene from lycopene cyclase epsilon. The γ-carotene acts as a vitamer of vitamin A in herbivores and omnivores. Burit and pitanga are some of the important sources of γ-carotene.

9.2.2.4 LYCOPENE

Lycopene is the bright red pigment found in the plant and animal sources. It has properties to fight against cancer and cardiovascular. It has an excellent antioxidant property. Fresh tomato and processed tomato products, watermelon, guava, pitanga, and papaya are good source of lycopene.

9.2.2.5 β-CRYPTOXANTHIN

It is a type of carotenoid, more specifically the xanthophyll. In the human body, beta-cryptoxanthin is synthesized and converted into vitamin A (retinol) and hence it is known as a pro-vitamin A. Beta-cryptoxanthin acts as a strong antioxidant and functions as preventing agent of cellular and DNA damage due to free radicals. It also reduces the risk of lung and colon cancers. It is a source of vitamin A, but about two times less than beta-carotene. These pigments are found in a wide range of fruits and vegetables such as nectarine, orange-fleshed papaya, tangerine, tree tomato, mango, and pitanga (*Eugenia uniflora*).

9.2.2.6 LUTEIN

It is basically a xanthophyll pigment found in vegetables. It belongs to the category of yellow to red pigments, but considered as yellow pigment; hence in high concentrations, it shows the orange-red color. In nature, this can absorb excess light energy from sunlight and prevents the damage (especially from high-energy light rays called blue light). Consumption of this compound keeps eyes healthy. Lutein is also anti-angiogenic and inhibits vascular endothelial growth factor. The richest source of lutein are green leafy vegetables and other green or yellow vegetables (like mustard green, turnip greens, collards, green peas, Artichoke, Asparagus, Broccoli, Brussels sprout, Cantaloupe, Cilantro, etc.).

9.2.2.7   ZEAXANTHIN

The name zeaxanthin is derived from the name of *Zea mays* in which it provides primary yellow color. It is one of the most commonly found carotenoid alcohols in the plant, animal, and microorganism sources. This pigment gives characteristic color like paprika, saffron, or yellow in corn, etc. It is found abundantly in the green leaves of plants, where it acts as nonphotosynthetic quenching agent and deals with exited form chlorophyll (Fig. 9.1). Zeaxanthin reduces the risk of age-related muscular degeneration and age-related eye problems.

9.2.2.8   CROCIN

These are hydrophilic carotenoids found in saffron stigma and gardenia flower; and responsible for the golden yellow color of saffron. They have medicinal value and highly priced pigment often savored for its color and aroma, usually used in the preparation of several dishes. It is one of the important natural drug carotenoid compounds that are found in flowers of crocus and gardenia. It acts as antioxidant and neural protective agent. It has a deep red color and it forms crystals after melting. It can dissolve in water and forms an orange solution.

9.2.2.9   BIXIN AND NORBIXIN

These are apo-carotenoids: the carotenoid derivatives. Bixin is soluble in fat but insoluble in water. However, upon hydrolysis bixin becomes norbixin, which is a water-soluble derivative. These pigments are found in annatto seeds (Fig. 9.1) and are responsible for reddish-orange color of seeds.

9.2.2.10   NEOXANTHIN

Neoxanthin is a carotene as well as xanthophyll and it is produced as inter-mediate product during the biosynthesis of plant hormone abscisic acid. It is found in both 9-cis and all-trans isomer forms. Green leafy vegetables and spinach are the major sources of this pigment.

9.2.2.11   DELTA-CAROTENE

δ-Carotene is a form of carotene with an ε-ring at its one end while the another end is uncyclized. It is synthesized as intermediate product during production of lycopene and α-carotene in some specific photosynthetic plants. It is fat soluble in nature and is found in fruit-like peach palm.

9.2.2.12   ZETA-CAROTENE

It is special type of carotenoids; and it especially differed from α-carotene and β-carotene as it is acyclic in nature. Passion fruit is rich source of zeta-carotene.

9.2.2.13   VIOLAXANTHIN

It is one of the important orange colored natural xanthophyll pigments found in a variety of plant sources. Violaxanthin is produced through special biosynthesis from zeaxanthin by epoxidation. As a food additive, it is repre-sented by code "E161e" as a food-coloring agent. Mango and mamey, and sapota are major source of violaxanthin.

9.2.3   ANTHOCYANINS

Anthocyanins are group of flavonoids that are water-soluble pigments, mainly containpunicalagin, punicalin, gallagic acids, and ellagic acids. They are responsible for a wide range of colors varying from red/orange to blue/purple. They are mostly found in flowers, apart from fruits and vegetables. Apple, bilberry, blackberry, blackcurrant, back carrot, blueberry, cherry, figs, grapes, gooseberry, litchi pericarp, mulberry, plum, pomegranate, raspberry, red currant, red cabbage, strawberry, and sweet potato are good sources of anthocyanins.

9.2.4   BETALAINS

Betalains are water-soluble nitrogen-containing pigments, which comprise the red-violet betacyanins and the yellow betaxanthins (Fig. 9.1). These pigments are found mostly in beetroot, also replaces anthocyanins in flowers and fruits of Caryophyllales families like and cacti *opuntia*, carnations, amaranths, iceplants, beets, and in some higher fungi.[46]

9.2.5   CURCUMINOIDS

The major pigments in this group are curcumins, which are the class of phenols (Fig. 9.1). It is insoluble in common solvent like water and ether but soluble in organic solvents like ethanol, acetone, and dimethyl sulfoxide. The curcumonids contain curcumin I, demethoxycurcumin (curcumin II), bisde-methoxycurcumin (curcumin III), and cyclocurcumin as their constituents. These curcumonids are responsible for the bright yellow color of turmeric and known for their antiseptic, anti-allergic, and anticancer properties. It is used in medicine and food coloring.

9.3   EXTRACTION METHODS

As discussed in the previous sections, some of the pigments are water soluble and some are fat soluble, and moreover, polarity of the pigment molecules varies significantly. Therefore, the extraction of the pigments requires different methods as their extractability varies in different systems. Anthocyanins are water soluble, whereas curcumins are fat soluble. Some of the pigments are soluble in polar solvents such as methanol, ethanol, and some are in ether, etc. In many cases, a number of pigments are present in plant but one of them is predominant in nature. Therefore, different methods are followed for extraction of the targeted pigment. Some of the methods used for extraction of plant pigments are discussed briefly in this section. Conventional method of extraction is shown in Figure 9.2.

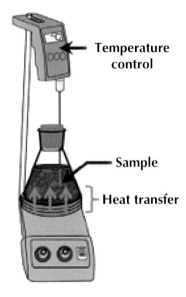

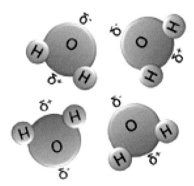

Water (dipole molecules)

FIGURE 9.2   (See color insert.) Laboratory setup for extraction of pigments using conventional method.
Source: http://bestmicrowave.hubspace.org/4325/microwave-assisted-extraction.html

9.3.1 AQUEOUS EXTRACTION

In this method, water-soluble pigments like anthocyanins are extracted using water as solvent. Though the pigments are water soluble, they are better extracted at slightly elevated temperature and in acidified water. Many research studies have optimized the conditions and 40°C was found to be optimum for many water-soluble pigments like betacyanin and betaxanthin.[31]

9.3.2 VEGETABLE OIL EXTRACTION

Fat-soluble pigments (like xanthophylls) are easily extracted using vegetable oil as solvent.[2] Research findings have reported the use of olive, sun flower seed, almond, canola, peanut, corn, soybean oils for extraction of chlorophyll, carotenoids, lycopene, bixin, crocin, lutein, etc.

9.3.3 SOLVENT EXTRACTION

Organic solvent extraction method is one of the widely used methods for extraction of colors and pigments from plant tissue. Most of these solvents have boiling point less than 65°C. The extraction efficiency is affected by operating temperature, solvent to solid ratio, extraction time, size of the raw material, and type of solvent. Despite having several disadvantages being chemical in nature, which has potential hazard of environmental and safety, high solvent, less recovery, energy requirement, and toxicological effect, it is still practiced as one of the most widely used method for extraction of pigments. Ethanol is one of the major polar solvents, which is generally recognized as safe (GRAS), but the efficiency is less than methanol, which is toxic in nature.

9.3.4 MICROWAVE-ASSISTED EXTRACTION

Among the novel methods of extraction, microwave-assisted extraction (MAE) is the most popular, for plant-based bioactive compounds. In MAE, the electromagnetic energy in the form of microwaves is utilized for extraction of targeted compound. Microwave is basically combination of two oscillating fields such as electric field and magnetic field and they propagated perpendicular to each other. During microwave exposure radiations were transmitted as waves, which can penetrate into cellular structure of the biomaterials where it interacts with polar solvent molecules such as water or other dielectric fluid (Fig. 9.3). The interaction of microwave with polar solvent creates spontaneous heat.

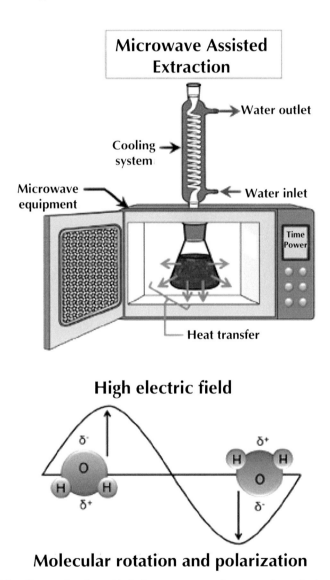

FIGURE 9.3 (See color insert.) Laboratory setup for extraction of pigments using microwave-assisted method. *Source*: http://bestmicrowave.hubspace.org/4325/microwave-assisted-extraction.html

During interaction, molecules of the dielectric liquid try to attain the vibrational frequency of electric field as well as magnetic field at same instant. Hence, there is always alternation in molecular movement direction. This change in alternate field movement at molecular level generates heat, which is the result of molecular collision and friction. This collision and friction of dipole molecule ruptures the cell wall, and cell matrix opens up for interaction with the solvent medium. As cell wall is the main barrier for leaching out the bioactive compounds and pigments, it is becoming vanished by this cellular damage. MAE can be carried out in two ways: (1) MAE with solvent (leaching based), and (2) solvent-free microwave-assisted extraction (SFMAE: distillation based). Selection of the method is fully based on the type of targeted compound, their volatility, boiling point, and thermostability.

During microwave-assisted leaching based extraction, the polar dielectric solvent penetrates into the cellular matrix of the feed stocks by diffusion, where the targeted solute pigment is dissolved selectively in solvent until equilibrium has reached. The solution of solute pigment and solvent then diffuses at the surface of the parent material. In MAE, the process of diffusion takes place at an accelerating manner with higher extraction yield as compared to conventional heat based extraction. This higher acceleration may be the result of a complex and spontaneous enhancement in heat and mass transfer that are taking place simultaneously, unidirectional, and inside-out manner. However in other heating based extraction methods, the heat is transferred from heating medium to the stock material and mass transfer occurs from center to the outside and heat transfer from outside to center.

SFMAE is the most popularized extraction technique especially for extraction of volatile bioactive pigments. Raw material, which is the source of targeted bioactive pigments should pass sufficient amount of dielectric liquid (most popularly water content). Here, vibrational and frictional heat created by molecules would directly evaporate the volatile compound, which can be recovered by condensation in later process.

Microwave power density, microwave exposure time, and solvent:solute ratio are main considerable parameters, which can significantly affect the yield and quality of the extracted compound. Some research also indicated that the particle size of feed stock, soaking time, pH, and initial moisture content show significant effect on extraction kinetics. MAE method have some important advantages over other methods such as (1) lowest time of extraction, (2) minimum requirement of polar solvent or solvent free extraction, (3) higher extraction yield, (4) quicker heating of stock material, and (5) reduced thermal gradient and reduced equipment size. MAE has distinct superiority when compared with other modern extraction techniques especially supercritical fluid extraction (SFE) with respect to its simplicity in process and lower processing cost. MAE is an economical, accepted and practical novel extraction technique for the extraction of color compounds from vegetative origin. Due to the reduced use of organic solvent, MAE is recognized as a green technology. Table 9.1 indicates results of MAE of plant pigments.

TABLE 9.1   Microwave-assisted Extraction (MAE) Used for Extraction of Plant Pigment.

Extracted compound	Plant source	Extraction system	Operating conditions	References
Anthraquinones	Roots of *Morinda citrifolia*	Closed vessel	80% ethanol in water MAE, 15 min	[18]
Coumarin and melilotic acid	Flowering tops of *Melilotus officinalis* L.	Closed vessel	50% aqueous ethanol for 10 min	[32]
Essential oil	Fresh stems and leaves of *Lippia alba*	Open vessel	Solvent free extraction for 30 min	[47]
Furanocoumarins	*Pastinaca sativa* fruits	Closed vessel	80% methanol for 31 min	[58]
Phenolic compounds	*Hypericum perforatum* and *Thymus vulgaris*	Closed vessel	Aqueous HCl, 30 min MAE	[48]
Pigments	*Capsicum annum* powders	Closed vessel	Acetone:water (1:1) 120 s MAE	[25]
Polyphenols and caffeine	Green tea leaves	Open vessel	50% ethanol–water mixture for 4 min	[37]

MAE of pigments works best when polar solvents like water, methanol, ethanol, etc. are used. However, scientists claimed that by using two different solvents like ethanol and benzene, pigments are better extracted in MAE system. The less polar solvents like benzene do not allow the temperature to rise up sharply, thereby minimizing the thermal damage to the pigments. Therefore, extraction of both polar and nonpolar pigments can be accomplished in a MAE system.

The system consists of a microwave cavity in which plants samples are immersed in polar, or polar + nonpolar solvents. When microwave energy is applied, it generates heat in the polar solvents and breaks the cell wall matrix of the plant tissue containing the pigments, thus aiding in dissolving of plant pigments into the solvents. In some cases, vacuum is also applied along with microwave radiation so that extraction can be done at a lower temperature. The later method is more suitable for heat-sensitive pigments. Later in case of mixed solvents, the solvents are separated by gravity, and pigments are recovered through filtration or by evaporation of the solvents under controlled conditions.

9.3.5   PULSE ELECTRIC FIELD EXTRACTION

Pulse electric field extraction (PEFE) is a nonthermal extraction technique in which feed stock material is positioned in a bulk manner between two electrodes; and for treatment, it is exposed to a pulsed high-voltage electric field (typically 20–80 kV/cm) for less than 1 s of treatment time, at multiple short duration pulses (less than 5 μs). The PEFE works on the principle of rupture of cell membrane structure and change in semipermeable nature of the cell wall to enhanced permeability due to formation of small holes (called electropores and phenomenon identifies as "electroporation"). This electroporation phenomenon enhances the release of bioactive pigments from the cell matrix to the solvent environment through pored cell wall.

During the extraction, electrical potential passes through the membrane during which it separates the molecules of target compounds. This molecular separation depends on charge of electric potential, target compound, and dipole nature of membrane molecules. Also in pulse electric field (PEF) treatment, drastic increase of permeability occurs after exceeding a critical threshold value of potential (about 1 V of transmembrane potential). The homogeneous change causes the repulsion, which is responsible for creation of pores in weak areas of the cellular membrane, and due to these, there is flow of color-pigmented bioactive cellular matrix material to outer solvent environment. There broken in the membrane is reversible only if the pores are small, while the pores formed across large areas of the membrane are responsible for permanent destruction of the cell wall. This phenomenon of pulsating electric field treatment can be effectively used for enhanced and nonthermal extraction of pigments from plants.

Typically, a setup for PEFE of bioactive compounds and pigments consists of a treatment chamber with two electrodes in which source raw materials can be placed. A simple electrical circuit is made in such a way that the exponential decay pulses can be generated and it is then utilized for PEF treatment of cellular material. PEF extraction can be either continuous or batch process, based on the design of treatment chamber and scale of the process. Various processes and material factors may affect the performance of PEFE. The efficiency of PEFE purely depends on the operating parameters such as electric field potential and strength, input specific energy density, pulse properties, stock temperature, and type of source materials.

PEF enhances the rate of mass transfer during extraction by creating complex changes on the cell wall, which vigorously increases extraction yield with minimum extraction time. PEF treatment at moderate range (500 and 1000 V/cm and for 10^{-4}–10^{-2} s) can give higher extraction yield with only little increase in temperature. Like other nonthermal extraction methods, PEF can also be applied as a pretreatment to plant materials before the conventional extraction or utilized as independent PEF-assisted leaching based extraction. Table 9.2 indicates research studies to extract plant pigments using PEF extraction method.

TABLE 9.2 Pulse Electric Field (PEF) Extraction Method Used for Extraction of Plant Pigments.

Extracted compound	Source	Operating conditions	Reference
Anthocyanin	Red cabbage	2.5 kV/cm electric field strength; 15 µs pulse width and 50 pulses, specific energy 15.63 J/g	[15]
Anthocyanin	Purple-fleshed potato	3.4 kV/cm and 105 µs (35 pulses of 3 µs), water and ethanol (48% and 96%) as solvents	[39]
Natural pigments	Red beetroot	Field strength 1 kV/cm, pulse length 10 µs, and interpulse interval of 20 ms	[4]
Red pigment	Red beetroot	270 rectangular pulses of 10 µs at 1 kV/cm field strength	[14]

9.3.6 ULTRASOUND-ASSISTED EXTRACTION

Ultrasound is the sound wave having frequency range of 20 kHz–100 MHz. According to the range of frequency, ultrasounds are classified as low and high-power ultrasounds. The high-power ultrasounds (less than 1 MHz) can be utilized for selective extraction of bioactive pigments from the material (Fig. 9.4). Low-frequency, high-power ultrasound waves have a potential to break intermolecular bonds at cellular level and enhance the semipermeable nature of the cell wall.

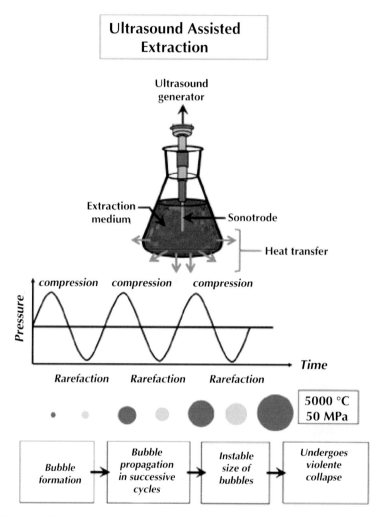

FIGURE 9.4 (See color insert.) Laboratory setup for extraction of pigments using US-assisted method.
Source: http://bestmicrowave.hubspace.org/4325/microwave-assisted-extraction.html

When ultrasonic waves travel through a medium, they create alternate expansion and compression of traveling medium. During expansion, there was pulling of molecules apart from each other, while pushing toward each other in compression. These mechanisms create bubbles in a liquid, which can grow and finally collapse, such phenomenon is called cavitation. During treatment, generated cavitation effects can alter the various physical properties and enhance many complex chemical reactions with specific modifications. During cavitation, bubbles collapse near the surface of the cell wall and produce temperatures up to 5000 K and pressure up to 1000 atm. These generated effects of temperature and pressure with cavitation have potential for enhancement of extraction yield at minimum of treatment time.

In ultrasound-assisted extraction (UAE), developed cavitations in the surrounding solvent environment have key role for accelerated extraction kinetics. The mechanical effects of ultrasound (i.e., friction, temperature, and pressure generated due to cavitation) enhance the rate of penetration of the solvent through cell membrane and improve mass transfer. During extraction, ultrasound waves can also disrupt the cell walls, which facilitated the release of pigments and bioactive compounds from cell matrix. Hence, efficient disruption of cell wall and effective mass transfer are most responsible factors for enhancement of ultrasonic-based extraction of bioactive compounds.

The extraction by ultrasound should pass through two main steps: (1) diffusion across the cell wall and (2) rinsing the contents of cell membrane and matrix after breaking the cell walls. The severity of extraction is affected by various parameters such as moisture content of feed stock, raw material preparation, particle size of feed stock and solvent type (i.e., polar or nonpolar) are the important variables. Additionally, temperature, pressure, frequency, and time of sonication can also be considered the important operating parameters during UAE.

This method of extraction can also be combined with conventional extraction techniques to increase the efficiency of the existing extraction methods. In the solvent extraction system, an ultrasound can be used as a pretreatment to enhance the extraction efficiency.[18] The main advantages of UAE are[5] (1) reduced extraction time (only a few minutes of treatment), (2) efficient energy and solvent use (polar/biosolvent), (3) uniform effective mixing, (4) quick energy transfer, (5) reduced extraction temperature and degradation, (6) selective extraction, (7) smaller equipment size, (8) faster response to process parameters, and (9) quick start-up. It has been reported that ultrasound have potential to enhance the extraction yield but it depends on the type of solvent and its interaction with the solute.

The use of UAE is most recommended method for thermosensitive bioactive compounds and pigments, which may degrade and alter their functional properties under Soxhlet operating conditions. However, developed heat due to ultrasound can accurately control the required range of extraction temperature. The sonication time is one of the most critical process factors, as an excess of sonication can deteriorate the quality of extracts. Selection of proper solvent and its selectivity for targeted compound are very important otherwise extraction yield gives mixture of different bioactive compounds and needs further complex downstream separation process. The results on the use of UAE for extraction of plant pigments are summarized in Table 9.3.

TABLE 9.3 Ultrasound-assisted Extraction (UAE) Used for Extraction of Plant Pigments.

Extracted compound	Source	Extraction system	Operating conditions	Reference
Anthocyanin	Red raspberry (*Rubus idaeus* L.)	UAP and UAE	Solvent:materials 4:1 (mL/g), ultrasonic power 400 W, time 200 s	[7]
Anthocyanin	Grape	UAE	52.35% ethanol, 55.13°C temp., and 29.49 min time	[16]
Chlorophyll-*a*	*Dunaliella salina*	UAE	*N,N*'-dimethyl-formamide and methanol, 3 min sonication	[29]
Lycopene	Tomato	UAE	Temp. 86.4°C, solvent:tomato paste 8.0:1 (V/W), 29.1 min	[27]
Natural pigments	Annatto seeds	UAE	Temp. 72.7°C, time 7.25 min, seed:solvent 14%, duty cycle 8 s	[60]

9.3.7   SUPERCRITICAL FLUID EXTRACTION

Supercritical stage is reached when a fluid is forced to a pressure and temperature above its critical limit. At this stage, the viscosity of the fluid is similar to that of gas and the density is that of liquid having intermediate diffusivity between liquid and gas. This renders it a better solvent status so that the organic compounds are easily soluble in it and can be extracted. The critical temperature (T_c) and pressure (P_c) varies widely for the fluids,[19] for example,

- carbon dioxide (Tc = 31.2°C, Pc = 7.39 MPa);
- methane, ethane, and propane (32.4°C, 4.87 MPa);
- methanol (−34.4°C, 79.9 MPa);

- ethylene, propylene, methanol, and ethanol (241°C, 62.18 MPa);
- acetone nitrous oxide (36.7°C, 71.7 MPa);
- N-butene (−139.9°C, 36 MPa);
- N-pentane (−76.5°C, 33.3 MPa);
- sulfur hexafluoride (45.8°C, 37.7 MPa); and
- water (374.1°C, 22.06 MPa).

Since most of the fluids have a low critical temperature, labile pigments can be easily extracted without much degradation. Because of these advantages, SFE process is being widely used for extraction of plant pigments and other phytochemicals.

Among the solvents, supercritical-carbon dioxide ($SC-CO_2$) has been extensively used for extraction process. It is nonpolar in nature and solubility of this polar substance can be increased by adding small quantities of cosolvents like ethanol, acetone, and ethyl acetate.[49] The advantages of using supercritical CO_2 for extraction process are (1) ideal for extraction of heat labile products; (2) as this process would be carried out in presence of CO_2, oxidation of the phytochemical can be avoided; (3) the supercritical CO_2 is gas at ambient conditions, hence no residue of it after the processes ends and the product is exposed to ambient conditions; (4) $SC-CO_2$ is a by-product of many industries, hence its use will not cause any additional impact on environment causing greenhouse effect; (5) as this process is carried out in darkness, photosensitive pigments such as chlorophyll is better retained; (6) leakage of the CO_2 gas can be effectively checked through reliable CO_2 detectors; (7) $SC-CO_2$ is odorless, colorless, chemically inert and stable, GRAS, nontoxic, noninflammable, nonexplosive, inexpensive, easily available, and easy-to-handle fluid. Research studies on use of supercritical fluid for extraction of plant pigments are described in Table 9.4.

TABLE 9.4   Supercritical Fluid Used for Extraction of Plant Pigments.

Extracted compound	Solvent	Plant source	Operating conditions	Reference
Bixin	$SCCO_2$ + 5% ethanol	*Bixa orellana* (annatto seeds)	40–60°C, 20–30 MPa	[36]
Chlorophyll-*a* and *b*	$SCCO_2$	*Nannochloropsis gaditana* microalgae	40–60°C, 10–50 MPa	[30]
Lycopene and β-carotene	$SCCO_2$	Pulp and skin of ripe tomato	40–80°C, 17.2–27.6 MPa	[3]

TABLE 9.4   *(Continued)*

Extracted compound	Solvent	Plant source	Operating conditions	Reference
β-carotene	SCCO$_2$ + 0%, 5%, 10% ethanol	Carrot	40, 55, and 70°C; 20.7, 27.6, and 34.5 MPa	[54]
β-carotene, capsorubin, capsanthin, zeaxanthin, β-cryptoxanthin	SCCO$_2$ + 1% ethanol/ acetone	Capsicum	13.7–48.3 MPa	[22]

9.3.8   SUBCRITICAL FLUID EXTRACTION

Subcritical fluid is the fluid whose temperature lies between the boiling point at atmospheric pressure and the critical point. For water, the subcritical temperature usually lies between 200 and 374°C. The pressurized water or superheated water has higher capacity of extracting the pigments from plant cells. Water and propane have been commonly used as subcritical fluids for extraction of various bioactive compounds from the bio-produce. The extractability depends on the solvent–solid ratio, pressure, temperature, and extraction time. In some cases, subcritical fluids are found to be more efficient than SC-CO$_2$ extraction such as di-esters of xanthophylls.[21,12] The dielectric constant value for subcritical water is less than 50 at >100°C; hence, there is an increased solubility of pigment like anthocyanin at this condition.[24] Examples of pigment extraction using this method are shown in Table 9.5.

TABLE 9.5   Pigment Extraction with Subcritical Fluids.

Pigment extracted	Subcritical solvent	Plant source	Operating condition	References
Anthocyanin	Water	Red grape skin	110°C	[23]
Anthocyanin	Hydroethanolic solvents	Red grape pomace	40, 60, 80, 100, 120, and 140°C, 6.8 MPa	[34]
Anthocyanin	Water	Elderberry, raspberry, bilberry, chokeberry; stems, skins, and pomaces	120–160°C, 4 MPa	[24]
Oleoresin	Propane	Paprika	25°C, 3–5 MPa	[12,21]

9.3.9 ENZYME-ASSISTED EXTRACTION

Enzymes are product specific; therefore, they act as catalyst in processing conditions to recover the pigments in aqueous condition. Because of their nature, they try to disintegrate the cellular structure thus aiding in recovery of pigments. Mostly, cellulose, hemicellulose, pectinase enzymes are used for extraction of pigments.[40] The enzymes can be of microbial, plant-tissue origin, or chemically synthesized sources and work at moderate temperature and optimum pH conditions.

The optimum pH for cellulose is 4.5 and 5.0 pectinase for extraction of lycopene from whole tomato and tomato peel.[9] The enzyme treatment time may vary between 24 h to few minutes. In many of the cases, enzyme application is used as a pretreatment prior to solvent extraction so that the recovery is faster and more efficient as the targeted pigments had already broken free from the cellular matrix. Examples of pigments using enzyme-assisted extraction are shown in Table 9.6.

TABLE 9.6   Enzyme-assisted Extraction of Plant Pigments.

Pigment extracted	Enzyme	Commodity	Observations	References
Lycopene	Cellulase Pectinase	Whole tomato Tomato peel	pH 4.5 cellulase, 55°C for 15 min pH 5.0, 60°C for 20 min for pectinase Increase in yield whole tomato by 198% (cellulase) 226% (pectinase) Tomato peel: 107% (cellulase) 206% (pectinase)	[9]
Anthocyanin	Pectinesterase, pectinlyase, polygalacturo-nase, hemicel-lulase, cellulase, pectinase	Red grape skin	37°C for 6 h	[35]
Carotenoid	Cellulase + pectinase	Orange peel, sweet potato, and carrot	24 h, RT	[10]

9.4   SUMMARY

Considering that there is a huge demand of natural pigments from plant sources for utilization in many industries including food coloring, textile, and paper industries, it is advisable to explore different possibilities of recovering these pigments from the waste or by-products of food processing industry. Depending on the polarity and chemical characteristics of individual pigment, a wide range of extraction methods are recommended by various researchers. Some of the modern extraction methods that require less solvent and comparatively utilize green energy, and are more efficient are microwave-assisted solvent extraction, SFE, PEF, and UAE.

ACKNOWLEDGMENT

The authors express their gratitude to Indian Council of Agricultural Research–Central Institute of Agricultural Engineering for providing facilities for preparation of this manuscript.

KEYWORDS

- aqueous extraction
- enzyme-assisted extraction
- microwave-assisted extraction
- pulse electric field extraction
- solvent extraction
- subcritical fluid extraction
- supercritical fluid extraction
- ultrasound-assisted extraction

REFERENCES

1. Bartley, G. E.; Scolnik, P. A. Plant Carotenoids: Pigments for Photoprotection, Visual Attraction, and Human Health. *Plant Cell* **1995,** *7* (7), 1027.
2. Bertouche, S.; Tomao, V.; Hellal, A.; Boutekedjiret, C.; Chemat, F. First Approach on Edible Oil Determination in Oilseeds Products Using Alpha-Pinene. *J. Essent. Oil Res.* **2013,** *25,* 439–443.

3. Cadoni, E.; De Giorgi, M. R.; Medda, E.; Poma, G. Supercritical CO2 Extraction of Lycopene and β-Carotene from Ripe Tomatoes. *Dyes Pigm.* **1999,** *44* (1), 27–32.

4. Chalermchat, Y.; Mustafa, F.; Petr, D. Pulsed Electric Field Treatment for Solid–Liquid Extraction of Red Beetroot Pigment: Mathematical Modelling of Mass Transfer. *J. Food Eng.* **2004,** *64* (2), 229–236.

5. Chemat, F.; Tomao, V.; Virot, M. Ultrasound Assisted Extraction in Food Analysis. In *Handbook of Food Analysis Instruments*; Otles, S., Ed.; CRC Press: Boca Raton, FL, 2008; pp 85–94.

6. Chemat, S.; Lagha, A.; Ait, A. H.; Bartels, P. V.; Chemat, F. Comparison of Conventional and Ultrasound-Assisted Extraction of Carvone and Limonene from Caraway Seeds. *Flavour Fragran. J.* **2004,** *19,* 188–195.

7. Chen, F.; Sun, Y.; Zhao, G.; Liao, X.; Hu, X.; Wu, J.; Wang, Z. Optimization of Ultrasound-Assisted Extraction of Anthocyanins in Red Raspberries and Identification of Anthocyanins in Extract Using High-Performance Liquid Chromatography–Mass Spectrometry. *Ultrason. Sonochem.* **2007,** *14* (6), 767–778.

8. Chengaiah, B.; Rao, K. M.; Kumar, K. M.; Alagusundaram, M.; Chetty, C. M. Medicinal Importance of Natural Dyes—A Review. *Int. J. PharmTech Res.* **2010,** *2* (1), 144–154.

9. Choudhari, S. M.; Ananthanarayan, L. Enzyme Aided Extraction of Lycopene from Tomato Tissues. *Food Chem.* **2007,** *102* (1), 77–81.

10. Çinar, I. Effects of Cellulase and Pectinase Concentrations on the Color Yield of Enzyme Extracted Plant Carotenoids. *Process Biochem.* **2005,** *40* (2), 945–949.

11. Curl, A. L. The Xanthophylls of Tomatoes. *J. Food Sci.* **1961,** *26* (2), 106–111.

12. Daood, H. G.; Illes, V.; Gnayfeed, M. H.; Meszaros, B.; Horvath, G.; Biacs, P. A. Extraction of Pungent Spice Paprika by Supercritical Carbon Dioxide and Subcritical Propane. *J. Supercrit. Fluids* **2002,** *23* (2), 143–152.

13. Ferruzzi, M. G.; Blakeslee, J. Digestion, Absorption, and Cancer Preventative Activity of Dietary Chlorophyll Derivatives. *Nutr. Res.* **2007,** *27* (1), 1–12.

14. Fincan, M.; Francesca, D. V.; Petr, D. Pulsed Electric Field Treatment for Solid–Liquid Extraction of Red Beetroot Pigment. *J. Food Eng.* **2004,** *64* (3), 381–388.

15. Gachovska, T.; Cassada, D.; Subbiah, J.; Hanna, M.; Thippareddi, H.; Snow, D. Enhanced Anthocyanin Extraction from Red Cabbage Using Pulsed Electric Field Processing. *J. Food Sci.* **2010,** *75* (6), E323–E329.

16. Ghafoor, K.; Yong, H. C.; Ju, Y. J.; In, H. J. Optimization of Ultrasound-Assisted Extraction of Phenolic Compounds, Antioxidants, and Anthocyanins from Grape (*Vitis vinifera*) Seeds. *J. Agric. Food Chem.* **2009,** *57* (11), 4988–4994.

17. Gong, W. Q.; Liu, R. Optimization of Microwave-Assisted Extraction of Pu er Tea Pigments via Response Surface Methodology. *J. Food Sci.* **2010,** *8,* 31.

18. Hemwimon, S.; Pavasant, P.; Shotipruk, A. Microwave-Assisted Extraction of Antioxidative Anthraquinones from Roots of *Morinda citrifolia. Sep. Purif. Technol.* **2007,** *54* (1), 44–50.

19. Herrera, M. C.; Luque de Castro, M. D. Ultrasound-Assisted Extraction of Phenolic Compounds from Strawberries Prior to Liquid Chromatographic Separation and Photodiode Array Ultraviolet Detection. *J. Chromatogr.* **2007,** *1100,* 1–7.

20. Herrero, M.; Cifuentes, A.; Ibanez, E. Sub- and Supercritical Fluid Extraction of Functional Ingredients from Different Natural Sources: Plants, Food-by-Products, Algae and Microalgae: A Review. *Food Chem.* **2006,** *98* (1), 136–148.

21. Illes, V.; Daood, H. G.; Biacs, P. A.; Gnayfeed, M. H.; Meszaros, B. Supercritical CO2 and Subcritical Propane Extraction of Spice Red Pepper Oil with Special Regard to Carotenoid and Tocopherol Content. *J. Chromatogr. Sci.* **1999,** *37* (9), 345–352.

22. Jaren, G. M.; Nienaber, U.; Schwartz, S. J. Paprika (*Capsicum annuum*) Oleoresin Extraction with Supercritical Carbon Dioxide. *J. Agric. Food Chem.* **1999,** *47* (9), 3558–3564.

23. Ju, Z.; Howard, L. R. Subcritical Water and Sulfured Water Extraction of Anthocyanins and Other Phenolics from Dried Red Grape Skin. *J. Food Sci.* **2005,** *70* (4), S270–S276.

24. King, J. W.; Grabiel, R. D.; Wightman, J. D. Subcritical Water Extraction of Anthocyanins from Fruit Berry Substrates. In *Proceedings of the 6th Intl. Symposium on Supercritical Fluids* at Versailles, France; International Society for the Advancement of Supercritical Fluids, Valence, France; April 2003; Vol 1, pp 28–30.

25. Kiss, G. A. C.; Forgacs, E.; Serati, T. C.; Mota, T.; Morais, H.; Ramos, A. Optimisation of the Microwave Assisted Extraction of Pigments from Paprika (*Capsicum annum* L.) powders. *J. Chromatogr. A* **2000,** *889,* 41–49.

26. Kuhl, M.; Chen, M.; Ralph, P. J.; Schreiber, U.; Larkum, A. W. Ecology: A Niche for Cyanobacteria Containing Chlorophyll *d. Nature* **2005,** *433* (7028), 820–820.

27. Lianfu, Z.; Liu, Z. Optimization and Comparison of Ultrasound/Microwave Assisted Extraction (UMAE) and Ultrasonic Assisted Extraction (UAE) of Lycopene from Tomatoes. *Ultrason. Sonochem.* **2008,** *15* (5), 731–737.

28. Luque, D.; Castro, M. D.; Garcia-Ayuso, L. E. Soxhlet Extraction of Solid Materials: An Outdated Technique with a Promising Innovative Future. *Anal. Chim. Acta* **1998,** *369,* 1–10.

29. Macias-Sanchez, M. D.; Mantell, C.; Rodriguez, M. D. L.; de la Ossa, E. M.; Lubian, L. M.; Montero, O. Comparison of Supercritical Fluid and Ultrasound-Assisted Extraction of Carotenoids and Chlorophyll *a* from *Dunaliella salina. Talanta* **2009,** *77* (3), 948–952.

30. Macias-Sanchez, M. D.; Mantell, C.; Rodriguez, M.; de La Ossa, E. M.; Lubian, L. M.; Montero, O. Supercritical Fluid Extraction of Carotenoids and Chlorophyll *a* from *Nannochloropsis gaditana. J. Food Eng.* **2005,** *66* (2), 245–251.

31. Maran, J. P.; Manikandan, S. Response Surface Modeling and Optimization of Process Parameters for Aqueous Extraction of Pigments from Prickly Pear (*Opuntiaficus indica*) Fruit. *Dyes Pigm.* **2012,** *95* (3), 465–472.

32. Martino, E.; Ramaiola, I.; Urbano, M.; Bracco, F.; Collina, S. Microwave-Assisted Extraction of Coumarin and Related Compounds from *Melilotus officinalis* (L.) *Pallas* as an Alternative to Soxhlet and Ultrasound-Assisted Extraction. *J. Chromatogr. A* **2006,** *1125* (2), 147–151.

33. Mason, T. J.; Paniwnyk, L.; Lorimer, J. P. The Uses of Ultrasound in Food Technology. *Ultrason. Sonochem.* **1996,** *3,* 253–260.

34. Monrad, J. K.; Howard, L. R.; King, J. W.; Srinivas, K.; Mauromoustakos, A. Subcritical Solvent Extraction of Anthocyanins from Dried Red Grape Pomace. *J. Agric. Food Chem.* **2010,** *58* (5), 2862–2868.

35. Munoz, O.; Sepulveda, M.; Schwartz, M. Effects of Enzymatic Treatment on Anthocyanic Pigments from Grapes Skin from Chilean Wine. *Food Chem.* **2004,** *87* (4), 487–490.

36. Nobre, B. P.; Mendes, R. L.; Queiroz, E. M.; Pessoa, F. L. P.; Coelho, J. P.; Palavra, A. F. Supercritical Carbon Dioxide Extraction of Pigments from *Bixa orellana* Seeds (Experiments and Modeling). *Braz. J. Chem. Eng.* **2006,** *23* (2), 251–258.

37. Pan, X.; Niu, G.; Liu, H. Microwave Assisted Extraction of Tea Polyphenols and Tea Caffeine from Green Tea Leaves. *Chem. Eng. Process*. **2003**, *42*, 129–133.

38. Perry, A.; Rasmussen, H.; Johnson, E. J. Xanthophyll (Lutein, Zeaxanthin) Content in Fruits, Vegetables and Corn and Egg Products. *J. Food Compos. Anal.* **2009**, *22* (1), 9–15.

39. Puertolas, E.; Cregenzan, O.; Luengo, E.; Alvarez, I.; Raso, J. Pulsed-Electric-Field-Assisted Extraction of Anthocyanins from Purple-Fleshed Potato. *Food Chem.* **2013**, *136* (3), 1330–1336.

40. Puri, M.; Sharma, D.; Barrow, C. J. Enzyme-Assisted Extraction of Bioactive from Plants. *Trends Biotechnol.* **2012**, *30* (1), 37–44.

41. Rodriguez-Amaya, D. B. Latin American Food Sources of Carotenoids. *Archiv. Latinoam. Nutr.* **1999**, *49* (3 Suppl 1), 74S–84S.

42. Romdhane, M.; Gourdon, C. Investigation in Solid–Liquid Extraction: Influence of Ultrasound. *Chem. Eng. J.* **2002**, *87*, 11–19.

43. Salisova, M.; Toma, S.; Mason, T. J. Comparison of Conventional and Ultrasonically Assisted Extractions of Pharmaceutically Active Compounds from *Salvia officinalis*. *Ultrason. Sonochem.* **1997**, *4*, 131–134.

44. Satoh, H.; Uchida, A.; Nakayama, K.; Okada, M. Water-Soluble Chlorophyll Protein in Brassicaceae Plants is a Stress-Induced Chlorophyll-Binding Protein. *Plant Cell Physiol.* **2001**, *42* (9), 906–911.

45. Siva, R. Status of Natural Dyes and Dye-Yielding Plants in India. *Curr. Sci.—Bangalore* **2007**, *92* (7), 916.

46. Strack, D.; Vogt, T.; Schliemann, W. Recent Advances in Betalain Research. *Phytochemistry* **2003**, *62* (3), 247–269.

47. Stashenko, E. E.; Jaramillo, B. E.; Martinez, J. R. Comparison of Different Extraction Methods for the Analysis of Volatile Secondary Metabolites of *Lippia alba* (Mill.) Grown in Columbia and Evaluation of its in Vitro Antioxidant Activity. *J. Chromatogr. A.* **2004**, *1025*, 93–103.

48. Sterbova, D.; Matejicek, D.; Vlcek, J.; Kuban, V. Combined Microwave-Assisted Isolation and Solid-Phase Purification Procedures Prior to the Chromatographic Determination of Phenolic Compounds in Plant Materials. *Anal. Chim. Acta* **2004**, *513* (2), 435–444.

49. Tanaka, Y.; Sasaki, N.; Ohmiya, A. Biosynthesis of Plant Pigments: Anthocyanins, Betalains and Carotenoids. *Plant J.* **2008**, *54* (4), 733–749.

50. Timberlake, C. F.; Henry, B. S. Plant Pigments as Natural Food Colors. *Endeavour* **1986**, *10* (1), 31–36.

51. Toma, M.; Vinatoru, M.; Paniwnyk, L.; Mason, T. J. Investigation of the Effects of Ultrasound on Vegetal Tissues during Solvent Extraction. *Ultrasonochemistry* **2001**, *8*, 137–142.

52. Tomes, M. L. Delta-Carotene in the Tomato. *Genetics* **1969**, *62* (4), 769.

53. Uquiche, E.; Jerez, M.; Ortiz, J. Effect of Pretreatment with Microwaves on Mechanical Extraction Yield and Quality of Vegetable Oil from Chilean Hazelnuts. *Innov. Food Sci. Emerg. Technol.* **2008**, *9*, 495–500.

54. Vega, P. J.; Balaban, M. O.; Sims, C. A.; Okeefe, S. F.; Cornell, J. A. Supercritical Carbon Dioxide Extraction Efficiency for Carotenes from Carrots by RSM. *J. Food Sci.* **1996**, *61* (4), 757–759.

55. Vinatoru, M.; Toma, M.; Filip, P.; Achim, T.; Stan, N.; Mason, T. J.; Mocanu, P.; Livezeanu, G.; Lazurca, D. Ultrasonic Reactor Dedicated to the Extraction of Active Principles from Plants. *Romanian Patent Number 98-01014*, 1998.

56. Vinatoru, M.; Toma, M.; Mason, T. J. Ultrasound-Assisted Extraction of Bioactive Principles from Plants and Their Constituents. *Adv. Sonochem.* **1999,** *5*, 209–247.

57. Wang, J. X.; Xiao, X. H.; Li, G. K. Study of Vacuum Microwave-Assisted Extraction of Polyphenolic Compounds and Pigment from Chinese Herbs. *J. Chromatogr. A* **2008,** *1198*, 45–53.

58. Waksmundzka-Hajnos, M.; Petruczynik, A.; Dragan, A.; Wianowska, D.; Dawidowicz, A. L.; Sowa, I. Influence of the Extraction Mode on the Yield of Some Furanocoumarins from *Pastinaca sativa* Fruits. *J. Chromatogr. B* **2004,** *800* (1), 181–187.

59. Yang, Z.; Zhai, W. Optimization of Microwave-Assisted Extraction of Anthocyanins from Purple Corn (*Zea mays* L.) Cob and Identification with HPLC–MS. *Innov. Food Sci. Emerg. Technol.* **2010,** *11* (3), 470–476.

60. Yolmesh, M.; Najafi, M. B. H.; Farhoosh, R. Optimization of Ultrasound-Assisted Extraction of Natural Pigment from Annatto Seeds by Response Surface Methodology (RSM). *Food Chem.* **2014,** *155*, 319–324.

PART III

Engineering Interventions in Fruits and Vegetables

CHAPTER 10

NONDESTRUCTIVE METHODS FOR SIZE DETERMINATION OF FRUITS AND VEGETABLES

AJITA TIWARI

ABSTRACT

In the present scenario, nondestructive methods applied in size determination of fruits and vegetables are gaining popularity as they provide quick and reliable approach. During many postharvest handling operations, determination of size parameters of horticultural commodities such as length, thickness, weight, volume, diameter, etc., are important to record. Size parameters of fruits and vegetables often perform a vital role during trading chain. Also size parameters are associated with the ripeness, maturity, as well as with overall quality of produce.

Different methods are compared in the present chapter, which is being used for estimation of various size properties of fruits/vegetables including length, width, volume, and density of horticultural produce. Electronic systems for size determination such as systems measuring volume of gap between equipment's outer casing of gauge and fruit/vegetable commodity, systems based on propagated waves, time of flight between fruit contour and radiation source, systems based on obstruction of light, two-dimensional- and three-dimensional-based machine-vision systems, computed tomography, and magnetic resonance imaging-based systems for size determination are also discussed.

10.1   INTRODUCTION

Postharvest practices applied in fresh fruits and vegetables (FAV) segment are always of great concern in trading point of view. This field is utmost important and is gaining focus due to continuous increase in demand for quality fresh produce. Today, variety of operations in food processing field can be accomplished by application of various machines/equipments in processing and in postharvest handling operations. This ultimately saves time and labor cost too, which lead to efficient processing operations.

Size parameters of FAV have played key role during marketing of these produce. Length, thickness, weight, volume, diameter, etc., are some important size parameters; these are generally followed for grading of produce in variety of quality fractions. Such grading has made it convenient for easy selection of graded produce.

In the developed and developing countries, the demand and popularity of FAV are escalating day-by-day in diets of inhabitants. Size parameters of FAV play a vital role during trading and consumers are interested in even-sized horticultural commodities while purchasing. Many times size parameters are linked with maturity, ripeness, and overall quality of produce. Usually such quality parameters are judged by visual examination of external appearance of the fruit. Many studies which focused on the use of machine vision for judging such quality parameters commented that machines are occupying the place of manual efforts for such tasks as machine working is assumed more consistent than human with virtue of following facts[2,13,54]:

- The insufficiency of labor in developed countries[63]
- The prospect to lessen labor costs[5]

For judging the quality of produce, numerous nondestructive techniques have been developed for uncovering various biochemical properties. In addition to internal quality aspects, quality of horticultural commodities is strongly associated with color of produce, gloss, texture characteristics, shape factor, firmness of intact fruit/vegetable, and absence of any type of surface defects too.

This chapter focuses on the various modern approaches, particularly electronic systems that are capable of determining the horticultural produce dimensional size nondestructively. It covers the necessities related to fruit size measurement, compares the sizers of mechanical and electronic equipment, provides some information related to the challenges related to volume determination using laboratory instruments, and lastly, classifies and compares various kinds of systems based on dimensional measurement.

10.2   SIGNIFICANCE OF SIZE DETERMINATION

For fruit size determination, the dimensions taken into consideration can be: volume, weight, and diameter. To predict optimum harvest time, the nondestructive tool used is fruit volume which can provide ripeness index,[23] yield prediction,[47] or to explain the linkage between rate of expansion of fruit and sensitivity to some physiological problems (e.g., cracking of fruits).[52] Fruit size determination is critical, in terms of many postharvest operations, because of the following reasons:

- Sorting of fresh fruits/vegetables into different size sets becomes possible. In case of pattern packaging with the definite size group, it is easy to allocate particular market and accordingly price of large sized, medium sized, and small sized produce is fixed, to match consumer preferences. The pattern packs, that is, of uniform sizes produce provide better protection over jumble packs.[55] Moreover, the advantage of pattern packages is the better utilization of the shipping box volume, as due to high density of packaging.[58]
- In case of modern online density sorting of fruit/vegetables, size determination is obligatory. In such cases, size-related parameters which are usually imperative are volume and weight.[1] Sorting according to density of fruits is suitable for following causes:

  a) In several fruit and vegetable species, density is correlated with either soluble solids content[36,59,61] or with starch content. Like in potato crop, densities of tuber are correlated with its starch content.[27]

  b) Determination of density is vital for separating the FAV into different classes, for example, citrus fruits that have been damaged during freezing,[46] fruits that have undergone puffy tangerines,[1,2] natural internal desiccation,[45] internally damaged fruits due to insects,[18] watermelons having high level of hollowness,[38] and apples having water core disease.[60] Therefore, in measurement of density of fruits/vegetables high level of accuracy is obligatory.

- Measurement of size parameters is vital for determining surface area of FAV. The importance of surface area is proved during calculating microbial load on particular food's surface[16] and it is also necessary for knowing heat, gas, and water vapor transfer.[11]
- Shape can be calculated independently or in combination with size measurements (e.g., roundness, compactness, aspect ratio,

and eccentricity.[15] Thus, sorting according to shape is possible by determining fruit/vegetable size parameters.

10.3   METHODS OF SIZE DETERMINATION

The size of horticultural commodities can be estimated according to various physical parameters, namely, length, width, diameter, volume and weight, circumference of fruit/vegetable, or combinations of these parameters.[55] Based on this, methods of size determination can be grouped in two categories, that is, methods based on dimensions and methods based on weight. In dimensional category, parameters such as volume, axes, perimeter measurement, and calculation of projected area are covered.

10.3.1   DIRECT DETERMINATION OF FRUIT WEIGHT

For weight sizing of FAV, there are basically two methods: direct and indirect. In case of indirect methods, fruit weight is estimated from dimensional parameters such as estimation of projected area of fruit/vegetable using equation or model.[12,33,35,62]

The main design difference between mechanical and an electronic weight sizer is that in case of mechanical sizer, the weight observation/measurement are carried out at the point of ejection.

10.3.2   ELECTRONIC SYSTEMS FOR SIZE DETERMINATION

Over the past few decades, for nondestructive nature-based size determination in fruits/vegetables, numerous electronic-based systems have evolved. According to their principle of measurement, the whole system has been categorized into six different groups[51]:

1. Systems that measure the volume of the space/gap between equipment's exterior casing of embracing meter/gauge and fruit/vegetable commodity.
2. Size measurement-based systems in which distance is calculated from propagated waves, time of flight (TOF) between fruit/vegetable contour and a radiation source.
3. Systems based on the hindrance of light.

4. Two-dimensional (2D)-based machine-vision systems.
5. Three-dimensional (3D)-based systems.
6. *Other systems*: This category comprises those systems which work on inside images, such as magnetic resonance imaging (MRI) and computed tomography (CT).

10.3.2.1 MEASURING GAP VOLUME BETWEEN OUTER CASING AND FRUIT/VEGETABLE SAMPLE

10.3.2.1.1 Principle

These work on 3D systems and are not based on the machine vision. The outer casing (surrounding) dimensions are fixed. The fruit is passed through it and the fruit size measurement is done.

10.3.2.1.2 Application

In agriculture, a large inconsistency of objects in a range of sizes and shapes is found, leading to size grading problems. For example, for a low-cost horticultural product like potatoes, the primarily used mechanical size-determining systems have errors of up to 30%. Use of this system can cause damage, but they are economical and have a throughput up to 20 t h^{-1}. Alternative methods are required, which can be an improvement in accuracy and profitable for the companies. The use of optical ring sensors (ORSs) was mentioned to estimate fruit volumes travelling at relatively high speed in case of cucumber or zucchini (elongated produce).[48]

The system creates an enveloping helix. This helix is useful in estimating fruit/vegetable's length, major as well as minor axes, and volume. Volume of fruit is measured by addition of individual products of cross section areas and its multiplication by helix pitch. A report on potatoes study confirmed coefficient of variance less than 2.5%.[21] For smaller objects, errors were found to be large due to the geometry of the system. This system is based on light-blocking principle,[48,51] which consists of alternative arrangement of infrared transmitters and receivers fitted in a circular frame.

When inner ring space clear, owing to the Lambertian response of all transducers, each receiver identifies emitted light by transmitter. As soon as a sample is established in ring corresponding to receivers blocked from activated transmitter, a shadow zone comes into view. Emitted light rays indicating chords in circumference and transmitters are turned on in sequence, on the ring. To approximate contour of the object, the two closest "tangential" chords (i.e., noninterrupted chords) are utilized.[49]

In case of tomato and kiwifruit the ORS is used. Using this sensor, the influence of sample orientation on precision of measurements was attained.[50] Comparison of controlled versus random orientations was done. Controlled orientation represented a lack of swinging movement of fruit.

The work has been done on the basis of relationship between the concentric double sphere's capacitance and the inner sphere radius. In case of watermelon, for measuring volume of fruit by electric method, use of precision instrument was referred.[38] In this system, capacitance between external electrode (i.e., outer sphere) and inner sphere (generated by watermelon fruit) symbolized by polygonal grounded tunnel. The polygonal tunnel was considered as outer casing in Kato's method and volume of fruit was assessed by gap between electrical capacitance.[52]

An online system for measuring volume was developed, which worked on the correlation between sample's volume in a Helmholtz resonator and an acoustic resonant frequency.[53] The Helmholtz resonator comprises a wide part known as chamber (cavity) and narrow part (throat). These both parts combined, look like the shape of a wine bottle. Blowing sound or hitting the throat lip creates which contains a resonant component. This resonant component is known as the Helmholtz frequency and it depends on cavity volume minus its content. The importance of this resonant frequency is in obtaining data on volume content. In this concept, the cavity wall has two openings, over which the conveyor belt passed.

Out of these three systems (electrical capacitance, ORS, and Helmholtz resonator) the most adaptable system is ORS. This is because of the fact that the ORS can compute fruit axes besides fruit volume. For volume assessments precision, ring sensor as well as electrical capacitance methods (ECM) are similar. The ORS (where conveying speed is in range of 100–200 cm/s) is superior in comparison of remaining two systems. Helmholtz resonator system does not however seem ready for viable use because of the throughput limitation; therefore, various comparisons have been confined to the other two systems. Regarding accuracy, ECM has more precise than ORS, but it has lower throughput.

10.3.2.2   TOF RANGE FINDING SYSTEMS

A prime attention regarding this system is that whatever the wave types comprised (mechanical or electromagnetic) in addition to a transmitter for generating the signal and a receiver for detecting the reflected signal, an accurate measurement of time is also needed. For rapid assessment of fruit size an apparatus has been developed. This apparatus consisted of three rods in the form of an inverted tripod.[39] On the base of the tripod an ultrasonic distance sensor was arranged.

10.3.2.2.1   Application

Ultrasonic distance sensors are useful in packing lines for fruits sizing. TOF with laser range finder is another technique which could be used for finding fruit size. Two different techniques that are called as laser scanners, used in laser range finders: TOF and triangulation. The triangulation is one accurate method for reconstructing the 3D surface of an object.

10.3.2.3   SYSTEMS BASED ON THE BLOCKING OF LIGHT

The three optoelectronic systems are based on hindering of light.[32] The first system computes width of fruit samples when passes through conveyor belt in horizontal position. It estimates the horizontal width of fruits in the direction of movement. This system has a set of transmitter and receiver fitted on either sides of conveyor belt in such a way that passing fruit is sensed by them. Receiver will detect the light emitted by the transmitter, provided the space between the transmitter and the receiver is clear.

The second type of system is also similar as above mentioned system. It comprises number of transmitters with subsequent receivers fitted as pair at conveyor's either sides in order to sense fruit movement.

The last type measures the height of a sample fruit during traveling through two arrays of optical transducers. The fruit has forward movement through passage between two lateral vertical arrays which are fitted in vertical position. One array is provided with LEDs while other has equipped with photodiodes.

10.3.2.3.1 Application

With use of sizer working on abovementioned principle, the lemons have been classified into three size groups with sorting accuracy of 94.7%.[10] The sizer throughput was of 0.8 fruits per lane. Average sorting accuracy for the first type of sizer was 92.8% as compared to the second type with 93.9%.[32] Based on the hindering laser light for jalapeno chilies a sizer was developed.[22] The fruits traveled on a conveyor belt. Then they were passed between a bar of photodetectors and laser line generator. The belt was running at a speed of 1 m/s and allowed sorting throughput of 15 fruits per lane. For precise measurements of fruits, precaution should be taken to orient fruit sample on conveyor belt in position where fruit's polar axes are made perpendicular to the line. With application of machine vision it is possible to increase the sorting accuracy.

Fruit horizontal width can be measured with both types of system sizers, whereas the vertical width or height measurement can be done with the use of type III sizer. Therefore, for oblate-shaped fruits such as tangerines, tomatoes, and grapefruits, system of type I and II would be better because such fruits have stable support for rest when moved on horizontal surface. Type III is superior than rest two as it can additionally compute projected area by integrating other height observations during the traveling of fruit samples. In case of prolate-shaped fruits such as lemon, type III sizers are preferred.

10.3.2.4 2D MACHINE-VISION SYSTEMS

Video cameras are used in 2D systems. To get the images of the fruits, these video cameras are generally equipped with CCD/CMOS (complementary metal oxide semiconductor) sensors. Especially developed computing software and hardware are used for analyzing the images. Some features that can be estimated from 2D images are diameter, projected area, and perimeter. Such systems comprise computer hardware, image-capturing board, and lighting devices (Fig. 10.1). Pictorial images are converted into numerical form with the help of the image-capturing board (digitizer).[8] In image preprocessing, segmentation followed by feature extraction are achieved with the system software. In order to increase number of views, sometimes lateral mirrors are installed along with the zenithal cameras.

Machine Vision

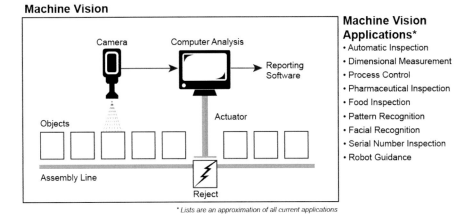

FIGURE 10.1   Flexible machine-vision system for nondestructive testing.
Source: http://www.sensoray.com/resources/blog.htm

10.3.2.4.1   Application

With the use of machine vision system, fruit size and shape can be investigated (Fig. 10.2). In addition to this, it also delivered the reflectance data which is required for grading on the basis of color and blemish grading. Description of digital image boundary using the chain code technique concluded that measuring area is more trustworthy than perimeter measurements. This is because of the fact that the digitization and thresholding processes lead to quantization errors which have no effect on final measurements.[64] The image analysis algorithm has been developed that works on fresh market tomatoes. This algorithm helps to classify fruits based on size, shape, surface defects, and color.[57] For assessment of shape and size of tomato, chain-coded boundary of tomato profile image was used. Strawberries were classified into three sized groups by using projected area of fruit image as fruit orientation angle and perimeter.[5]

On citrus fruits, the machine vision is used to define the fruit image boundary and to calculate the coordinates of centroid.[1,2] Distance from centroid to each boundary pixel was then worked out. The major axis was traced in the maximum distance direction, whereas the minor axis was passing through the centroid and kept as perpendicular to the major axis. An online machine-vision system has been developed for automatic quality grading of oranges, peaches, and apples. This system acquired four images per fruit.[7] A repeatability test was conducted for sizing performance evaluation of the system, using apples.

FIGURE 10.2 **(See color insert.)** Sorting system using *machine-vision system* are installed to sort fruit into different grades based on weight, dimensional size, color, shape, limited defects, and other parameters.
Source: http://tao.umd.edu/html/industry_exp_.html

Fruit volume was determined in some 2D machine-vision system by extracting some features of material images. Volume of agricultural produce such as lemons, eggs, and peaches was determined by using image processing.[56] Dimensions of specific fruits were estimated from digitized picture of the product. These pictures were obtained by a CCD camera.

For a fixed shape solid having variable size scale, using dimensional analysis, size compute is proportional to volume.[43] The separation of potatoes in various size fractions is performed by trays placed between two consecutive rollers in parallel rows. As adjacent tubers in a row would usually touch, for single tuber identification a *blob-splitting* algorithm was developed.[43]

Although the root mean squared percentage error (RMSPE) value was not mentioned, sometimes a rough guess of the RMSPE is obtained by dividing the RMSE by mean. As Merchant system assumes hypothesis of constant tuber density, RMSPEs are not essential. Therefore, density-based sorting could not be performed. From 2D digital images to estimate fruit volume, a machine-vision algorithm was developed which included a neural network.[19]

During working on pears from a set of four images, RMSPE reported of 3% and 1.9% when the volume of single image was calculated. The RMSPEs reported were of 1.6% for lemons, 1.3% for orange, and 5.3% for potatoes when all four images used for volume estimation.[17]

10.3.2.5   3D SYSTEMS FOR MACHINE VISION

Vast number of techniques is included in this group of systems. In the near future, this system will probably supplement in many applications of 2D vision systems. Due to deficient information comprised in the images, the capability of 2D visual systems may be limited.[9] Based on the working of citrus fruits for volumetric inaccuracy, it was observed that the two factors were responsible. One of which was for volume estimation of the 2D measurements extension.[46] In case of automatic inspection systems, 3D information is needed in order to improve geometry characterization of samples products.[24,25,34,54]

With support of third dimension, errors during measuring volume of product reduced by 41% in case of oyster pieces.[41] A range image is defined as huge collection of distance measurements from an identified reference coordinate system to points on surface of the sample.[6]

These methods can be classified into two: active and passive.

Passive 3D methods are also called as range or shape from X techniques, where X indicates various 3D signals such as shading, stereo, motion, occluding contours, shadow of darkness, and texture. Many passive 3D methods are monocular systems, except some the important exemptions of stereovision and shape from silhouettes monocular systems only need a view of the inspected object from one direction.[3] In active methods group indicates energy projected on surface of sample product.

10.3.2.5.1   APPLICATION

By using circular cross section of fruit with the stereo vision desktop system, appropriate results were obtained.[29] In case of elliptical produce in cross section, namely, mangoes, apple, kiwi, etc., the systems performance was less efficient. The preferred system should be developed keeping cost and speed in mind.[30] Based on online system the 3D system was developed.[26,40,41]

From the industrial applicability point of view, active 3D machine-vision system is vigorous than its 2D and passive 3D counterparts.[40] One of the foremost applications of these 3D machine-vision systems in near future will be its use in estimation of surface area and assessment of 3D shapes besides of estimation of volume.

10.3.3 MISCELLANEOUS TECHNIQUES

In this chapter, many different size-sensing techniques have been covered and discussed in detail. Of all the systems described in this chapter, systems based on 2D machine vision are most extensive, as these are practically used not only for classifying samples according to shape and size, but also has potential to categorize samples according to color as well as external defects. Generally, electronic weighing technologies and 2D machine vision are appropriate for large number of horticultural produce. In case of 2D machine-vision technique, the fruit rotation is generated to examine the fruit's whole surface, which may damage to the skin of fruits by abrasion mainly at high speeds. The judgment of quality parameters (such as surface bruises, firmness, density, color, etc.) while selecting the food materials (especially FAV) by the consumers is done by visual inspection. Numerous efforts have been made to find out these properties.

In case of Satsuma mandarins, the fluorescence technique has been used to sense the unseen surface bruises on the surface.[62] This method has been successfully used to evaluate the freshness of eggs and cucumbers and found that it is extremely helpful in knowing the freshness of agricultural produce. Spectral radiometer system was used to measure eggplant's gloss and considered as a feasible parameter in judging the freshness.[44]

Use of gloss meter (Fig. 10.3) was mentioned for measuring gloss of curved surfaces of bananas. A conventional gloss meter was additionally used to measure ripening banana peel gloss[65] as color of peel is an indicator of ripened banana. By this new gloss meter, correct measurement of peel color is possible which facilitates prediction of the right time and ripening level. It is also capable to evaluate the gloss of orange, tomato, green bell pepper, eggplant, and onion.[45]

FIGURE 10.3   (See color insert.) Dual and triple angle gloss meter (Elcometer 407L). *Courtesy*: http://www.elcometerusa.com/Laboratory/Elcometer-407L-Dual-and-Triple-Angle-Glossmeter/.

For the assessment of horticultural produce in conventional digital image analysis techniques, first the video camera is used to record the image followed by video signal's digitization for analysis of digital images. However, with use of artificial retina system it is possible to transmit acquired information directly to processing algorithm.

Many active 3D machine-vision methods are highly depend on digital image analysis (noise reduction, image enhancement operations such as feature extraction, edge detection, etc.). Digital image analysis is not required in case of artificial retina with the charge simulation method. This can be said an advantage, as many times digital image analysis needs specific use and high-cost software for processing of images. Another preferred standpoint of the artificial retina concept is that the photosensors used in this device are comparatively cheaper than other such arrangements.

The major drawback of artificial retina concept is in its inability to find out diameter of fruits and limitation to assess fruit volume only.[21] The prototype of artificial retina machine-vision system conveyed the data acquired to a retina model, which was similar to the "hardware" prototype.[37] This prototype consisted of a hemispherical chamber (artificial retina) fitted with number of photosensors. One of these elements was located at the Fresnel lens center placed at the hemisphere base, another was kept at the hemisphere pole, and the remainder was evenly distributed on circumferences of hemisphere at different heights from the base.[51]

In order to estimate tubers volume, use of optoelectronic device was mentioned to sort potato tubers calculated from density measurements according to their starch content.[28] For online weight estimation of apples and oranges, the microwave technology was used.[14] The method was based on the effect of fruit moisture content on the resonant frequency of a resonant wave guide cavity.[51] Due to higher percentage of moisture (more than 80%) in apples and oranges, estimation of the water volume represented an accurate measure of fruit volume, and therefore the weight.

MRI was performed to evaluate wine grapes, and the reconstruction techniques for 3D representation of berry clusters was developed.[4] CT was used to determine the 3D structure of sugar beet seeds.[42] The inspection system setup consist of a sealed microfocal X-ray tube, a flat panel X-ray detector, a turntable with a sample holder, and a network of computers for data acquisition, reconstruction, and image processing.

10.4 SUMMARY

Many nondestructive methods for analysis of FAV has made available with the development of technology. In this chapter, different sensing techniques for size measurement have been discussed. Out of so many techniques in the field, for nondestructive measurements the portable ultrasonic instrument seems a good option for volume estimation. It has the advantage of higher speed over the air pycnometer system developed.[31]

The ORS system shows good performance on horticultural produce which is elongated.[20] It can work without correctly oriented fruits in order to estimate volume, but correct orientation is required for axis determination. The basic requirement for precise volume estimation is that sample fruit should cross the ring without swinging. The ECM[38] precisely deliberated the volume of large fruits like watermelons, but it required the fruit orientation. 2D machine-vision systems[2,7,19] can be a better preference as it does not demand the fruit orientation. Among 3D machine-vision systems, the active triangulation systems have been implemented online, whereas passive type (stereovision, VI method) have not yet been. 3D shape recovery is important, in applications where with volume estimation, surface area is of also importance. The 2D machine-vision method is a good option if only volume measurement is needed.[17]

The different systems have been compared on basis of accuracy, adaptability, and throughput. Task of online determination of projected area, diameters, and perimeter are possible to judge by using machine vision systems. A better method for 3D reconstruction of fruit surface is the mixed approach of height profiles attained from range sensor with 2D object image boundary obtained by camera. Above all, the important and promising factor in this area is 3D multispectral scanning which utilizes mixed approach of multispectral data and 3D surface reconstruction.

KEYWORDS

- size determination
- machine vision
- neural network
- nondestructive
- photosensors

REFERENCES

1. Aleixos, N. *Desarrollo de Tecnicas de Vision Artificial, Utilizando Procesadores Digitales de Serial: Aplicacion a la Detection de Defectos en Frutas en Tiempo Real* (Development of Machine Vision Techniques, Using Digital Signal Processors: Application to Real-time Fruit Defect Detection). PhD Dissertation, Universidad Politecnica de Valencia, Spain, 1999; p 210.
2. Aleixos, N.; Blasco, J.; Navarron, R.; Molto, E. Multispectral Inspection of Citrus in Real-time Using Machine Vision and Digital Signal Processors. *Comput. Electron. Agric.* **2002,** *33* (2), 121–137.
3. Aloimonos, J. Visual Shape Computation. *Proc. IEEE* **1988,** *76* (8), 899–916.
4. Andaur, J. E.; Guesalaga, A. R.; Agosin, E. E.; Guarini, M. W.; Irarrazaval, P. Magnetic Resonance Imaging for Non-destructive Analysis of Wine Grapes. *J. Agric. Food Chem.* **2004,** *52* (2), 165–170.
5. Bato, P. M.; Nagata, M.; Cao, Q.; Hiyoshi, K.; Kitahara, T. Study on Sorting System for Strawberry Using Machine Vision (Part 2). *J. Jpn. Soc. Agric. Mach.* **2000,** *62* (2), 101–110.
6. Besl, P. J. Active Optical Range Imaging Sensors (Chapter 1). In *Advances in Machine Vision*; Sanz, J. L. C., Ed.; Springer-Verlag Inc.: New York, 1989; pp 1–63.
7. Blasco, J.; Aleixos, N.; Molto, E. Machine Vision System for Automatic Quality Grading of Fruit. *Biosyst. Eng.* **2003,** *85* (4), 415–423.
8. Brosnan, T.; Sun, D. W. Improving Quality Inspection of Food Products by Computer Vision: A Review. *J. Food Eng.* **2004,** *61* (1), 3–16.
9. Chen, C.; Chiang, Y. P.; Pomeranz, Y. Image Analysis and Characterization of Cereal Grains with a Laser Range Finder and Camera Contour Extractor. *Cereal Chem.* **1989,** *66* (6), 466–470.
10. Chen, S.; Fon, D. S.; Hong, S. T.; Wu, C. J.; Leu, K. C.; Tien, B. T. Electro-optical Citrus Sorter. ASAE Paper No. 923520; ASAE: St. Joseph, Michgan, USA, 1992; p 10.

11. Clayton, M.; Amos, N. D.; Banks, N. H.; Morton, R. H. Estimation of Apple Fruit Surface Area. *J. Crop Hortic. Sci.* **1995,** *23*, 345–349.

12. Davenel, A.; Guizard, Ch.; Labarre, T.; Sevila, F. Automatic Detection of Surface Defects on Fruit by Using a Vision System. *J. Agric. Eng. Res.* **1988,** *41* (1), 1–9.

13. Deck, S. H.; Morrow, C. T.; Heinemann, P. H.; Sommer III, H. J. Comparison of a Neural Network and Traditional Classifier for Machine Vision Inspection of Potatoes. *Appl. Eng. Agric.* **1995,** *11* (2), 319–326.

14. De Waal, A.; Mercer, S.; Downing, B. J. On Line Fruit Weighing Using a 500 MHz Waveguide Cavity. *IEEE Electron. Lett.* **1988,** *24* (4), 212–213.

15. Du, C. J.; Sun, D. W. Recent Developments in the Applications of Image Processing Techniques for Food Quality Evaluation. *Trends Food Sci. Technol.* **2004,** *15* (5), 230–249.

16. Eifert, J. D.; Sanglay, G. C.; Lee, D. J.; Sumner, S. S.; Pierson, M. D. Prediction of Raw Produce Surface Area from Weight Measurement. *J. Food Eng.* **2006,** *74* (4), 552–556.

17. Forbes, K. A. Volume Estimation of Fruit from Digital Profile Images. M.Sc. Thesis, Department of Electrical Engineering, University of Cape Town, Cape Town, South Africa, 2000; www.dip.ee.uct.ac.za/(kforbes/Publications/ msckaf.pdf> (accessed Aug 20, 2017).

18. Forbes, K. A.; Tattersfield, G. M. Volumetric Determination of Apples Using Machine Vision Techniques. *Elektron J. South Afr. Inst. Electr. Eng.* **1999,** *16* (3), 14–17.

19. Forbes, K. A.; Tattersfield, G. M. In *Estimating Fruit Volume from Digital Images*. Fifth Africon Conference in Africa: AfriCon; IEEE: Los Alamitos, CA, USA, 1999; Vol. 1, pp 107–112.

20. Gall, H. A Ring Sensor System Using a Modified Polar Coordinate System to Describe the Shape of Irregular Objects. *Meas. Sci. Technol.* **1997,** *8* (11), 1228–1235.

21. Gall, H.; Muir, A.; Fleming, J.; Pohlmann, R.; Gocke, L.; Hossack, W. A Ring Sensor System for the Determination of Volume and Axis Measurements of Irregular Objects. *Meas. Sci. Technol.* **1998,** *9* (11), 1809–1820.

22. Hahn, F. Automatic Jalapeno Chilli Grading by Width. *Biosyst. Eng.* **2002,** *83* (4), 433–440.

23. Hahn, F.; Sanchez, S. Carrot Volume Evaluation Using Imaging Algorithms. *J. Agric. Eng. Res.* **2000,** *75* (3), 243–249.

24. Hall-Holt, O.; Rusinkiewicz, S. In *Stripe Boundary Codes for Real-time Structured-light Range Scanning of Moving Objects*, Proceedings of the 8th IEEE International Conference on Computer Vision (ICCV 2001), Vancouver, BC, Canada, July 07–14, 2001; Vol. 2; pp 359–366.

25. Hardin, W. 3D Advanced Technologies Make Food Inspection Palatable, 2006. www.machinevisiononline.org/public/articles/articlesdetails.cfm?id=2827 (accessed Aug 15, 2017).

26. Hatou, K.; Morimoto, T.; De Jager, J.; Hashimoto, Y. In *Measurement and Recognition of 3-D Body in Intelligent Plant Factory*, Abstracts of the International Conference on Agricultural Engineering (AgEng), Madrid, Sept 23–26,1996; Vol. 2; pp 861–862.

27. Hoffmann, T.; Fiirll, C; Ludwig, J. A. In *System for the On-line Starch Determination at Potato Tubers*, Proceedings of the International Conference on Agricultural Engineering (AgEng), Leuven, Belgium, Sept 12–16, 2004; p 215.

28. Hoffmann, T.; Wormans, G.; Fiirll, C; Poller, J. A System for Determining Starch in Potatoes, 2005 (online). http://vddb-dt.library.lt/fedora/get/

LT-eLABa-0001:J.04-2005-ISSN_1392-1134.V_37.N_2.PG_34-43/DS.002.1.01. ARTIC.

29. Hryniewicz, M.; Sotome, I.; Anthonis, J.; Ramon, H.; De Baerdemaeker, J. 3-D Surface Modeling with Stereovision. In *ISHS Acta Horticulturae: Proceedings of the 3rd International Symposium on Applications of Modelling as an Innovative Technology in the Agri-Food Chain*; Hertog, M. L. A. T. M., Nicolai, B. M., Eds.; ActaHort (CD-rom format): Leuven, Belgium, 2005; p 674.

30. Imou, K.; Kaizu, Y.; Morita, M.; Yokoyama, S. Three-dimensional Shape Measurement of Strawberries by Volume Intersection Method. *Trans. ASABE* **2006,** *49* (2), 449–456.

31. Iraguen, V.; Guesalaga, A.; Agosin, E. A Portable Non-destructive Volume Meter for Wine Grape Clusters. *Meas. Sci. Technol.* **2006,** *17* (12), N92–N96.

32. Iwamoto, M.; Chuma, Y. Recent Studies on Development in Automated Citrus Packinghouse Facility in Japan. *Proc. Int. Soc. Citric.* **1981,** *2*, 831–834.

33. Jahns, G.; Nielsen, H. M.; Paul, W. Measuring Image Analysis Attributes and Modelling Fuzzy Consumer Aspects for Tomato Quality Grading. *Comput. Electron. Agric.* **2001,** *31* (1), 17–29.

34. Jain, S. 2003. A Survey of Laser Range Finding. http://awargi.org/ee236a.pdf (accessed Mar 15, 2016).

35. Jarimopas, B.; Siriratchatapong, P.; Chaiyaboonyathanit, T.; Niemhom, S. Image-processed Mango Sizing Machine. *Kasetsart J.* **1991,** *25* (5), 131–139.

36. Jordan, R. B.; Clark, C. J. Sorting of Kiwifruit for Quality Using Drop Velocity in Water. *Trans. ASAE* **2004,** *47* (6), 1991–1998.

37. Kanali, C.; Murase, H.; Honami, N. Three-dimensional Shape Recognition Using a Charge-simulation Method to Process Primary Image Features. *J. Agric. Eng. Res.* **1998,** *70* (2), 195–208.

38. Kato, K. Electrical Density Sorting and Estimation of Soluble Solids Content of Watermelon. *J. Agric. Eng. Res.* **1997,** *67* (2), 161–170.

39. Laing, A.; Smit, O.; Mortimer, B. J. P.; Tapson, J. Ultrasonic Fruit Sizing Device. *J. South Afr. Acoust. Inst.* **1995,** *6*, 60–65.

40. Lee, D. J.; Eifert, J.; Zhan, P.; Westover, P. Fast Surface Approximation for Volume and Surface Area Measurements Using Distance Transform. *Optic. Eng.* **2003,** *42* (10), 2947–2955.

41. Lee, D. J.; Lane, R. M.; Chang, G. H. *Three-dimensional Reconstruction for High-speed Volume Measurement, SPIE Machine Vision and Three-dimensional Imaging Systems for Inspection and Metrology*; 2001; Vol. 4189, pp 258–267.

42. Maisl, M.; Kasperl, S.; Oeckl, S.; Wolff, A. In *Process Monitoring Using Three Dimensional Computed Tomography and Automatic Image Processing*, Proceedings of the 9th European Conference on Non-Destructive Testing, 2006. <www.ndt.net/article/ ecndt2006/doc/We.3.7.1.pdf> (accessed Jan 23, 2016).

43. Marchant, J. A. A Mechatronic Approach to Produce Grading (Mechatronics: Designing Intelligent Machines), Proceedings of an International Conference, Robinson College, Cambridge, UK, Sept 12–13, 1990; pp 159–164.

44. Matsuoka, T.; Miyauchi, K.; Yano, T. Basic Studies on the Quality Evaluation of Agricultural Products (Part 2): The Evaluation of Color and Gloss Decrease on the Surface of Eggplants. *J. Jpn. Sci. Agric. Mach.* **1996,** *58*, 69–77.

45. Missinovitch, A.; Ward, G.; Mey-Tal, E. Gloss of Fruits and Vegetables. *Food Sci. Technol.* **1996,** *29*, 184–186.

46. Miller, W. M.; Peleg, K.; Briggs, P. Automated Density Separation for Freeze-damaged Citrus. *Appl. Eng. Agric.* **1988,** *4* (4), 344–348.

47. Mitchell, P. D. Pear Fruit Growth and the Use of Diameter to Estimate Fruit Volume and Weight. *Hortic. Sci.* **1986,** *21* (4), 1003–1005.

48. Moreda, G. P. *Disefio y Evaluation de un Sistema Para la Determination en Linea del Tamano de Frutas y Hortalizas Mediante la Utilization de un Anillo Optico* (Design and Assessment of a System for On-line Size Determination of Fruits and Vegetables, Using an Optical Ring Sensor). PhD Dissertation, Universidad Politecnica de Madrid, Spain, 2004; p 219.

49. Moreda, G. P.; Ortiz-Cafiavate, J.; Garcia-Ramos, F. J.; Homer, I. R.; Ruiz-Altisent, M. Optimal Operating Conditions for an Optical Ring Sensor System to Size Fruits and Vegetables. *Appl. Eng. Agric.* **2005,** *21* (4), 661–670.

50. Moreda, G. P.; Ortiz-Canavate, J.; Garcia-Ramos, F. J.; Ruiz-Altisent, M. Effect of Orientation on the Fruit On-line Size Determination Performed by an Optical Ring Sensor. *J. Food Eng.* **2007,** *81* (2), 388–398.

51. Moreda, G. P.; Ortiz-Canavate, J.; Garcia-Ramos, F. J.; Ruiz-Altisent, M. Non-destructive Technologies for Fruits and Vegetables Size Determination — A Review. *J. Food Eng.* **2009,** *92*, 119–136.

52. Ngouajio, M.; Kirk, W.; Goldy, R. A Simple Model for Rapid and Nondestructive Estimation of Bell Pepper Fruit Volume. *Hortic. Sci.* **2003,** *38* (4), 509–511.

53. Nishizu, T.; Ikeda, Y.; Torikata, Y.; Manmoto, S.; Umehara, T.; Mizukami, T. Automatic, Continuous Food Volume Measurement with a Helmholtz Resonator. *CIGR J. Scie. Res. Dev.* 2001 (e-journal). <http://cigr-ejournal.tamu.edu/submissions/volume3/FP%20 01%20004.pdf> (accessed Sept 03, 2016).

54. Njoroge, J. B.; Ninomiya, K.; Kondo, N.; Toita, H. Automated Fruit Grading System Using Image Processing. In *SICE 2002*; IEEE: Los Alamitos, CA, USA, 2002; p 310.

55. Peleg, K. Sorting Operations (Chapter 5). In *Produce Handling, Packaging and Distribution*; AVI Publishing Co.: Westport, CT, 1985; pp 53–87.

56. Sabliov, C. M.; Boldor, D.; Keener, K. M.; Farkas, B. E. Image Processing Method to Determine Surface Area and Volume of Axis-symmetric Agricultural Products. *Int. J. Food Prop.* **2002,** *5* (3), 641–653.

57. Sarkar, N.; Wolfe, R. R. Feature Extraction Techniques for Sorting Tomatoes by Computer Vision. *Trans. ASAE* **1985,** *28* (3), 970–979.

58. Studman, C. J. Fruits and Vegetables (Chapter 3, Section 3.3): Handling Systems and Packaging. In *CIGR Handbook of Agricultural Engineering*; ASAE: St. Joseph, Michigan, USA, 1999; Vol. IV, pp 291–339.

59. Sugiura, T.; Kuroda, H.; Ito, D.; Honjo, H. Correlations Between Specific Gravity and Soluble Solids Concentration in Grape Berries. *J. Jpn Soc. Hortic. Sci.* **2001,** *70* (3), 380–384.

60. Throop, J. A.; Rehkugler, G. E.; Upchurch, B. L. Application of Computer Vision for Detecting Watercore in Apples. *Trans. ASAE* **1989,** *32* (6), 2087–2092.

61. Ting, S. V.; Blair, J. G. The Relation of Specific Gravity of Whole Fruit to the Internal Quality of Oranges. *Proc. Fla. State Hort. Soc.* **1965,** *78*, 251–260.

62. Uozumi, J.; Kawano, S.; Iwamoto, M.; Nishinari, K. Spectrophotometric System for the Quality Evaluation of Unevenly Colored Food. *Nippon Shokuhin Kogyo Gakkaishi* **1987,** *34*, 163–170 (in Japanese).

63. Walsh, K. B. In *Commercial Adoption of Technologies for Fruit Grading, with Emphasis on NIRS*, Frutic '05 (Information and Technologies for Sustainable Fruit and Vegetable Production), Montpellier, France, 2005; pp 399–408.

64. Wechsler, H. A New and Fast Algorithm for Estimating the Perimeter of Objects for Industrial Vision Tasks. *Comput. Graph. Image Proc.* **1981,** *17* (4), 375–385.

65. Wryde, G.; Nussinovitch, A. Peel as a Potential Indicator of Banana Ripeness. *Food Sci. Technol.* **1996,** *29*, 289–294.

CHAPTER 11

ROLE OF ARTIFICIAL SENSORS FOR MEASUREMENT OF PHYSICAL AND BIOCHEMICAL PROPERTIES OF FRUITS AND VEGETABLES

ANWESA SARKAR, KHURSHEED ALAM KHAN, ANUPAMA SINGH, and NAVIN CHANDRA SHAHI

ABSTRACT

Food safety is undoubtedly the most critical public health concern which is equally important for consumers as well as for food industry. Certain features such as freshness, firmness, color, texture, odor, etc., are commonly measured in the quality control of horticultural commodity and their processed food products to ensure food safety. Earlier analytical techniques had been used for quality analysis of fruits and vegetables. However, such techniques require proper sampling from the entire lot of raw materials. Also, absolute damage to the tested samples is inevitable. Therefore, in recent years research has been focused on progress and development of nondestructive approaches/instruments for determining the organoleptic properties of food. The development of sensors has great potential as it could be used in many areas of applications of food industry. The sensation of smell, taste, and vision resulting from a chain of particular molecular recognitions are being used as a tool for quantification of food samples, beverages, and chemical products in these devices (electronictongue, e-tongue; electronicnose, e-nose; and electronicvision, e-vision). These systems, frequently referred as artificial senses, are getting popularity due to their simplicity, rapidity, and objectivity. E-tongues are applied to analysis of liquid samples, whereas e-noses—for gases, and e-visions for solids. This technology provides many advantages such as impossibility of online monitoring, individual variability, subjectivity, adaptation, harmful exposure to dangerous compounds, etc.

These features of sensors will completely change the scenario for quality control of fruits and vegetables and their processed products. With correct operation, calibration, validation, and training, thesesensors can contribute significantly for the maintenance and improvement of product quality.

11.1   INTRODUCTION

Today, food safety is prime health issue applicable for consumer as well as food industry. Physical and biochemical features such as firmness, color, texture, microbial content, etc., are routinely measured as a tool for quality control of horticultural produce or processed products. Earlier analytical techniques were used for food quality control which required proper sampling of food component of interest as it was to be inevitably destroyed in the experimental procedure. Therefore, recent approaches are focused on developing nondestructive approaches/instruments, for example, electronic sensors that could detect and recognize odors, tastes, and appearances for food quality evaluation. Sensor is a device that sense or measures a physical property/quantity or input quantity and records the information or indicate the measured quantity/property and sometimes respond to it. In simple words, sensor can be said converting device which converts measured/quantity into an output signal which may be in analog/digital form.[12] Color, odor, and taste are crucial characteristics in the processed food and beverage industries. With the increasing urbanization and enhanced purchasing powers, consumers are now demanding quality product. Therefore, the food and beverage industries are facing huge pressure due to competition of market share and also emphasison rigorous quality analysis and quality control to meet consumer expectations. Therefore, to deliver the best product at reasonable price and achieve maximum consumer satisfaction, raw material quality is equally important as finished products for manufactures.

The organoleptic characteristics of products (such as: odor, mouthfeel, taste, texture, etc.) are measured to assure the quality of the product. Nowadays, a new era of odor, aroma, and taste control evaluation has been emerged, that is, analysis by gas and liquid sensor array system. They are gaining attraction as simple and rapid working and have wide application in many food commodities, cosmetics, as well as packaging businesses. The sensation of smell, taste, and vision resulting from a series of particular molecular recognition is used as base tool which analyze quality of food, beverage, and chemical products.[22] Sensors array with partial overlapping selectivity perform this function and treat the obtained data with multivariate methods.[21]

These systems are also mentioned as artificial senses due to their ability to perform similar function as human. To mimic the nonspecific recognition, electronic nose (e-nose), electronic tongues (e-tongue), and electronic vision (e-vision) have been developed that consist of a number of solid-state sensor arrays.[3] Calibration is done in most sensors against known standards for accurate results. Sensors are very useful in day-to-day life for instance touch sensation (tactile sensor). Sensitivity of sensor is the property, which tell the amount of output change when measured quantity changes. Sensor, that can successfully determine even very small changes, has very high sensitivity.

The idea of the e-tongue and taste sensor is growing rapidly due to emerging potential.[4,5] Number of applications in the industry has been reported, as mentioned in this chapter that focuses on the role of artificial sensors for measurement of physical and biochemical properties of fruits and vegetables.

11.2 THE CONCEPT OF ELECTRONIC SENSING

The basic idea of human-based sensing is to develop electronic device, which can logically compensate human perception. The human perception is deeply related to human safety aspects as it can identify hazardous situations in its close proximity, functioning as a warning system using the information of knowledge and earlier experience. Therefore, the concept of electronic sensor is to develop a complementary human-based perception facility to notify about the conditions of significance. An artificial sensing system only becomes effective when it is simple and considers that mainly nonexperts are interacting with this type of system. Another motivation behind developing a new generation of artificial sensing systems is for individual local use that can provide fast and accurate indications.

The sensor is an analytical device that converts a response from sample to detectable or measurable signal, which is usually in the form of graph. The overall features of a good sensor are[1]:

- *Selectivity*: The sensors used in any application should be highly selective for the focused task.
- *Reproducibility of signal response*: Sensor should give same response when tested for samples having similar concentration when analyzed many times.

- *Quick response and recovery time*: Time required for sensor to response to analyte should be fast enough. Then recovery time for sensor should be small as possible for reusability.
- *Stability and operating life*: Sensors should produce stable results as most of the biological compounds may show unstability in different biochemical and environmental conditions.

11.3   E-NOSE TECHNOLOGY

E-nose can be defined as an array of chemical sensors, which control and analyze electronically, in order to identify any odor (simple or complex) and attempts to distinguish gas or gas mixtures by mimicking the mechanism of mammalian nose for identifying the specific pattern of vapor response. Compared to other analysis techniques such as gas chromatography, the so called e-nose are easy to construct and it can offer a selective analysis with high sensitivity in short time.[24] Differences between traditional nose and e-nose are described in Figure 11.1 and Table 11.1.

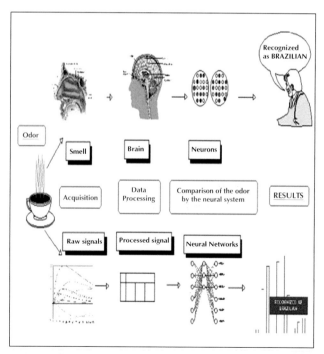

FIGURE 11.1   (See color insert.) Differences between traditional nose and e-nose.
Source: http://www.enose.nl/rd/technology/

TABLE 11.1   Differences Between Traditional Nose and E-nose.

Bio-nose	E-nose
Bio-nose has mucus membrane on inside layer and having hairs which act as filter for dust, etc.	E-nose has inlet sampling system provided with filtration facility
Bio-nose uses lungs system to transfer odor to epithelium layer	It uses a pump for smelling the odor
The human nose contains the olfactoryepithelium, which contains millions ofrecognizing cells which interact with odor	E-nose has a number of sensors that interact differently with a group of odorous molecules
Human nose receptors in nose translate the chemical response after sensing any odor to electronic nerve impulses, then unique patterns of these impulses are propagated by neurons through a complex network, and then reach to higher brain for interpretation	Similarly, the chemical sensors provided in e-nose interact with the smell/odor and generate electrical signals. A personal computer reads these unique signal patterns and interprets them with pattern classification algorithms

Sensor array is used instead of single sensor. This array is specific to a particular type of vapor so that identification of gas or gas mixture is possible by response pattern of array. An e-nose comprises two mechanisms:

- Chemical detection mechanism (e.g., electronic sensors array)
- Pattern recognition mechanism (e.g., neural network)

E-noses have been commercialized for some time; but being typically larger in size and expensive its application was not so popular. Therefore, to overcome these disputes of e-nose technology, research hadfocused on making smaller devices, having less cost, and with more sensitivity. As a result, recently a smaller version as nose-on-a-chip has been developed (Fig. 11.2). It is a single computer chip constituted with sensors and processing components.

Odor is composed of nothing but a number of molecules having specific shape and size and every molecule has a corresponding shape and size receptor of human nose.

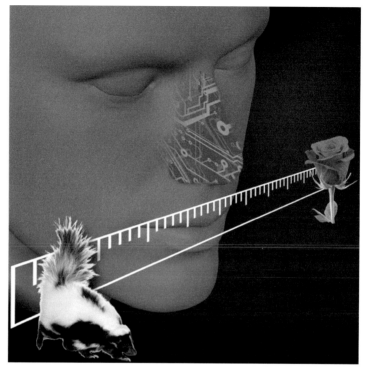

FIGURE 11.2   (See color insert.) E-chip on a nose.
Source: Brodsky, G.; Sobel, N. Electronic Nose was Tuned to the Perceptual Axis of Odorant Pleasantness. *PLoS Comput. Biol. Issue Image* **2010,** *6* (4), ev06.i04 (Open Access): Creative Commons Attribution 2.5 Genericlicense. https://commons.wikimedia.org/wiki/File:An_Electronic_Nose_Estimates_Odor_Pleasantness.png

When any molecule of odor is received by specific receptor, signal is sent to brain and then brain identifies particular smell which is associated with that molecule. E-nose based on biological model follows a similar concept (Fig. 11.1). The receptors are replaced by albeit sensors and it transmits signal to the program for processing. Research areas like e-noses are called biomimetics, or biomimicry, in which human-made applications copies the patterned of natural phenomena. Earlier the use of e-noses was limited to quality control applications in food and beverage and also in cosmetics industries. However, current application of e-nose is being harnessed in other fields such as in medical field for detection of odors related to specific diseases, for detection of pollutants in environment and detection of gas leakage.

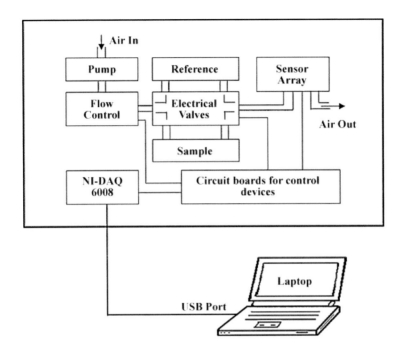

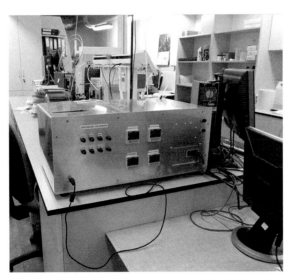

FIGURE 11.3 **(See color insert.)** Principle of e-nose.

Source: https://www.elprocus.com/electronic-nose-work/; Katlewa, L.AnElectronic Nose Prototype. This file is Licensed Underthe Creative Commons Attribution-Share Alike 4.0 International License, 2013. https://commons.wikimedia.org/wiki/File:Enose_prototype_ Analytical_Dept_Chemical_Faculty_GUT_Gdansk.jpg

11.3.1 WORKING PRINCIPLE OF E-NOSE

An array, which consists of 32 polymer–carbon black composite sensors, is being used in e-noses nowadays. They are being developed and tested by selecting numerous chemical molecules from alcohols to aromatics for identification and quantification of broad range of target compounds. E-noses consist of three major segments: (1) sample delivery mechanism, (2) detection section, and (3) computing system (Fig. 11.3). First system allows head space generation for volatile components of provided sample, which is then injected to detection system. The sample delivery system guarantees constant operating conditions. This detection system has a sensor set. When the sensors come in contact with volatile compounds, its electrical properties changes. Every sensor responds to different volatile components in different specific way.

When the volatile compounds are absorbed by surface of sensor they cause a physical change of sensor.[13] The electronic interface records specific response and transform it in a digital value. Most frequently used sensors nowadays are conducting polymers, metal-oxide semiconductors (MOS), surface acoustic wave, quartz crystal microbalance, and MOS with field effect transistors. Another type of e-noses have also been developed which utilizes ultrafast gas chromatography, mass spectrometry, or as a detection system. Computing system combines the responses received by all sensors, which represent the input for data and also perform fingerprint analysis and give output in interpretable format.

11.4   E-TONGUE

Human tongue primarilydetects five different tastes: sweet, sour, salty, umami, and bitter, by use of gustatory receptor cells. After gustatory cell's perception, this taste information is conveyed via cranial nerves to brain stem nuclei. The theory of e-tongue originates from human gustatory system mechanism. E-tongues are analytical instruments which artificially reproduce the taste sensation by chemometric processing. The information is evaluated in cerebral cortex, and thus different tastes are perceived. The e-tongue has widespread applications in food, processed food products, and beverage industries for monitoring the product quality. For example, to analyze flavor ageing in beverages, to quantify spicy or bitterness levels in drinks, and to the stability of medicines in reference to taste.

Sensitivity of e-tongue does not decrease with time due to exposure, which is one of the important advantages of the e-tongue. Also as the device is portable and compact in nature, in situ measurements are possible. It also prevents the exposure of human to awkward tastes and toxic substances. The e-tongue system uses a sensor's array made of ultrathin films and composite films of several polymers. Sensors made of different materials generated different electric responses, and this variation is required to produce "finger-print" of tested samples.[13] There are several applications of e-tongue in food industry:

- In wine industry, custom-designed e-tongueprovided with array of hybrid sensors consisting of voltammetric electrodes is used to discriminate and recognize different Spanish red wines based on origin, grape variety, denomination, and vintage.
- This sensor system is used for beer to predict beer, sensory attributes including sweet, bitter, sour, caramel, fruity, artificial, and burnt.
- It is also possible to determine contents of caffeine and catechins in green tea.
- Adulterations in milk, fruit juices, or any other beverages can be detected efficiently by e-tongue.
- Microelectrode is used as an e-tongue for discriminating oils according to their geographic origin and quality.
- E-tongue is also being used for predicting sodium nitrate ($NaNO_2$), sodium chloride ($NaCl$), and potassium nitrate (KNO_3) in meat products.

11.4.1   FEATURES OF AN E-TONGUE

There are two types of e-tongue systems frequently used in food industries. Both of them measure variation in electronic potential of liquid samples but they have different sensor technologies. Lipid membrane is used in some sensing system, whereas some uses chemical field effect transistor tech-nology.[23] Chemically adapted polymer beads are organized on a tiny silicon wafer called as a sensor chip. These beads change color according to the type and quantity of certain chemicals which is recorded by a digital camera and then converted into data by a video capture board attached to a computer. In a nutshell, it can identify the presence of analytes, bacteria, and/or toxins in food and beverage industry; environmental solutions; and in medical field. The e-tongue is made of following components:

- *Working electrode* can be made of inert material such as gold, platinum, glassy carbon, iridium, rhodium, etc., that is used as working surface on which the electrochemical reaction takes place. Surface area should be very less (few mm^2) to limit current flow.
- *Reference electrode* measures the electrode potential. A reference electrode always has a constant electrochemical potential even when no current is flowing through it.
- *Auxiliary electrode* is a stainless steel electrode that serves as inert conductor completing the cell circuit.

11.5 E-VISION

In recent years, e-vision technology has imparted wide range of applications in agricultural and horticultural sectors for instance: remote sensing for precision farming, natural resources assessments, postharvest quality of produce, and sorting applications. E-vision systems provide an idea about quality attributes such as size and shape, texture, and color of sample with its numerical attributes.

It has a great prospect in agricultural industry because the technology is simple and of low operational cost compared to destructive analytical techniques. Also it provides opportunity to analyze entire commodity, that is, each particle, which is not possible in other analytical techniques.[8,16] Even of its high potential for defect and disease detection in agricultural products and food, its application is still limited due to high initial cost of concerned equipments.

11.5.1 COMPONENTS OF E-VISION SYSTEM

E-vision system measures the reflectance, fluorescence, and transmittance properties of the investigating material under UV or NIR illumination. The components of an e-vision system are: camera, computing system equipped with lighting system, and an image-acquisition board. Software is essential to transmit the electronic signal to computer and also to perform storage and processing of images (Fig. 11.4).

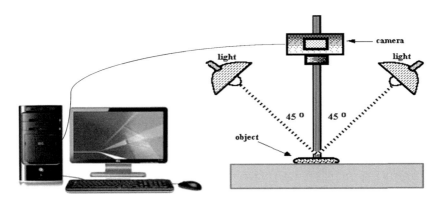

FIGURE 11.4   Components of an e-vision system.
Source: Own work.

11.5.1.1   LIGHTING

The ranges of light used in e-vision system are 200–400 nm for ultraviolet imaging, 400–700 nm for visible imaging region, and 700–2500 nm for near-infrared imaging. Thermal imaging, where wavelength lies above 2500 nm has also some applications for food and agricultural products. When sample produce is illuminated by light it is either transmitted or reflected, or absorbed due to its optical property. Due to variation in wavelength, angle of incident light, and physical and chemical structure of the sample, the optical properties and fluorescence emission from the sample also differs. Therefore, recognizing or analyzing objects or scenes is possible with a lighting system of known characteristics, by measuring the change in optical properties. The selection and configuration of a lighting unit applied in a machine-vision system is largely based on its application.

11.5.1.2   FRAME GRABBER

Frame grabber is used for image acquisition, controlling camera, and also for preprocessing of images. The frame grabber analyzes analogue as well as digital images depending on the type of camera being used. Advanced frame grabber card communicate signals with computer through software for application in agriculture and food processing.

11.5.1.3   IMAGE PROCESSING AND ANALYSIS SOFTWARE

A computer is used for digital-image processing for manipulation of information within an image. Image processing consists of following steps: (1)
image enhancing, (2) extraction of image feature, and (3) classification of
image features. Image enhancement operation comprises morphological
operations, use of filters, and operations on pixel level are done to adjust
errors, for example, poor contrast or removal of noise in image caused by
inadequate illumination. Statistical procedures of basic image statistics such
as mean, variance, and standard deviation is used for feature extraction from
digital images. After identification of image features, different numerical
techniques (e.g., neural networks and fuzzy inference systems) are applied
for image feature classification.

11.5.1.4   CHARGE-COUPLED DEVICECAMERAS

Different types of imaging cameras are being utilized in e-vision systems
ranging from monochrome cameras for recognition of simple shape and size
to common aperture multispectral cameras for detecting defects present on
the surface of fruits and diseases on grains, meat, vegetables, and fruits.
Charge-coupled device (CCD) image sensors translate light into electrons. A
CCD is sensitive photon detector. CCD has numerous small sections known
as pixels, which helps in building an image of focused sample.

11.5.1.5   MONOCHROME IMAGING

Monochrome imaging sense visual and near-infrared using a single-chip
CCD. Monochrome imaging technique is useful for determination of food
product quality.[4,10] It can successfully detect the blemishes and bruises on
apples.[2,13,14,20] This imaging technology was used in detecting cracks, scars,
and asparagus spreading tips.[15,17] Online grading of apples using e-vision
technology is also being commercialized.[11,18] The major challenges for
online checkup is accessibility of good quality images which in turn give
distinct features of defects in sample. Addition to the efficient hardware
and software are required which can process the images enough fast on
real time.

11.5.1.6   COLOR IMAGING

A single-chip CCD camera is generally used for color imaging by alternating the pixels in the CCD camera for red, green, and blue colors acquisition in the array for simulation of colors as seen by human eye. However, this method is not adequate for complicated e-vision applications. Despite of that, research has been done in color-imaging technique for evaluation of agricultural product quality. Color difference was used to differentiate bruised and nonbruised regions of Golden Delicious apples.[20] Color-imaging techniques were applied in online poultry quality grading.[2]

11.5.1.7   MULTISPECTRAL IMAGING

Such images are obtained by taking a spectral image series at narrow-band wavelength by using a liquid crystal tunable filter or by change of filters fitted at front of the camera. The advantage of multispectral imaging is that it can acquire multiple spectral images simultaneously and process these images of different spectral bands. A multispectral imaging system provided with rotating wheel which can hold six filters was used for discrimination between healthy and unhealthy leaf and for comparison of chlorophyll distribution on leaf surface.[19]

Muir et al.[11] has mentioned the use of spatial information at eight wavelengths, in order to spot the blemishes on the surface of potato. Use of intensified multispectral imaging system was mentioned to differentiate wholesome poultry carcasses from unwholesome carcasses.[12] Multispectral images were used to characterize chicken heart images for disease detection.[2] Multispectral fluorescence imaging was used in observation of diffusion of herbicide within leaves, after herbicide application.[6] The technique can provide additional information when compared to other techniques as every pixel of image possess a unique spectral signature of sample at respective pixel.

Spectroscopic data analysis method can also be useful for knowing chemical composition at every pixel or an aggregate of pixels. Due to these features, hyperspectral imaging can be more useful and has potential to identify materials and can detect minor features of sample. Martinsen and Shaare[9] mentioned use of hyperspectral imaging in measuring distribution of soluble solids in kiwifruit. A hyperspectral imaging system has also been developed which is capable of ground/laboratory and airborne data acquisition.[7]

11.6 SUMMARY

Biomimetic measurement methods (demonstrated by the e-nose, e-tongue, and the e-vision) are increasingly being introduced in various sectors of food and agriculture. It is an emerging novel technology where both hardware and software functions together for better performance of the sensors in order to analyze the sample characteristics. E-tongue, e-nose, and e-vision are systems for auto-recognition and classification of liquids/gases or solids using nonspecific sensors array, data collectors, and data analysis tools. E-tongues have application in analysis of liquid samples, whereas e-noses for analysis of gases and e-visions for analysis of solids are used. The output or result of e-tongue/e-nose/e-vision analysis can be in terms of identification of the sample or estimation of its characteristic properties like concentration. This technology has numerous advantages when compared to manual efforts. Human capacity may not utilize consistently. E-tongue/e-nose/e-vision can be used where chances of exposure to harmful substances. Moreover individual variability, inability to monitor online subjectivity, adaptation are another limitations of human observer. With correct operation, calibration, validation, and training process this technology can contribute significantly for the maintenance and improvement of product quality. These features of artificial sensing have proved them useful in food and beverage industry, environmental solutions, and public security and such an approach will completely change the scenario of quality control in food product and processing market.

KEYWORDS

- artificial sensing
- auxiliary electrode
- charge-coupled device cameras
- electronic nose
- electronic tongues
- electronic vision

REFERENCES

1. Adley, C. C. Past, Present and Future of Sensors in Food Production. *Foods* **2014**, *3*, 491–510.
2. Davenel, A.; Guizard, C.; Labarre, T.; Sevila, F. Automatic Detection of Surface Defects on Fruit Using a Vision System. *J. Agric. Eng. Res.* **1988**, *41*, 1–9.
3. Gardner, J. W; Bartlett, P. N. *Electronic Noses: Principles and Applications*; Oxford University Press: New York, 1999; Vol. 2, p 33.
4. Hwang, H.; Park, B.; Nguyen, M.; Chen, Y. R. Hybrid Image Processing for Robust Extraction of Lean Tissue on Beef Cut Surfaces. *Comput. Electron. Agric.* **1997**, *17*, 281–294.
5. Kim, M. S.; Chen, Y. R.; Mehl, P. M. Hyperspectral Reflectance and Fluorescence Imaging System for Food Quality and Safety. *Trans. Am. Soc. Agric. Eng.* **2001**, *3*, 721–729.
6. Kim, M. S.; Mc. Murtrey, J. E.; Mulchi, C. L.; Daughtry, C. S. T.; Chappelle, E. W.; Chen, Y. R. Steady-state Multispectral Fluorescence Imaging System for Plant Leaves. *Appl. Optics* **2001**, *401*, 157–166.
7. Mao, C., Heitschmidt, J; Meyer, G. E., De, Shazer, J. A. Hyperspectral Imaging with Liquid-crystal Tunable Filter for Biological and Agricultural Assessment. In *Precision Agriculture and Biological Quality–Proc*; SPIE, Bellingham, WA, 1999; pp 172–181.
8. Marks, J. S.; Schmidt, J.; Morgan, M. T.; Nyenhuis, J. A.; Stroshine, R. L. *Nuclear Magnetic Resonance for Poultry Meat Fat Analysis and Bone Chip Detection.* Industry Summary for US Poultry and Egg Association: Tucker, GA, 1998; p 21.
9. Martinsen, P.; Shaare, P. Measuring Soluble Solids Distribution in Kiwifruit Using Near-infrared Imaging Spectroscopy. *Postharv. Biol. Technol.* **1998**, *14*, 271–281.
10. Mc. Donald, T. P.; Chen, Y. R. Visual Characterization of Marbling in Beef Ribeyes and Its Relationship to Taste Parameters. *Trans. Am. Soc. Agric. Eng.* **1991**, *346*, 2499–2504.
11. Muir, A. Y.; Porteous, R. L.; Wastie, R. L. Experiments in the Detection of Incipient Diseases in Potato Tubers by Optical Methods. *J. Agric. Eng. Res.* **1982**, *27*, 131–138.
12. Otto, M.; Thomas, J. D. R. Model Studies on Multiple Channel Analysis of Free Magnesium, Calcium, Sodium, and Potassium at Physiological Concentration Levels with Ion-selective Electrodes. *Anal. Chem.* **1985**, *57*, 47–51.
13. Ramamoorthy, H. V.; Mohamed, S. N.; Devi, D. S. E-nose and E-tongue: Applications and Advances in Sensor Technology. *J. Nano Sci. Nanotechnol.* **2014**, *3*, 370–376.
14. Rehkugler, G. E.; Throop, J. A. Image Processing Algorithm for Apple Defect Detection. *Trans. Am. Soc. Agric. Eng.* **1989**, *32*, 267–272.
15. Roberts, L. G.; Borkowitz, D. A.; Clapp, L. C.; Koester, C. J. *Machine Perception of Three-dimensional Solids in Optical and Electro-optical Information Processing*; Tippett MIT Press: Cambridge, MA, USA, 1965; pp 159–197.
16. Schatzki, T. F.; Haff, R. P.; Young, R.; Can, I.; Le, L. C.; Toyofuku, N. Defect Detection in Apples by Means of X-ray Imaging. *Trans. Am. Soc. Agric. Eng.* **1997**, *40*, 1407–1415.
17. Singh, N.; Delwiche, M. J., Machine Vision Methods for Defect Sorting Stonefruit. *Trans. Am. Soc. Agric. Eng.* **1994**, *37*, 1989–1997.
18. Sliwinska, M.; Wisniewska, P.; Dymerski, T.; Namiesnik, J. and Wardencki, W. Food Analysis Using Artificial Senses. *J. Agric. Food Chem.* **2000**, *11*, 345−351.

19. Taylor, S. K.; McCure, W. F. In *NIR Imaging Spectroscopy: Measuring the Distribution of Chemical Components*, Proceedings of the Second International NIR Conference, Tsukuba, Japan, July 2, 1989; pp 393–404.

20. Throop, J. A.; Aneshansley, D. J.; Upchurch, B. L. Near-IR and Color Imaging for Bruise Detection on Golden Delicious Apples. *Proc. Int. Soc. Optic. Eng.* **1993**, *36*, 33–44.

21. Trinchi, A.; Li, Y. X.; Wlodarski, W.; Kaciulis, S.; Pandolfi, L.; Russo, S. P.; Duplessis, J.; Viticoli, S. Investigation of Sol-gel Prepared Ga-Zn Oxide Thin Films for Oxygen Gas Sensing. *Sens. Actuators B* **2003**, *108*, 263–270.

22. Vlasov, Y.; Legin, A.; Fresenius, J. Non-selective Chemical Sensors in Analytical Chemistry: From "Electronic Nose" to "Electronic Tongue". *Anal. Chem.* **1998**, *361*, 255–260.

23. Winquist F.; Rulcker C.; Krantz, Y.; Lundstrom, I. *Electronic Tongues and Combination of Artificial Senses.* The Swedish Sensor Centre and the Division of Applied Physics, Department of Physics and Measurement Technology, Linkoping University; 2015; p 112.

24. Yingchang, Z.; Hao, W.; Xi, Z.; Wang, D. H. Electronic Nose and Electronic Tongue. *Anal. Chim. Acta* **2000**, *403*, 273–277.

PART IV

Challenges and Solutions:
The Food Industry Outlook

CHAPTER 12

WASTE-REDUCTION TECHNIQUES IN FRESH PRODUCE

KUMARI S. BANGA, SUNIL KUMAR, and KHURSHEED ALAM KHAN

ABSTRACT

Fruits and vegetables are sources of micronutrients and essential in human diet. Waste of food has always remained a matter of concern, as it requires resources and labor to produce. In addition, it has potential to feed big chunk of population as well. Existing technologies with skilled and supervised manpower have potential to reduce the surfeit losses and waste of the fruits and vegetables. This chapter includes conventional technologies as well as recent technologies being used across the world. Farmers, wholesalers, retailers, and consumers all are the stakeholders in postharvest chain of fresh produces. Each section of chapter has managerial suggestions and information for all stakeholders to improve quality and shelf life of agricultural fresh produces. Lack of consumer awareness and unskilled labor in postharvest chain are the major issues in menace of postharvest losses. Emphasis has also been given for awareness program, training, and skill development to inculcate a right protocol to follow by stakeholders under postharvest chain of fresh produces.

12.1 INTRODUCTION

Globally, one-third of the produced food for human consumption is wasted or lost according to FAO study.[6] Food waste is acknowledged as a global problem that needs to be solved to achieve sustainable development. It increases greenhouse gases emission through production, processing and its disposal and contributes to world hunger as food is wasted instead of being delivered to the ones in need. India held the second rank in producing 235.85 million tons of fruits and vegetables during 2012–2013. Based on independent study on postharvest losses, annual loss of fruits and vegetables was 6.7–15.8% and 4.58–12.44%, respectively, during various unit operations.[28] Postharvest losses of foods are not restricted to developing countries only. Many of the other countries are still struggling to prevent losses by modernizing and upgrading the supply chains.

There are two terms associated with the waste reduction, namely, food waste and food loss. Food waste is the unused surplus cooked food that occurs at the consumer end of the postharvest chain. Food loss takes place during harvesting, handling, distribution, and processing operations. Biotic and abiotic components are responsible for losses and quality deterioration of horticultural crops.[48] In case of fresh produces, the waste starts from the field and continues up to consumer plate. Pests, natural ripening process, mold, and other microbe's infestation are biotic factors,[31] whereas temperature, relative humidity, poor transportation, etc. are the abiotic factors (Table 12.1).

Harvesting, handling, storage, and distribution are major unit operations associated with postharvest chain of fruits and vegetables. Maintaining sound quality and quantity of fresh produces are the major challenge during this chain.[34] The fresh produces come into highly perishable category of foods due to high respiration rate and moisture content. It needs more attention in comparison to the cereals, pulses, and oilseeds throughout the period of harvesting to consumption, to restrict a chemical, biochemical, and physiological changes to a minimum by control of temperature and humidity in a closed space. The controlled environment will decrease water loss and respiration rate as well as lower the sensitivity to ethylene to prolong the shelf life of fresh produces.

This chapter explores various methods and techniques to reduce the postharvest losses in fruits and vegetables. It includes factors responsible for the deterioration of fresh produces as well as the quality parameters to measure the impact of deterioration. The protocols and recommendations of each method to save food from being waste are also discussed.

12.2   REASONS TO REDUCE FOOD LOSS AND FOOD WASTE

- Food is the output of a long chain of activities from field to the consumer plate. It requires energy and money in terms of labor, water, seed, fertilizer, pesticides, transport, processing, etc. to get food. Therefore, waste of the food is indirectly a waste of money and resources. Food waste has a worth of 580 billion rupees for perishable and nonperishable food items. It is associated with production, processing, supply, and preparation of food. In India, about 20–40% of food production gets spoiled before reaching consumers.

- *Monetary loss*: Inedible food can be used for animal feed, compost, and biogas generation. Else, there are many nongovernment organizations (NGOs) in the cities that took surplus food of parties, conferences, meetings, etc., and handover it to the needy one or to the animal shelters.
- The carbon footprint of food development has environmental impacts such as pollution, greenhouse effect, etc. The wasted food is generally used for landfill or incineration for waste management. Rotting of food in landfill, rice cultivation in field, incineration, processing, and transportation produce greenhouse gases (methane and carbon dioxide).

TABLE 12.1 Biotic and Abiotic Components Involved in Food Deterioration.

Biotic factors	Insects, pests, natural ripening process, mold, and other microbe's infestation, genetic disorders, senescence
Abiotic factors	Environmental factors: temperature, relative humidity, air-velocity physical factors: poor transportation, stacking height

12.3 QUALITY PARAMETERS OF FRUITS AND VEGETABLES

The quality of fruits and vegetables is defined as "the combination of attributes or properties that give them value in term of human food."[30] It denotes the degree of acceptability and bundle of the properties liked by wholesalers, retailers, and consumers. Handlers and users during postharvest operations have a different understanding of quality. Accordingly, there are two types of quality orientation: product- and consumer-oriented quality. Product-oriented quality is the shelf life or storability in which fruits and vegetables have its characteristic attributes. Consumer-oriented quality is the degree of acceptance of the product by the consumer. A high-quality product may not be acceptable to the consumers due to food habits, taste, living locations, age, religion, etc.[10,24,27] The ultimate objective of the postharvest losses control is to preserve the parameters (Table 12.2) which in combination turns into the quality.

TABLE 12.2 Classification of Quality Parameters.

Sr. No.	Type	Quality parameters
	Safety and hygiene	Mycotoxins, contaminant, adulterant, natural toxicant, pesticide residues, microbial contamination, foreign matter, antinutrients, pH, titratable acidity, aflatoxin, pathogens, parasites, insect infestation
2.	Nutritional availability	Content and bioavailability of vitamins (C, E, K), folate, protein content, protein efficiency ratio, fibers, carbohydrate, flavonoids, carotenoids, anthocyanins, coumarins, cancer inhibitor, the soluble solids, etc.
		Texture: softening, drying up, development, firmness, juiciness, fibrousness
3.	Sensory attributes	*Flavor*: ratio acid/sugar, phenols, change in aroma content, and off-odors appearance
		Development: stipe elongation (mushroom), head opening (endive)
		Taste: sweetness, saltiness, umami, sourness, bitterness
		Color: discoloration, uniformity, intensity, spectral, gloss
		Surface texture: smoothness, waxiness, gloss, wrinkling, damage
		Size: weight, volume, and dimensions
4.	Cosmetic appearance	*Shape*: regularity, length, sphericity, geometric mean diameter
		Physiological: browning, genetic effects, germination, defects, ripe/unripe
		Convenience: 100% usable product, trimmed, peeled, washed, and cut-fresh produces

12.4 TECHNIQUES TO REDUCE THE WASTE IN FRESH PRODUCES

The waste is started from the field and continues up to the consumption. Therefore, the techniques and methods employed for reducing waste and loss used in field, transportation, storage, and processing are discussed here.

12.4.1 IN-FIELD PRACTICES

The inherent variability of fruits and vegetables is the major challenge in the designing of safe-storage atmosphere as well as a package. It depends on soil conditions, climate conditions, seed, variety, weather, harvesting method, and agricultural practices of grown places. First-class quality of raw material yields first-class products.[3] In-field practices play an important role in the longevity of storage with the highest quality. The harvesting of the crop in morning reduces the postharvest losses as well as the cost of maintaining the quality of the products.[63] Field heat in agricultural produces is at its minimum in morning, therefore, less cooling will be needed to reach a safe temperature. It increases the shelf life of the product by reducing high rate of transpiration and respiration.[34]

Respiration rate depends on the temperature, growing area and climatic condition, variety, and injury during harvesting and handling. This recommendation of morning harvesting has exceptions in case of citrus fruits and long-transportation routes to reach market with fresh produce. Citrus fruits can be damaged during handling in the morning due to turgid state.[19] Established condition of preharvesting resulted in a greater tolerance for chilling injury in stress and unstressed tomato.[51] In the case of long routes, the farmers have to harvest in the evening to reach market in the coolest time of the day. The use of hybrid varieties also helps in reducing postharvest losses. Hybrid varieties have longer shelf life, higher yield, slow ripening process, uniform firmness and color, insect resistance, etc. in comparison to normal varieties.[16] Mutation develops resistance to bear the high temperature, which is a major factor to affect the postharvest life of fruits and vegetables.[52]

All fruits and vegetables can be classified as climacteric or nonclimacteric on the basis of the biochemical process of ripening. The climacteric continues to ripen even after the harvest because of the ethylene gas and burst of respiration. But in the case of the nonclimacteric, there is small or no production of the ethylene gas. Therefore, ripening process is terminated at the time of harvest of the produce. Table 12.3 indicates classification of fruits and vegetables as climacteric and nonclimacteric. The harvesting of the climacteric fruits and vegetables can be accomplished before ripening and stored in closed space with control of temperature, gases, and RH.[58] At early harvesting, the climacteric produces are firm and green. They are able to withstand the stress and injury during transportation and handling; and late ripening provides sufficient time to reach market with higher quality produces. It helps in fetching good price of agro produces as well as controls the supply in a surplus of arrival in market.[58] The fruits harvested from trees of poor canopy ventilation and lower branches will have more diseases and rot during transportation, storage, and marketing.

TABLE 12.3   Classification of Fruits and Vegetables on the Basis of Respiratory Behavior.

Climacteric fruits	Apple, papaya, apricot, passion fruit, avocado, peach, banana, pear, cherry, biriba, persimmon, breadfruit, plum, cherimoya, sapote, feijoa, soursop, fig, tomato, guava, watermelon, kiwifruit, mango, muskmelon, nectarine, jackfruit, peach
Nonclimacteric fruits	Blueberry cocoa, cashew, cherry, cucumber, grape, grapefruit, lemon, lime, olive, orange, pepper, pineapple, strawberry, tamarillo, litchi, mousambi, kinnow

Lower branched fruits have more exposure to pathogens. The chemical composition and nutritional status of the fruits and vegetables depends on the fertilizers used in agricultural lands. The amount of boron and potash used in the field causes postharvest browning and internal breakdown during CA storage, respectively. Use of optimum composition of micro and macronutrients yields higher quality and quantity of produces with long shelf life.[64] The micro-irrigation methods are efficient in delivering the nutrient requirement to a particular plant in comparison to the conventional irrigation methods. It saves the water as well as increase the productivity with high-quality farm produces. Drip irrigation in grapefruits increases the percentage of edible grape by 3.88–5.78%, vitamin C content, and ratio of total soluble solid concentration/titrated acid without any detrimental effect on fruit yield.[18]

12.4.2   PRECOOLING

Farmers used to travel several kilometers to reach the market or cold storage to sell or store their produces. Field heat increases the rate of biological metabolism and hence reduces the shelf life due to their highly perishable nature.[38] The precooling is the process to remove the field heat using the cool fluids prior to storage and transportation. Therefore, quality remains comparatively intact during traveling and fetches higher prices to farmers. The time lag between harvesting and precooling should be kept to a minimum. Precooling retards the production of ethylene that reduces the rate of ripening and senescence in case of the climacteric fruits and vegetables.[12] There are five methods of precooling used for perishable agricultural produces: Hydrocooling, room cooling, vacuum cooling, forced air cooling, and evaporating cooling. Generally, the commodity and the cost–benefit ratio are basis for the selection of cooling method.[63] Moreover, the choice of cooling method depends on product flow, packaging requirement, nature of product, and economic constraints. Most of the methods are carried out in the packed room or central cooling facilities.[12]

12.4.2.1   EVAPORATING COOLING

Evaporating cooling is a well-known short-term preservation method of fruits and vegetables soon after harvest. It reduces the storage temperature as well as increases relative humidity of the enclosure to maintain freshness of horticultural produces. Evaporative cooling is also known as zero-energy cooling system due to adiabatic exchange of the heat. It has been practiced in some tropical and subtropical countries as a method to remove field heat.[38]

12.4.2.2   HYDROCOOLING

Water is used as a potential cool fluid to remove field heat. It is a better method of heat removal than the forced air since water has highest specific heat. Hydrocooling cleans the product and reduces field heat and microbial load. At the field, irrigation water is also used for the fruits and vegetables to remove field heat as well as soil particles. No moisture loss from the product happens in this cooling method. The treated water with gaseous chlorine or sodium hypochlorite is used to prevent the proliferation of decay microorganisms.[65]

12.4.2.3   VACUUM COOLING

Vacuum cooling is the method of fast removal of the field heat. It is used for the leafy vegetables, which lose water vapor very rapidly, that is, broccoli. The product with the packed crushed ice is kept in a strong steel container under a low pressure of 4.6 mmHg. Vacuum depresses the boiling point of water. Water evaporates from the surface of the product and leaves a cooled product. It causes 4% moisture loss of its weight during the precooling process. Moisture loss is recovered by spraying the water on the surface of the product.[65] It is a costlier cooling method than the other methods.[12]

12.4.2.4 FORCED AIR COOLING

It is a forced convection method of field heat removal using the refrigerated air. A batch tunnel cooler with packed produces in boxes or bins is used for forced air cooling. Product quantity, packaging material, airflow rate, and dimensions of the product are the deciding factors for the cooling time. Material is wrapped in a packaging film to avoid the extent of the moisture loss from the product except for the citrus fruits. The severity of moisture loss is depended on the difference between initial and final temperature of precooling.[65]

12.4.2.5 ROOM COOLING

In room cooling, harvested produces are placed in a refrigerated room. Produce packed in the wooden boxes, bulk containers, and other packages are exposed to the cold air. It is a slow heat-removal method and requires 1 m/s velocity for necessary turbulence. Cold air enters the room from the ceiling and goes to condenser from the bottom of room. Room cooling is used for the medium- and low-perishable fruits and vegetables at the cooling rate of 0.5°C/h.[12]

12.4.2.6 CRYOGENIC COOLING

It is the conveying of the produce through a tunnel of the cryogens for heat removal. The solid CO_2 and liquid nitrogen absorb the heat from the produce and utilize it for the phase change. It is a rapid method, relatively cheap to install but costly to run. The product flow rate and evaporation rate of cryogens are the considerable points to supervise during use of cryogenic cooling.[12]

12.4.3 PRETREATMENTS OF FRESH PRODUCE

Pretreatments are generally applied to the fresh produces prior to further processing operations or storage. It is an optional treatment to improve the quality of fruits and vegetables. Osmotic dehydration, chemical applications, blanching, etc. are pretreatment methods that are practiced before the drying, canning, and storage. Pretreatments of fresh produces offer following advantages:

- Prevent discoloration and off-flavor.
- Minimize nutrient loss (retention of vitamin A and C).
- Prevent enzymatic action.
- Increased rate of drying.
- Extend shelf life.
- Clean the product's surface.
- Reduce microbial load.
- Softening of the tissues.
- Improve heat transfer in heat processing.

12.4.3.1 OSMOTIC DEHYDRATION

It is a preservation as well as pretreatment technique given to cellular materials prior to the dehydration without a phase change. The hypertonic solution is used for mass transfer through a semipermeable membrane for partial removal of the water. Diffusion is the principle mechanism for mass transfer under the effect of the concentration gradient. It reduces the harshness of high temperature and prevents the major physical, chemical, and biological changes.[35] Pretreatment of osmotic dehydration prior to microwave drying of the apple cubes increased the overall quality. Rehydration of dehydrated apple cubes had higher firmness compared to other methods.[54] The osmotic dehydration of aonla (Indian gooseberry, *Phyllanthus emblica*) slices increases the overall acceptability[9] and may reduce the extent of discoloration.[71] Osmotic pretreatment using NaCl solution of sliced mushroom before combined hot air and microwave drying improved the structure, rehydration ratio, porosity, shortened drying time, and bulk volume of dried slices.[66]

12.4.3.2 CURING

Curing is the application of high temperature and relative humidity to fruits and vegetables. During harvesting and transportation, the outer layer named as cork layer may be damaged. Curing repairs the damaged cork layer over the surfaces of the root crops, namely, potato, yam, and sweet potato and reduces the postharvest losses during storage. Cork layer protects the root crops from microbial infection and excessive water loss. Curing is also used in citrus fruits but the mechanism of repair is different. It heals the wounds over the surface and reduces the disease level. Cured fruits had higher soluble solids: acid ratio, firm texture, and a lower incidence of superficial scald in comparison to not heat treated fruits.[64]

12.4.3.3 FREEZING

Freezing as a pretreatment increased the drying rate of apples in combination with the microwave and convective drying. Slight heating converts ice to the vapor using the sublimation principle of phase change. Pretreatment of apples with ethanol helps in maintaining the rehydration capacity and shrinkage of dried product.[21]

12.4.3.4 CHEMICAL PRETREATMENT

The hardness and firmness of the dehydrated products were increased with the calcium pretreatment at different temperatures.[2] Pretreatment with SO_2 helped to retard the discoloration and oxidation of grapes, fresh apples, radish, and potatoes during preparation, storage, and distribution. Pretreatment with sulfites had a preservative and antibrowning effect on a cut or shredded fruits and vegetables.[29] Pretreatments using honey, 1-methylcyclopropene, ethanol, ascorbic acid, citric acid, high acid fruit juices, and syrups are being used in fruits to improve shelf life and flavor.[10]

12.4.3.5 BLANCHING

It is a pretreatment used to destroy the peroxidase of fruits and vegetables. Peroxidase is a deteriorating enzyme responsible for quality degradation during storage. Blanching prevents browning, off flavor, and from being bitter. It reduces the microbial load of the produces and softens the tissues due to involved heat in the process. Fruits with higher sugar content and acid prevent enzymatic action. Retention of vitamins A and C and rehydration capacity were improved in the blanched dried product. Boiling water or saturated steam is used for blanching. The sliced or whole fresh produces are submerged in boiling water for proper blanching. Blanching time depends on the altitude of location, quantity of product, type and texture of product, and thickness of shredded slices. Nutrient losses in minimal processed agro-produces were lesser with water blanching than steam blanching.[4]

12.4.4 STORAGE OF FRESH PRODUCES IN COLD STORAGE

Cold storage is a confined place, capable of maintaining the recommended temperature and relative humidity with the adjusted mixture of the gases. The gas mixture is preadjusted before the storage and maintained throughout the period of storage. The recommended temperature, relative humidity, and gas mixture (Tables 12.4–12.8) depend on the stage of ripeness, variety, the level of infestation, microbial load, injury, etc.[31,38,41,45,50,51,57,58,64,67] It is a cost-intensive process of storage for a substantial period without loss of freshness of stored commodities. Controlled atmosphere (CA) conditions for fruits and vegetables are presented in Tables 12.7 and 12.8, respectively. CA decreases the severity of insect infestation, microbe infection, lower the ethylene production and respiration rate.[64] It has the provision of the ethylene control to restrain the rate of ripening and color change. Abusing of the recommended temperature and relative humidity caused ripening and chilling injury in papaya. Postharvest losses are increased due to shortage of the cold storage facility at farms, markets, and ports and lack of knowledge and training about the role of temperature, ethylene, and sanitation in storage.[56]

TABLE 12.4 Gases Used in Controlled and Modified Storage of Fruits and Vegetables.

Gases	Description
Carbon dioxide[50,64]	Colorless, fungi-static, bacteriostatic and odorless gas, water, and lipid soluble
	Delayed yellowing and loss of both chlorophyll and ascorbic acid in broccoli. Solubility of CO_2 decreases with increasing temperature
	Increased fruit firmness after storage
	Lower incidence of physiological disorders
	Problem of pack collapse
	Reduced chilling injury symptoms significantly
	Reduction in enzyme activity at supercritical stage
	Sensitive to dairy products

TABLE 12.4   *(Continued)*

Carbon monoxide[50,64]	Colorless, fungi-static, and odorless gas
	Flammable and explosive in air, toxic
	Delaying ripening and maintaining good quality of mature green tomatoes Reduce fungal load strawberries
	Inhibit discoloration of lettuce
	Little inhibitory effect on microorganisms
	Reduce the level of chilling injury symptoms in capsicum and tomatoes Maintaining the red color in fresh meat
Ethylene[41]	Naturally occurring organic compound, plant hormone, colorless, effective at ppm and ppb concentrations
	Active in presence of O_2 and low levels of CO_2
	Control flowering, flower development, and growth
	Destruct chlorophyll, stimulates seed germination and senescence
	Inhibit ethylene synthesis in nonclimacteric fruit
	Necessary for ripening and yellowing of climacteric fruit
	Postharvest coatings and packaging influence synthesis of gas
Nitrogen[50,64]	Colorless, inert, tasteless, and odorless gas
	Inhibited ripening in plums, tomato, and avocados
	Little or no effect on flavor
	No antimicrobial activity
	Reduced chilling injury symptoms significantly
	Retard fruit softening and ethylene production, brown discoloration
	Solve problem of pack collapse
	Used as cushion gas and filler gas
Oxygen[50]	Colorless, odorless gas
	Responsible for aerobic spoilage of vitamins and flavor, oxidative deterioration, enzymatic reactions, color retention in red meat
	Avoid anaerobic fermentation in fish
Other gases[50]	Chlorine, nitrogen dioxide, ozone, propylene oxide, and sulfur dioxide

TABLE 12.5 Difference Between Controlled Atmosphere Storage and Modified Atmosphere Storage.

Controlled atmosphere	Modified atmosphere
Bulk storage	Reduction in retail waste (packaging of minimal processed fruits and vegetables)
Continuous control and regulation on the basis of produce requirement	Initial modification depends on respiration rate and packaging material
Created by modification of gas composition by adding gases	Created by either actively (addition or removal of gas) or passively (produce generated)
High degree of control over gas concentration	Low degree
Longer storage life of produces	Less
More expensive technology	Less
Specific temperature, RH, and gas composition should maintain	May or may not be maintained

TABLE 12.6 Optimum Storage Conditions for Selected Fruits and Vegetables.

Fruit type	Optimum temperature	RH	Storage life (weeks)	Relative perishability	Respiration rate at 5°C (mg CO_2/kg/h)
Apple	−1 to −4	90–95	30–365 days	–	Low
Banana	13–14	90–95	2–4	High	Moderate (10–20)
Cabbage			2–4	High	Moderate (10–20)
Carrot	0	90–95	28–180 days	–	Moderate (10–20)
Custard apple	7–10	85–90	1–2	Very high	–
Dry onion			8–16	Low	Low
Grape	−1	90–95	4–8	Moderate	Low (5–10)
Guava	5–10	90	2–3	High	–
Jackfruit	11–12.8	85–90	3–5	High	–
Litchi	2	90–95	21–35 days	–	–
Mango	13	90–95	2–3	High	–

TABLE 12.6 *(Continued)*

Fruit type	Optimum temperature	RH	Storage life (weeks)	Relative perishability	Respiration rate at 5°C (mg CO_2/kg/h)
Mushroom	–		1–2	Very high	Extremely high (>60)
Orange	3		4–8	Moderate	Low
Papaya	7	85–90	2–4	High	–
Pineapple	7–13	85–90	2–4	High	–
Pomegranate	0–4	90–95	2–3 months	Moderate	–
Potato	3	95	8–16	Low	Low
Spinach	–		1–2	Very high	Extremely high (>60)
Strawberry	0	90–95	1–2	Very high	High (20–40)
Sweet corn	0	90–95	4–8 days	Very high	Extremely high (>60)
Tomato (ripe)	7–10		1–2	Very high	Moderate (10–20)

TABLE 12.7 Controlled Atmosphere (CA) Recommendations for Fruits.

Commodity	Temperature range (°C)	CA	
		% O_2	% CO_2
Apricot	0–5	2–3	2–3
Avocado	5–13	2–5	3–10
Banana	12–16	2–5	2–5
Blackberry	0–5	5–10	15–20
Blueberry	0–5	2–5	12–20
Cactus pear	5–10	2–3	2–5
Cherimoya	8–15	3–5	5–10
Cherry	0–5	3–10	10–15
Cranberry	2–5	1–2	0–5
Durian	12–20	3–5	5–15
Fig	0–5	5–10	15–20
Grape	0–5 or 5–10	2–5 or 15–20	1–3
Grapefruit	10–15	3–10	5–10

TABLE 12.7 *(Continued)*

Commodity	Temperature range (°C)	CA	
		% O_2	% CO_2
Guava	5–15	2–5	0–1
Kiwifruit	0–5	1–2	3–5
Lemon	10–15	5–10	0–10
Lime	10–15	5–10	0–10
Loquat	0–5	2–4	0–1
Lychee (litchi)	5–12	3–5	3–5
Mango	10–15	3–7	5–8
Olive	5–10	2–3	0–1
Orange	5–10	5–10	0–5
Papaya	10–15	2–5	5–8
Peach, clingstone	0–5	1–2	3–5
Peach, freestone	0–5 or 4–6	1–2 or 15–17	3–5
Persimmon	0–5	3–5	5–8
Pineapple	8–13	2–5	5–10
Plum	0–5	1–2	0–5
Pomegranate	5–10	3–5	5–10
Raspberry	0–5	5–10	15–20
Strawberry	0–5	5–10	15–20
Sweetsop (custard apple)	12–20	3–5	5–10

TABLE 12.8 Controlled Atmosphere (CA) Recommendations for Vegetables.

Commodity	Temperature range (°C)	CA	
		% O_2	% CO_2
Artichokes	0–5	2–3	2–3
Asparagus	1–5	Air	10–14
Beans, green snap	5–10	1–3	3–7
Processing	5–10	8–10	20–30
Broccoli	0–5	1–2	5–10
Brussels sprouts	0–5	1–2	5–7
Cabbage	0–5	2–3	3–6
Chinese cabbage	0–5	1–2	0–5
Cantaloupes	2–7	3–5	10–20

TABLE 12.8 *(Continued)*

Commodity	Temperature range (°C)	CA	
		% O_2	% CO_2
Cauliflower	0–5	2–3	3–4
Celeriac	0–5	2–4	2–3
Celery	0–5	1–4	3–5
Cucumbers, fresh	8–12	1–4	0
Pickling	1–4	3–5	3–5
Herbs	0–5	5–10	4–6
Leeks	0–5	1–2	2–5
Lettuce (crisphead)	0–5	1–3	0
Lettuce (cut or shredded)	0–5	1–5	5–20
Lettuce (leaf)	0–5	1–3	0
Mushrooms	0–5	3–21	5–15
Okra	7–12	Air	4–10
Onions (bulb)	0–5	1–2	0–10
Onions (bunching)	0–5	2–3	0–5
Parsley	0–5	8–10	8–10
Pepper (bell)	5–12	2–5	2–5
Pepper (chili)	5–12	3–5	0–5
Processing	5–10	3–5	10–20
Radish (topped)	0–5	1–2	2–3
Spinach	0–5	7–10	5–10
Sugar peas	0–10	2–3	2–3
Sweet corn	0–5	2–4	5–10
Tomatoes (green)	12–20	3–5	2–3
Tomatoes (ripe)	10–15	3–5	3–5
Witloof chicory	0–5	3–4	4–5

Higher storage temperature causes rapid loss of the vitamin C in acid fruits. Stored tangerines at 7–13°C for 8 weeks have 40% loss of vitamin C. Long-term storage and chilling injury of tangerines lead to off-flavor development due to ethanol and acetaldehyde. In the case of lemon, significant loss of vitamin C was reported at 24°C.[51] At 5°C, loss of vitamin C was 5% in mango, watermelon, and strawberry pieces; 10% in pineapple pieces; and 12% in kiwifruit slices in 6 days' storage.[23]

Potato tubers had a slight loss of vitamin B_1 and niacin at 10°C after 30 weeks of storage. Volatile losses of isopentyl, butyl, hexyl acetates, and alcohols were reported on account of storage of "Jonathan" apples at 10°C. Storage at 10°C or less repressed the toughening of mushroom due to wall thickening and it was associated with the browning. The loss of firmness in broccoli was observed due to storage at 5°C or higher. The firmness increases with the storage time and temperature. In case of apples and avocado, the indirect effect of temperature was observed as ripening of fruits. Ripening may lead to loss of firmness and an increase in color of fruits. The varietal difference of recommended atmosphere condition is observed in the apples.[51]

Minimally processed kiwifruit had the shortest postcutting life, whereas the fresh-cut mango and watermelon had postcutting life of 9 days at 5°C. The temperature was observed as most detrimental factor for the postcutting life of the fresh-cut pineapple than low pressure, combinations of gases, and relative humidity. Fresh-cut pineapple had a shelf life of 2 weeks at 5°C or lower.[10] No significant losses in carotenoids were observed in watermelon cubes and kiwifruit slices. Pineapples had highest carotenoids (25%) loss at 5°C; followed by 10–15% in mango, strawberry, and cantaloupe pieces after storage of 6 days. Fresh-cut exposes the interior parts of the fruits more to the oxygen and light. Fresh-cut of pineapples, mangoes, cantaloupes, watermelons, strawberries, and kiwifruits had no significant effect of light exposure on the degrading quality and nutrient content.[23] The categories of the cold storages generally used are cold storage for warm material, short-term and long-term cold storage for precooled multimaterials, and controlled cold storage.

Codes and standards for safe cold storage of fruits and vegetables are formulated by the Bureau of Indian Standards (BIS) and International Organization for Standardization (ISO). When ISO and BIS codes are not available, then relevant codes of ASHRAE, IARW, ASME, and other international codes are to be followed.[5] Static (SCA) and dynamic controlled atmosphere (DCA) are two types of CAs storage. In SCA, the composition of gases, temperature, and RH is altered in the start of storage and maintained throughout the storage period. The composition of gases, temperature, and RH will depend on the type of processed or raw fruits and vegetables. The respiring fruits and vegetables are stored under the DCA. The dynamic term denotes the variable requirement of the gas composition, temperature, and RH based on stages and age of respiring produces.[64] Solar-driven cold stores, hypobaric and hyperbaric storage, jacketed storage, etc. are other forms of the CA storage.[50]

12.4.5   COLD CHAIN

Cold chain is the logistics system that is integrated with the refrigeration system to maintain a safe temperature in transit, storage, and distribution of the fruit and vegetable (Table 12.6). The temperature fluctuation is one of the most deteriorating factors for the perishable commodity. It delivers the foods in their fresh form by restricting the chemical, biochemical, and physiological changes to a minimum. As a rule of thumb, "each 10°C increase in temperature roughly doubles the reaction rates." It may also be noted that each 10°C increase in temperature roughly reduces shelf life of fresh produces to half. The Van't Hoff rule states that the "velocity of a biological reaction increases 2 to 3 folds for every 10°C rise in temperature."[58] Higher temperatures result in an exponential rise in biological respiration to keep the produces alive. Metabolism produces heat and water vapor, which increase the deterioration rate by ripening, senescence (aging), shrinkage, etc.[38] As shown below, 686 kcal/mol is produced during the reaction.

$$C_6H_{12}O_6 + 6O_2 \rightarrow 6CO_2 + 6H_2O + 686 \text{ kcal/mol}$$

The ultimate objective of cold chain is to maintain the safe temperature and relative humidity depending on the requirements of a commodity throughout the distribution and storage phase. Table 12.6 presents the optimum storage conditions and classification as per respiration rate and relative perishability. Farmers, wholesaler, retailer, fruits/vegetables market staff, buying agents, and consumers all are the actors in the cold chain. More than 90% of total production is sold to commission agents or wholesalers by farmers in India. The need for cold chain starts from the field after crop harvesting and continues up to the consumption, that is, farm to fork. Countries like India have warm tropical climate (high temperature and relative humidity) most of the year and therefore need cold chain.

Low-temperature storage and refrigerated transportation system are mandatory in countries like India to prevent senescence, microbial, and enzymatic spoilage. India has the cold stores facility for only 10% of the total perishable produces. More than 95% of the cold stores are in private hands. High charges of the cold stores divert most of the fruits and vegetables to markets in open carts and trucks. Development of cold-chain infrastructure (grading, sorting, packaging, storage, processing, and transportation facilities) between the farm and table reduces the waste and losses.[44]

Short shelf life and perishable nature of fruits and vegetables are crucial for the quality delivery to the consumer. Fruits and vegetables are sources of vitamins and minerals necessary for the growth of human body. Health awareness in consumers increases the demand of fresh agricultural produces. Moreover, increased purchasing power, changing lifestyle, urbanization, and globalization are also reasons to increase the demand of fresh agricultural produces. Cold-chain management is defined as "a set of approaches utilized to efficiently integrate suppliers, manufacturers, warehouses and stores, so that merchandise is produced and distributed in the right quantities to the right location and at the right time, in order to minimize system-wide costs while satisfying service level requirements."[60]

India is the second largest producer of fruits and vegetables in the world after China. The export potential of the country is comparatively low due to poor cold-chain infrastructure (Fig. 12.1), which consists of supply procurement, transport, storage, transport, and end consumer.[44] About 30% of horticultural produces get wasted due to inadequate cold chain thus reduction in remuneration of farmers. The declination in consumption of cereals (1–2%) and consumption of fruits and vegetables (2–3%) has been reported to increase annually. The development of cold-chain infrastructure can reduce the postharvest losses and prolong the shelf life. The employment opportunities in export and production of fruits and vegetables can improve the income and bring health and happiness to farmer and workers.[44]

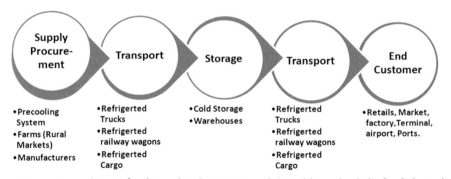

FIGURE 12.1 **(See color insert.)** Infrastructures of the cold supply chain for fruits and vegetables.

Advantages of efficient cold-chain management in transit of fruits and vegetables are[43,44]:

- Affordable right prices for high-quality produces.
- Balance the gap of demand and supply.
- Development of business and employment opportunities.
- Export development.

- Health and happiness of consumer.
- Income improvement of farmers.
- Increase profitability.
- Increase shelf life of fresh produces.
- Off-season availability of fresh produces.
- Price stabilization.
- Reduction in food waste and loss.
- Value addition to stakeholder, consolidators, and customers.

12.4.6   PROCESSING OF AGRICULTURAL PRODUCES

Lack of cold storage of sufficient capacity and cold-chain management are major bottlenecks in the postharvest chain of fresh produces. Processing of the surplus production is the method to save the produce being wasted or lost. In India, only 2% of total produced fruits and vegetables are being processed compared to 23%, 65%, and 78% in China, the United States, and Philippines, respectively.[43] Processing is the process of converting the agricultural produces into end products of comparatively longer shelf life. Pickles, juices, jam, jelly, squash, drying, evaporation, and dehydration are examples of the processing. Microbial and enzymatic stability increase the shelf life of the product. Processing has potential to convert the nonmarketable or surplus fruits and vegetables into useful and economic products, such as juice, sauce, vinegar, etc.[11]

According to Aguilera and Stanley,[1] the processing is viewed as a "controlled effort to preserve, transform, destroy, and/or create structure/ texture changes." Processing is the sequence of unit operations to transform a raw material to an end product. Cleaning, grading, size reduction, material handling, storage, drying, etc. are unit operations that are generally found in a typical processing plant. The recipe of the abovementioned products from fruits and vegetables is beyond the scope of this chapter; however, selected unit operations are described in this section to understand the processing procedure.

12.4.6.1   CLEANING OR WASHING

Cleaning or washing is a prior operation before the execution of the processing. Raw material comes with the soil, sand, insects, or mud from the field. End quality of the processed product and safety of the processing equipment depends on the raw material. Water is generally used in cleaning of the fruits and vegetables. Washing water of temperature less than 5°C with recommended quantity 5–101 L/kg should be used before peeling or cutting of the fruits and vegetables. Some preservatives like ozone, chlorine and citric acid can be used with washing water to reduce microbial load and to retard enzymatic activity. Preservatives can improve the shelf life and sensory quality of the products.[3]

12.4.6.2   PEELING, CUTTING, AND SHREDDING

Peeling, cutting, and shredding are the operation of removing the outer cover of fruits and vegetables before further processing. For example, potato, apples, banana, orange, etc. require peeling before further processing. Mechanical peeler, chemical peeler, or high-pressure steam peeler are used on industrial scale to remove the outer cover. Peeling should be gentle as much as possible. Therefore, hand peeling with the sharp knife is the ideal method of peeling. Microbial growth, browning, and respiration rate of abrasion peeling were observed higher in comparison to hand peeling. The knife or blades used for peeling, cutting, or shredding should be sharp and made of stainless steel. Sensory quality and fluid retention of sharp knife cut produce are comparatively higher than other peeling methods.[3]

12.4.6.3   CANNING

Canning is preservation method of perishables in which peeled or raw fruits and vegetables are filled in cans or glass containers followed by exhausting, sealing, and steaming at 100°C (acid fruits). Heat destroys the microorganisms present in food and container and prevents recontamination. The sealed cans are immediately cooled after heat processing to prevent overcooking and stored or transported after labeling and casing. Typically, the canned products are apples, beets, carrots, peas, tomatoes, beans (cut and whole), spinach, pineapple, pears, corn, apricots, peaches, cranberries, etc.

12.4.7   PACKAGING

Packaging is prevalent and imperative, which encompasses, magnify, and protects the product during processing and manufacturing, handling, and storage. Packaging plays a vital role in protection, containment, communication and convenience of fresh produce and processed food between the farm and fork. Consumer demands fresh, healthy, and convenient food. Different packaging methods have potential to meet various requirements of consumer. It works as a safeguard against the deteriorating factors to reduce food losses and waste.[6] Packaging is not only the wrapping of the material; rather, it also conveys the information about the packed product to the consumer. It provides information about nutritional value, shelf life, ingredients, brand name, how to use, price, etc. on the film or containers. Glass, paper, metal, wood, cloth, and plastics are materials that are used in the manufacturing of the convenient shapes and sizes of packages. Glass bottles, tin can, trays, pouches, films are designed using the basic ergonomic principles for human convenience. Famers still use the jute sacks, leaves, and branches of plants, ropes, etc. in the fabrication of baskets and bags to handle the fresh produces.[34] Packaging is divided into three categories:

- *Primary packaging* is used to designate the layer of packaging in immediate contact with the product. It keeps the product in storage for long periods of time. In this case, it is crucial that it will keep the product absolutely sealed-off from its environment.
- *Secondary packaging* is used to pack various primary packed units in a comparatively bigger package. It eases the handling, protection, transport, and storage of fresh produces. Secondary packaging displays the brand name and information about the primary packed produces.
- *Tertiary packaging* is used to convey the huge quantity of the secondary packaged products to the retailers. Brown corrugated fiberboard, shipping cargo container, and refrigerated container on trucks or train are used for the tertiary packaging. It is very common among export firms in transit to send the produces for long distances without quality deterioration.

Packaging reduces the rate of spoilage and waste produced in transit of fruits and vegetables. Protection of fresh produce needs to be the primary goal for packaging sustainability and also desires trade-offs between packaging and food waste. Selection of proper packaging requires information about characteristics of fruit and vegetable, marketing/consumer requirements, logistics, economics, and environmental impact. Investment in packaging to reduce food losses and waste is a wise idea. It maintains the freshness, nutritional value, safety, taste, and prolongs the shelf life.

Packaging provides a barrier against oxygen and moisture to reduce the weight loss and metabolism of fruits and vegetables. Packaging can deliver 100% edible fresh cut and other ready-to-eat (RTE) products made from fruits and vegetables to the customers. The concept of edible packaging yield zero waste at consumer end.[14] The collected by-products and waste in the manufacturing of fresh cut, RTE, edible food at a single place can be utilized more efficiently in the production of animal feed, biogas, compost, etc. Jute sacks were widely used for almost all types of horticultural crops primarily due to its lowest price, good strength, and easy availability. Used packing materials must have the property of protection of respiring fruits and vegetables from compression, collapse, and mechanical damage, vibrations.[34]

Packaging techniques to ensure microbiological and nutritional quality of minimal processed or whole fruits and vegetables are discussed in this section.

12.4.7.1 MODIFIED ATMOSPHERE PACKAGING

Modified atmosphere packaging (MAP) is the alteration of gaseous environment of package according to the requirement of the produces.[50] Gas packaging is also used synonymously with the MAP. It may be active or passive type. Active MAP is displacement or replacement of the gaseous environment with a desired gas mixture during filling of material. However, passive MAP is used after some specified time due to respiration of stored living commodity and for fresh and fresh-cut fruits and vegetables. MAP retards rate of respiration as well as ripening process to save produces against senescence and deterioration. Marginal increase in shelf life, cultivar, and location of cultivation and packaging material properties restrict the potential use of this technology.

The design of MAP depends on the filled weight, respiration rate of the product, and film surface area for gas exchange. Gaseous environment was modified after a certain period to reduce quality deterioration due to enzymatic action, transpiration, insect infestation, microbial infection, and respiration.[13] Laminates, co-extrusion, perforation, diced holes, and mixing of certain additives with virgin material during the manufacturing of the plastic films are technologies to alter the properties of the packaging films. Gas flushing and compensating vacuum are two techniques extensively used for gas replacement for MAP. There are three types of gas mixtures used in the MAP based on the product, packaging material, and storage conditions[50]:

- Inert blanketing (N_2).
- Semireactive blanketing (CO_2/N_2 and $CO_2/N_2/O_2$).
- Fully reactive blanketing (CO_2 or CO_2/O_2).

12.4.7.2   EDIBLE COATING AND BIODEGRADABLE PACKAGING

Edible coating and biodegradable packaging of fresh-cut fruits and vegetables preserve the microbiological and physicochemical quality and nutritional supply. It forms a layer on the surface of food material to improve the barrier and mechanical properties. Edible coating is safe to eat and biodegradable in nature. Polysaccharides, lipids/fats, and composites are used in preparation of film in addition of plasticizers. Edible coating on the minimal processed food prolongs the shelf life and keeps freshness and nutrition intact. Popularity of minimal processed food is increasing since it takes minimum time in preparation and reduces retail waste. Certain additives, mineral, vitamins can be mixed in edible coating to improve the nutritional value and flavor of base produce.[22,47]

12.4.7.3   SMART PACKAGING

Smart packaging is combination of intelligent and active packaging. Active packaging is the incorporation of active agents in the surrounding environment of food to improve shelf life and conditions in package. Active compounds absorb or emit substances to counteract against deteriorating factors. Ethylene absorber, carbon dioxide emitter, ethanol emitter, moister absorber are examples of active compounds. Intelligent packaging is associated with the communication about the condition of stored product to the consumers. Smart packaging is able to detect, sense, record, track, communicating, and applying science logic to facilitate decision making, to extend shelf life, enhance safety, improve quality, provide information, and warn for possible problems. Time–temperature indicator, fresh check, biosensor, RFID tags, etc. are examples of smart packaging.[33]

12.4.7.4   VACUUM PACKAGING

Vacuum packaging or hypobaric packaging is the earliest form of MAP and still extensively used commercially for meat, cheese, coffee, etc. It is not suitable for the soft and bakery products because of irreversible deformation due to negative pressure.[50] The low pressure reduces the partial pressure of the oxygen in the interstitial space of respiring commodity. Small amount of O_2 slows down the respiration rate and inhibits growth of spoilage microorganisms.[25] Moderate vacuum packaging (300 mmHg) improved the sensory and microbial quality of mung bean sprouts, sliced apples, tomatoes, and mixture of cut vegetables. It inhibited the growth of pathogens quickly.[3]

12.4.7.5   HURDLE TECHNOLOGY

Hurdle technology or combined technology is not a packaging method but rather a combination of a number hurdles to save the produces from abiotic and biotic deterioration.[40] Hurdles are factors applied on product to inactivate or delay the microbial growth for food preservation. Hurdles can be classified into physical, physicochemical, microbiological, and miscellaneous (Table 12.9). About 50 hurdles are being used in food preservation. Hurdles disturb the homeostasis reactions of microorganisms and cause metabolic exhaustion. Pickling is the traditional hurdle technology to preserve the fruits and vegetables.[40] Shelf life and stability of minimally processed fruits and vegetables (such as carrot, papaya, mango, banana, and pineapple) were enhanced under preservation by hurdle technology.[55]

TABLE 12.9   Potential Hurdles for Food Preservation.

Hurdle type	Parameter	Application
Physical	High temperature	Heating, blanching, pasteurization, evaporation, sterilization, baking, frying, extrusion
	Low temperature	Chilling, cooling, freezing
	High pressure	High-pressure processing, hyperbaric packaging
	Low pressure	Hypobaric packaging, vacuum hermetic fumigation
	Electromagnetic energy	Surface disinfectants, microwave heating, radiofrequency heating, pulsed magnetic field, high-intensity electric field
	Ionic radiation	Irradiation, surface disinfestations
	Packaging	MAP, edible coatings, active and passive packaging, aseptic packaging

TABLE 12.9 *(Continued)*

Hurdle type	Parameter	Application
	Ultrasonication	Microbial deactivation
Physicochemical	Reduced water activity	Drying, curing, conserving
	Increased acidity	Acid addition or formation
	Reduced redox potential	Removal of oxygen or addition of ascorbate
	Preservatives	Sorbate, sulfite, nitrite, ethanol, carbon dioxide, ozone, salt, sugar, smoking, spices, and herbs
Microbiological	Competitive flora	Microbial fermentations, bacteriocin, antibiotics, protective cultures

12.4.7.6   SHRINK PACKAGING

Shrink packaging is a form of passive MAP. The flexible polymer film is shrinking over the commodity due to blow of hot air or heat. It increases the shelf life and maintains the freshness of fruits and vegetables. It also helps in reducing weight loss, fruit deformation, and chilling injury and prevents secondary infections. It is used for the packaging of apple, cucumber, citrus, papaya, tomato, melon, etc.[17]

12.5   ROLE OF ENGINEERING INTERVENTIONS IN WASTE REDUCTION

The engineering interventions start from the field and keep influencing up to the consumption of fruits and vegetables. It saves time and help to reduce the postharvest losses, waste, and human drudgery. The harvesting and handling equipment ease the work of farmers and minimize the injury to horticultural produces. More than 50% of fresh produces are transported to the market in unrefrigerated open trucks or bullock carts without packaging or packaged in gunny bags. External injury determines the further storability of the harvested horticultural crops. Hand picking or collecting with equipment such as sickle, spade, hoe, axe, and ladder is being used by farmers.[34] This equipment needs to be upgraded time to time by ergonomically well designs with suitable material of construction as per requirement of farmer and horticultural produces.

The hydraulic platform, crane, and fruit harvester for mango, guava, lime, sapota, etc. developed by the Indian Institute of Horticultural Research—Bangalore are examples to use on farms in place of conventional equipment. However, the cost, maintenance, and small land holdings restrict the full replacement by these technologies. The cooling and washing units for the fruits and vegetables based on refrigeration or evaporation cooling has direct effect on shelf life of produces. Soft brushing and infrared lamps are used for cleaning and grading. The transport vehicles should have sufficient provision to minimize the vibrations, stacking, and heat stresses. Drop of fruits and vegetables from a height is also injurious for shelf life. Surfaces of conveying and handling equipment are generally kept softer and have proper pathways to guide the fruits and vegetables.

The hand-peeling gives the peeled product of longer shelf life and this method is resource efficient.[28] The use of small vending refrigeration units has provision for displaying the produces in a portable thermally insulted cabinet. It maintains the environmental and hygiene requirement of produces and makes it suitable for small and supermarkets.[20] Central Institute of Post-Harvest Engineering and Technology has developed affordable mobile cool chamber (refrigerated transportation rickshaw) for the peddler.

The quality evaluation of fruits and vegetables are generally carried out by manual/visual or destructive methods. These methods are time consuming, labor intensive, and sample destructive and suffer from inconsistent and inaccurate grading. These methods can be replaced by nondestructive methods of quality evaluation and defect detection.

12.5.1 MACHINE VISION

Machine vision using X-ray, hyperspectral imaging, MRI, computed tomography, ultrasound, etc. has potential to replace existing destructive technologies. The machines for continuous grading and sorting of mango, peach, orange, date, potato, pomegranate, apples, based on image processing techniques has been developed. It offers consistency, uniformity, noncontact, speed of work, automation, accuracy in grading, quality evaluation, and monitoring during storage.[42]

12.5.2   HERMETIC STORAGE

Hermetic storage of agriculture produce in flexible bag is a pesticide free and environment-benign technology. It can be used in three manners: organic storage, fumigation with inert gas (GHF), and vacuum hermetic fumigation (VHF). It helps in retardation of respiration rate and disinfestations of insect infestation of stored commodities. Combinations of CO_2, pressure, essential oils VHF and higher temperature with hermetic environment may decrease exposure time for disinfestations.[37] VHF storage ensures maintenance of eating quality and the shelf life of the horticultural produces.[15]

12.5.3   MODERN INTELLIGENT SENSOR SYSTEM

Integration of modern intelligent sensor system with conventional logistics methods also help in reduction of postharvest losses and quality deterioration. Intelligent cold-chain infrastructure with sensors and actuators has been developed for monitoring and collection of data of commodities in transit. These systems are accurate, rapid action, and of low cost.[26]

12.5.4   IRRADIATION

Irradiation of horticultural produces reduces or eliminates the microbiological load and also increases the shelf life up to 3–5 times. Low doses of irradiation in conjunction with other technology furnish better quality. It restrains the sprouting in the potatoes and ripening in the papaya and tomatoes. The international organizations WHO, FAO, etc. have accepted irradiation as a safe and effective nonchemical technology for fruits and vegetables. Irradiation has potential to meet the requirement of modern consumer for safe, healthy, fresh, and residue-free food.[7]

Product quality is always a concern in adopting technologies. The conventional technologies of drying produce product of low quality having case-hardening and dark color. Sometimes, case-hardening and dark color is desirable.

12.5.5   MICROWAVE-ASSISTED DRYING

Microwave-assisted drying (MWAD) is a rapid dehydration method particularly for fruits and vegetables. Short drying time, improved food quality, instantaneous on-off, etc. are advantages of MWAD. MWAD-dried banana slices had lighter color and higher rehydration ratio and 64% shortened drying time in comparison to the traditional airflow drying.[71]

12.5.6   VACUUM/STEAM/VACUUM PROCESS

Vacuum/steam/vacuum process is a surface disinfecting process used in raw fruits and vegetables. Air, dirt, and water work as an insulator on food surface and cause food loss due to thermal damage. Vacuum removes air and water from surface, which is followed by steam outburst. Steam burst destroys the biofilm of bacteria and mold within 1–2 s of process time. The condensed steam is removed from surface by second exposure of vacuum. It has been applied on papayas, mangoes, avocados, kiwis, carrots, cucumbers, and peaches.[36]

12.5.7   OHMIC HEATING

Ohmic heating is also known as electrical resistance heating, joule heating, or electroconductive heating. Rapid and uniform heat is generated internally due to electric resistance of food. It inactivates the enzymes and microorganism and reduces the process time of blanching. Electroporation by ohmic heating improves the juice extraction of apples. It improves the shelf life and quality of apricots. It degrades the heat-sensitive compound in food but degradation is comparatively less than the conventional method.[32]

12.5.8   HIGH-PRESSURE PROCESSING

High-pressure processing is a technique of applying high pressure on submerged fruits and vegetables in medium. It can be used in combination with heat to reduce exposure time. HPP inactivate and eliminate the pathogenic and spoilage microorganism with minimum modification in sensory and nutritional quality of fruits and vegetables.[46]

12.6   UTILIZATION OF WASTE PRODUCT AND BY-PRODUCT

In a whole postharvest chain, it is not possible to eliminate food loss and waste 100%. Fruit and vegetable processing, packing, distribution, and consumption generate a huge quantity of fruit and vegetable wastes. Food loss may happen due to inappropriate postharvest management, lack of equipment, and shortage of basic facility to store the foods. Food waste also happens due to lack of consumer awareness and education regarding handling, processing, storage, cooking, etc. of food. The processing of fruits and vegetables also yield the by-products, that is, peels of fruit and vegetable, stones from mango, etc. These by-products can be used in the development of costly products. Fruits and vegetables wastes are rich source of micronutrients. It is mixed in urea and molasses for the development of multinutrient animal feed.[68] Anaerobic digestion instead of landfill is also a disposal method to generate energy from food waste to recover input used in producing fruits and vegetables.

Solid and liquid horticultural waste with cattle dung can be treated in a biogas plant to produce methane gas to meet energy needs in rural and urban areas. The nitrogen-containing stabilized sludge is used as fertilizer on the farm. Moreover, anaerobic digestion sanitizes the solid and liquid waste efficiently and reduces the public health hazards.[59] Fruits and vegetables waste/by-products are good source of the polyphenols, flavors, dietary fibers, dyes, gelling agents, and antioxidants. Since polyphenols and antioxidants have anti-inflammatory properties, they help in prevention of cancer, cardiovascular, and degenerative diseases. Therefore, the extracted bioactive compounds from apple pomace, kiwifruits, grapefruits, cauliflower, cabbage, broccoli, etc. can be utilized in manufacturing of food ingredients.[70] The expensive activated carbon was used in waste-water treatment so far. Bio-absorbent developed from vegetable waste is a cheaper and simple solution in treating aromatic waste water.

The phytochemicals developed from food waste are used to protect the crops. Nematode control by olive pomace, fungistatic property of citrus waste, and insect antifeedant compound in citrus peel and seeds are examples of phytochemicals. The development of innovative products using food waste or by-products reduces the waste quantity as well as adds new market segments. It is a long-term solution for waste utilization.[39] Production of ethanol is an alternative to the fossil fuels using the wastes of fruits. It disposes of citrus waste within 7 days by simultaneous saccharification and fermentation (SSF) process. The dried residue from SSF is utilized as a cattle feed.[53]

Wine production, organic acids, limonene, pectin, butanol, volatile compounds, enzymes, and single cell protein can be extracted or made by using the food waste or by-products. Mushroom cultivation on paddy straw in combination with the fruits and vegetables waste is a solid-state waste-management strategy. The protein content of mushroom grown on fruit and paddy substrate is higher than vegetable substrate. Moreover, the fruits and vegetables waste with paddy straw can increase the amount of six amino acids in mushroom than paddy straw alone.[61] Composting of food waste can convert organic matter into readily usable plant nutrients in the presence of air, water, insects, and microorganisms. By segregating, recycling, and composting, a family of four members can reduce their waste from 1000 kg to less than 100 kg every year.[62]

12.7   AWARENESS AND EDUCATION FOR WASTE REDUCTION

Wastage means bigger environmental impact. It can increase the green-house gases emission through production, processing, and its disposal and contributes to world hunger as food is wasted instead of delivering to the needy. Food waste is associated with production, processing, supply, and preparation of food. In India, about 20–40% of food grown gets spoiled before reaching to the consumers. Prevention of food loss and waste is the responsibility of consumer, seller, and farmer during the journey of food from farm to fork. Awareness programs are arranged by national and international institutes to spread knowledge of good agricultural practice among the fruits and vegetables growers.

The objectives of such programs are to grow high-quality farm produces through optimization of farm input; understanding of environmental impact in agriculture and farming operations. The success of RRR (reduce, reuse, and recycle) concept depends on the skill, education, training, and awareness of all actors of postharvest chain of fruits and vegetables. Reduce the food loss from farm to fork and reuse of leftover food on next time or handover to needy followed by the recycle of the valuable components from being food waste. RRR is a strategy to form zero or low waste society; and it needs an active and responsible participation of all.

Loss of the aroma has been reported due to longer storage of the horticultural produces, irrespective of storage temperature. First in first out (FIFO) is the principle to save the product quality characteristics. FIFO means to supply or sell the old product first from the storage. It will help the farmer, shopkeeper, retailer, or the wholesaler to keep the material for a long time to find suitable time to sell.

Awareness and education regarding the use of dustbins in houses, cold stores, shops, and fruits market for waste collection. It helps in cleanliness as well as biodegradable trash can be used in the production of the methane gas, compost, etc. The training and the skill development of the fruits and vegetables market workers and staff about protocols of safe handling of the agricultural produces. Loss of germination and storability caused by impact damage during free-fall has been observed in the soybean seeds from different heights. Free-fall on cement floor results greater loss in quality than galvanized iron floor.[49]

Warm fruits are more susceptible to vibration injury but are able to withstand the impact injury in comparison to cold fruits. Cold fruits are generally less plastic than warm fruits.[58] In cold storage, there should be regular checkups and monitoring of the machine like refrigeration etc. for optimum cooling. One important aspect in storage of the fruits and vegetables is the stacking height. During transport and storage, a particular stacking height should be maintained to save horticultural produces from bruising, injury, and external stresses. Use of paper carton provides cushioning as well as help to absorb the dripping to the fruits and vegetables.

In case of plastic and wooden cartons, the chips of papers are used for the cushioning and absorption of dripping. Handling of fruits and vegetables plays a major role in maintaining the shelf life. Softness and high water content make it more vulnerable for external injury. Drop from small height makes invisible injury to fruits which reduce and increase the probability of spoilage of other adjoined fruits too.

Inventory management and dynamic pricing are technologies to reduce food waste in a retailer shop. Real-time inventory monitor maintains the stock in shop based on the previous demand, price of the commodities, and future trends.[8] Dynamic pricing helps to sell about-to-expire products first using the price changing by discounts, loyalty points, coupons, and promotions. Dynamic pricing is the flexi price system, which changes the price based on available stock, sales forecasts, and the shelf life of the produces. Both technologies work to decrease the waste as well as increase the profit of the retailer. Sometimes, technologies have failed to maintain the gap of demand and supply. Retailers donate the surplus food material to biogas plants or to animal shelters.[69]

Government organizations like FAO, UNEP, etc. run campaigns to increase public awareness for the food loss and waste. Campaigns help public to realize their social responsibility toward the environment and resources. Wastage of food in marriages, meetings, conferences, etc. has also come under the campaign. Innovative ideas such as smartphone applications are used to get information of surplus food. NGOs seek out the venue and handover the food to poor. Optimization of shopping behavior, home visitor frequency, diet of a person, storage capacity, etc. could be an effective way to avoid food waste. Change in recipes in food-making also helps in utilizing the about to expire fruits and vegetables.[11]

The agriculture is emerging as business destination. Increase in per capita income, changed life style, and lack of time to prepare food has evolved the food industries. Food industry targets to make luxuries RTE food for convenience of consumer. Trained and educated professional in the agribusiness markets are appointed to control, planning, and production. They make the strategies and execute accordingly. Management helps in allocating the resources for maximum profit and minimum input. Efficient Customer Response is joint body of trade and industry working to meet the customer needs more efficiently by assortment, replenishment, promotions, and new product development. It depends upon space management, range, and pricing of products in retail outlets.[43]

12.8 SUMMARY

Each method or technique used in reduction of waste and loss in fresh produces has certain benefits and limitations. The cost involved, economic status, education, awareness, skill, etc. are major reasons that prevent adoption of such techniques. Rules and protocols for zero waste for all sectors of food production are also needed to develop and applied with public–private partnership. Zero waste saves the energy used in producing food and helps in securing food for future. The shortening of the postharvest chain and using the techniques in this chapter will disburse the cost of input and profit to farmers. All techniques and methods ultimately depend on the awareness, skill, and education of the all actors of the postharvest chain. It is required to arrange awareness programs, trainings, and skill development programs to spread the knowledge about waste reduction.

KEYWORDS

- **canning**
- **curing**
- **gas flushing**
- **high-pressure processing**
- **hyperbaric packaging**
- **microwave-assisted drying**
- **shrink packaging**

REFERENCES

1. Aguilera, J. M.; Stanley, D. W. Microstructural Principles of Food Processing and Engineering; Springer Science & Business Media, Aspen Publishers Inc.: Gaitherburg, MD, 1999; p 2.
2. Ahrne, L.; Prothon, F.; Funebo, T. Comparison of Drying Kinetics and Texture Effects of Two Calcium Pretreatments before Microwave-Assisted Dehydration of Apple and Potato. *Int. J. Food Sci. Technol.* **2003,** 38 (4), 411–420.
3. Ahvenainen, R. New Approaches in Improving the Shelf-Life of Minimally Processed Fruit and Vegetables. Trends Food Sci. Technol. **1996,** 7 (6), 179–187.
4. Anonymous. Drying Fruits and Vegetables. Food Preservation and Storage; OSU Extension Catalogue PNW397, 2009; p 11. http://extension.oregonstate.edu/lane/sites/default/files/images/pnw0397.pdf.
5. Anonymous. Technical Standards and Protocols for the Cold Chain in India; National Horticultural Board, Department of Agriculture and Cooperation, Ministry of Agriculture, Government of India, 2010. http://nhb.gov.in/documents/cs2.pdf.
6. Anonymous. Consolidated Final Report: Post-Harvest Losses in Selected Fruits and Vegetables in India. Technical Bulletin 41; Publication Committee, the Indian Institute of Horticultural Research (IIHR): Bangalore; 2013–2014; p 15.
7. Arvanitoyannis, I. S.; Stratakos, A. C.; Tsarouhas, P. Irradiation Applications in Vegetables and Fruits: A Review. Crit. Rev. Food Sci. Nutr. **2009,** 49 (5), 427–462.
8. Bahinipati, B. K. The Procurement Perspectives of Fruits and Vegetables Supply Chain Planning. Int. J. Supply Chain Manage. **2014,** 3 (2), 111.
9. Banga, S. Km.; Kumar, S.; Johry, P.; Singh, G. R. Influence of Operating Parameters on Process Kinetics of Aonla (Emblica officinalis) during Tray Drying. Green Farm. **2016,** 7 (1), 226–229.
10. Barrett, D. M.; Beaulieu, J. C.; Shewfelt, R. Color, Flavor, Texture, and Nutritional Quality of Fresh-Cut Fruits and Vegetables: Desirable Levels, Instrumental and Sensory Measurement, and the Effects of Processing. Crit. Rev. Food Sci. Nutr. **2010,** 50 (5), 369–389.
11. Blanke, M. Challenges of Reducing Fresh Produce Waste in Europe—From Farm to Fork. Agriculture **2015,** 5 (3), 389–399.

12. Brosnan, T.; Sun, D. W. Precooling Techniques and Applications for Horticultural Products—A Review. Int. J. Refrig. **2001,** 24 (2), 154–170.

13. Caleb, O. J.; Mahajan, P. V.; Al-Said, F. A. J.; Opara, U. L. Modified Atmosphere Packaging Technology of Fresh and Fresh-Cut Produce and the Microbial Consequences—A Review. Food Bioprocess Technol. **2013,** 6 (2), 303–329.

14. Corbo, M. R.; Campaniello, D.; Speranza, B.; Bevilacqua, A.; Sinigaglia, M. Non-Conventional Tools to Preserve and Prolong the Quality of Minimally-Processed Fruits and Vegetables. Coatings **2015,** 5 (4), 931–961.

15. Davenport, T. L.; White, T. L.; Burg, S. Optimal Low-Pressure Conditions for Long-Term Storage of Fresh Commodities Kill Caribbean Fruit Fly Eggs and Larvae. Hortic. Technol. **2006,** 16 (1), 98–104.

16. Deppe, C. The Tao of Vegetable Gardening: Cultivating Tomatoes, Greens, Peas, Beans, Squash, Joy, and Serenity; Chelsea Green Publishing: White River Junction, VT, 2015; pp 74–75.

17. Dhall, R. K.; Sharma, S. R.; Mahajan, B. V. C. Effect of Shrink Wrap Packaging for Maintaining Quality of Cucumber during Storage. J. Food Sci. Technol. **2012,** 49 (4), 495–499.

18. Du, T.; Kang, S.; Zhang, J.; Li, F.; Yan, B. Water Use Efficiency and Fruit Quality of Table Grape under Alternate Partial Root-Zone Drip Irrigation. Agric. Water Manage. **2008,** 95 (6), 659–668.

19. Eckert, J. L.; Eaks, I. L. Postharvest Disorders and Diseases of Citrus Fruits. In The Citrus Industry; Ruther, W., Ed.; Department of Agriculture and Natural Resources, University of California: Davis, CA, 1989; vol 5 (3326), pp 179–260.

20. Evans, J. A.; Foster, A. M., Eds. Sustainable Retail Refrigeration; John Wiley & Sons: Hoboken, NJ; 2015; pages 352.

21. Funebo, T.; Ahrné, L.; Prothon, F.; Kidman, S.; Langton, M.; Skjöldebrand, C. Microwave and Convective Dehydration of Ethanol Treated and Frozen Apple–Physical Properties and Drying Kinetics. Int. J. Food Sci. Technol. **2002,** 37 (6), 603–614.

22. Galgano, F.; Condelli, N.; Favati, F.; Di Bianco, V.; Perretti, G.; Caruso, M. C. Biodegradable Packaging and Edible Coating for Fresh-Cut Fruits and Vegetables. Ital. J. Food Sci. **2015,** 27 (1), 1A.

23. Gil, M. I.; Aguayo, E.; Kader, A. A. Quality Changes and Nutrient Retention in Fresh-Cut versus Whole Fruits during Storage. J. Agric. Food Chem. **2006,** 54 (12), 4284–4296.

24. Gontard, N.; Guillaume, C. Packaging and the Shelf-Life of Fruits and Vegetables. Food Packag. Shelf-life **2010,** 297, 297–303.

25. Gorris, L. G. M.; Peppelenbos, H. W. Modified Atmosphere and Vacuum Packaging to Extend the Shelf-Life of Respiring Food Products. Hortic. Technol. **1992,** 2 (3), 303–309.

26. Guo, B.; Qian, J.; Zhang, T.; Yang, X. Zigbee-Based Information Collection System for the Environment of Cold-Chain Logistics of Fruits and Vegetables. Trans. Chin. Soc. Agric. Eng. **2011,** 27 (6), 208–213.

27. Jha, S. N. Nondestructive Evaluation of Food Quality. Heidelberg: Springer, 2010; pp 375.

28. Jha, S. N.; Vishwakarma, R. K.; Ahmad, T.; Rai, A.; Dixit, A. K. Assessment of Quantitative Harvest and Post-Harvest Losses of Major Crops/Commodities in India. ICAR-All India Coordinated Research Project on Post-Harvest Technology, ICAR-CIPHET 2015: P.O. PAU, Ludhiana, 2015; p 112.

29. Joslyn, M. A.; Braverman, J. B. S. The Chemistry and Technology of the Pretreatment and Preservation of Fruit and Vegetable Products with Sulfur Dioxide and Sulfites. Adv. Food Res. **1954,** 5, 97–160.

30. Kader, A. A. A Summary of CA Requirements and Recommendations for Fruits Other than Apples and Pears; In VIII International Controlled Atmosphere Research Conference 600, 2001; pp 737–740.

31. Kader, A. A. Postharvest Technology of Horticultural Crops; UCANR Publications: California; 2002; p 3311.

32. Kaur, R.; Gul, K.; Singh, A. K. Nutritional Impact of Ohmic Heating on Fruits and Vegetables—A Review. Cogent. Food Agric. **2016,** 2 (1), 1159000.

33. Kerry, J.; Butler, P. Smart Packaging Technologies for Fast Moving Consumer Goods; John Wiley: England, 2008; p 356.

34. Kasso, M.; Bekele, A. Post-Harvest Loss and Quality Deterioration of Horticultural Crops in Dire Dawa Region, Ethiopia. J. Saudi Soc. Agric. Sci. **2016.** DOI:10.1016/j.jssas.2016.01.005.

35. Kowalska, H.; Lenart, A. Mass Exchange during Osmotic Pretreatment of Vegetables. J. Food Eng. **2001,** 49 (2), 137–140.

36. Kozempel, M.; Radewonuk, E. R.; Scullen, O. J.; Goldberg, N. Application of the Vacuum/Steam/Vacuum Surface Intervention Process to Reduce Bacteria on the Surface of Fruits and Vegetables. Innov. Food Sci. Emerg. Technol. **2002,** 3 (1), 63–72.

37. Kumar, s.; Mohapatra, D.; Kotwaliwale, N.; Singh, K. K. Vacuum Hermetic Fumigation for Food-Grain Storage. In Proceedings of 10th International Conference on Controlled Atmosphere and Fumigation in Stored Products, 2016; p 210.

38. Lal Basediya, A.; Samuel, D. V. K.; Beera, V. Evaporative Cooling System for Storage of Fruits and Vegetables—A Review. J. Food Sci. Technol. **2013,** 50 (3), 429–442.

39. Laufenberg, G.; Kunz, B.; Nystroem, M. Transformation of Vegetable Waste into Value Added Products: (A) The Upgrading Concept; (B) Practical Implementations. Bioresour. Technol. **2003,** 87 (2), 167–198.

40. Lee, S. Y. Microbial Safety of Pickled Fruits and Vegetables and Hurdle Technology. Int. J. Food Saf. **2004,** 4, 21–32.

41. Mir, N.; Beaudry, R. M. Modified Atmosphere Packaging. In The Commercial Storage of Fruits, Vegetables, and Florist and Nursery Stocks; Gross, K.C., Wang, C.Y., Saltveit, M., Eds.; Agriculture Handbook Number 66. USDA-ARS: Washington DC, 2004; p 42.

42. Nandi, C. S.; Tudu, B.; Koley, C. Machine Vision Based Techniques for Automatic Mango Fruit Sorting and Grading Based on Maturity Level and Size. In Sensing Technology: Current Status and Future Trends II; Springer International Publishing: Switzerland, 2004; pp 27–46.

43. Negi, S.; Anand, N. Supply Chain Efficiency: An Insight from Fruits and Vegetables Sector in India. J. Oper. Supply Chain Manage. **2014,** 7 (2), 154–167.

44. Negi, S.; Anand, N. Cold chain: A Weak Link in the Fruits and Vegetables Supply Chain in India. IUP J. Supply Chain Manage. **2015,** 12 (1), 48.

45. Nelson, K. E. Harvesting and Handling California Table Grapes for Market. UCANR Publications: California; 1979; pp 1913.

46. Oey, I.; Lille, M.; Van Loey, A.; Hendrickx, M. Effect of High-Pressure Processing on Color, Texture and Flavor of Fruit- and Vegetable-Based Food Products: A Review. Trends Food Sci. Technol. **2008,** 19 (6), 320–328.

47. Olivas, G. I.; Barbosa-Cánovas, G. V. Edible Coatings for Fresh-Cut Fruits. Crit. Rev. Food Sci. Nutr. **2005,** 45 (7–8), 657–670.

48. Parfitt, J.; Barthel, M.; Macnaughton, S. Food Waste within Food Supply Chains: Quantification and Potential for Change to 2050. Philos. Trans. R. Soc. **2010,** 365, 3065–3081.

49. Parde, S. R.; Kausal, R. T.; Jayas, D. S.; White, N. D. Mechanical Damage to Soybean Seed during Processing. In 2001 ASAE Annual Meeting; American Society of Agricultural and Biological Engineers, 1998; p 10.

50. Parry, R. T., Ed. Principles and Applications of Modified Atmosphere Packaging of Foods; Springer Science & Business Media: Suffolk, England, 2012; p 305. DOI: 10.1007/978-1-4615-2137-2.

51. Paull, R. Effect of Temperature and Relative Humidity on Fresh Commodity Quality. Postharv. Biol. Technol. **1999,** 15 (3), 263–277.

52. Pereira, M. A. B.; Tavares, A. T.; Silva, E. H. C.; Alves, A. F.; Azevedo, S. M.; Nascimento, I. R. Postharvest Conservation of Structural Long Shelf-Life Tomato Fruits and with the Mutant Rin Produced, in Edaphic Climatic Conditions of the Southern State of Tocantins. Ciên. Agrotecnol. **2015,** 39 (3), 225–231.

53. Pourbafrani, M.; Forgács, G.; Horváth, I. S.; Niklasson, C.; Taherzadeh, M. J. Production of Biofuels, Limonene and Pectin from Citrus Wastes. Bioresour. Technol. **2010,** 101 (11), 4246–4250.

54. Prothon, F.; Ahrné, L. M.; Funebo, T.; Kidman, S.; Langton, M.; Sjoholm, I. Effects of Combined Osmotic and Microwave Dehydration of Apple on Texture, Microstructure and Rehydration Characteristics. LWT—Food Sci. Technol. **2001,** 34 (2), 95–101.

55. Pundhir, A.; Murtaza, N. Hurdle Technology—An Approach towards Food Preservation. Int. J. Curr. Microbiol. Appl. Sci. **2015,** 4 (7), 802–809.

56. Rosa, S. Postharvest Management of Fruit and Vegetables in the Asia-Pacific Region/ Asian Productivity Organization; Food and Agricultural Organization (FAO): Rome, 2006; p 115.

57. Saltveit, M. E. In Summary of CA Requirements and Recommendations for Vegetables, VIII International Controlled Atmosphere Research Conference 600, 2001; pp 723–727.

58. Saltveit, M. E. Respiratory Metabolism. The Commercial Storage of Fruits, Vegetables, and Florist and Nursery Stocks. Agriculture Handbook; USDA-ARS Food Quality Laboratory: Beltsville, MD, 2004; p 66.

59. Sagagi, B.; Garba, B.; Usman, N. Studies on Biogas Production from Fruits and Vegetable Waste. Bay. J. Pure Appl. Sci. **2009,** 2 (1), 115–118.

60. Simchi-Levi, D.; Kaminsky, P.; Simchi-Levi, E.; Shankar, R. Designing and Managing the Supply Chains—Concepts, Strategies and Case Studies. Tata McGraw-Hill: New Delhi, 2008; p 208.

61. Singh, M. P.; Singh, V. K. Biodegradation of Vegetable and Agrowastes by Pleurotus sapidus: Novel Strategy to Produce Mushroom with Enhanced Yield and Nutrition. Cell. Mol. Biol. **2012,** 1, 1–7.

62. The Hindu. How to Compost Kitchen Waste in 6.5 Steps, 2013. http://www.thehindu.com/news/cities/chennai/how-to-compost-kitchen-waste-in-65-steps/article5016625.ece (accessed on July 31, 2017).

63. Thompson, A. K. Pre-Cooling and Storage Facilities. In USDA Agriculture Handbook; USDA: Washington DC, 2004; p 66.

64. Thompson, A. K. Controlled Atmosphere Storage of Fruits and Vegetables; CABI: Wallingford, Oxfordshire, UK, 2010 (online).

65. Thompson, J. F.; Mitchell, F. G.; Rumsay, T. R. Commercial Cooling of Fruits, Vegetables and Flowers; UCANR Publications: Richmond, CA, 2008; p 61.

66. Torringa, E.; Esveld, E.; Scheewe, I.; van den Berg, R.; Bartels, P. Osmotic Dehydration as a Pre-Treatment before Combined Microwave-Hot-Air Drying of Mushrooms. J. Food Eng. **2001,** 49 (2), 185–191.

67. Voss, R. E.; Baghott, K. G.; Timm, H. Proper Environment for Potato Storage. Communication by the Vegetable Research and Information Center, University of California: Davis, CA, 2004; p 110.

68. Wadhwa, M.; Bakshi, M. P. S. Utilization of Fruit and Vegetable Wastes as Livestock Feed and as Substrates for Generation of Other Value-Added Products; RAP Publication: Thailand, 2013; p 4.

69. Wang, X.; Li, D. A Dynamic Product Quality Evaluation Based Pricing Model for Perishable Food Supply Chains. Omega **2012,** 40 (6), 906–917.

70. Wijngaard, H. H.; Roble, C.; Brunton, N. A Survey of Irish Fruit and Vegetable Waste and By-Products as a Source of Polyphenolic Antioxidants. Food Chem. **2009,** 116 (1), 202–207.

71. Zhang, M.; Tang, J.; Mujumdar, A. S.; Wang, S. Trends in Microwave-Related Drying of Fruits and Vegetables. Trends Food Sci. Technol. **2006,** 17 (10), 524–534.

CHAPTER 13

NEGATIVE EFFECTS OF PROCESSING ON FRUITS AND VEGETABLES

TANYA LUVA SWER, SAVITA RANI, and KHALID BASHIR

ABSTRACT

Different processing techniques such as drying, curing, pickling, fermentation, etc., have been practiced worldwide for preserving the quality, improving the shelf life, and for ensuring the availability of fruits and vegetables beyond the season. Despite of these advantages, processing such as peeling, trimming, blanching, and thermal methods have been observed to have a deleterious effects on the nutritive value, sensory properties, and on the phytochemicals content of processed food products compared with their fresh counterparts. Vitamins, minerals, antioxidants, and color profile of foods especially vegetables were reduced significantly after the processing treatments. Therefore, this chapter creates awareness about negative effects of different processing methods on fruits and vegetables commodities.

13.1 INTRODUCTION

Availability of various fruits and vegetables is restricted to seasonal and regional factors. Besides, these produce are highly perishable and susceptible to spoilage, which renders their limited utilization. Therefore, effective processing techniques are required to preserve the quality and extend the shelf life of these produce[56] and to make them available worldwide throughout the year either in fresh form or in the form of processed product. One of the major aspects of food processing is that it facilitates the stabilization, preservation, transportation, and availability of a variety of fruits and vegetables in different regions for human consumption.[17]

There are various preservation techniques that are focused to deliver safe processed food products to consumers. However, processing can considerably alter these sensory properties and nutritive value of the food product in comparison to their fresh counter parts. For instance, peeling and trimming of fruits and vegetables prior to cooking could result in significant reduction of nutrient content of the resultant product and formation of some harmful chemicals.[3,32] Similarly, thermal or nonthermal processing at household or industrial levels have been reported to cause reduction in levels of phytochemicals in processed food products.[48]

Therefore, this chapter describes the negative effects of processing operations on the nutritional profile of fruits and vegetables.

13.2 FOOD PROCESSING METHODS AND THEIR IMPACTS ON NUTRITIVE VALUE OF FRUITS AND VEGETABLES

The significant development in food preservation started after the accidental discovery of fire by prehistoric human. Since then, many processing techniques have been developed and adopted throughout the centuries. Processing (such as drying, curing, pickling, fermentation, etc.) has been carried out for generations throughout the world. Though wide range of novel processes have been studied over the last 100 years, some of the earlier or conventional techniques are still in use and are available in several commercial formats. Processing techniques despite their usefulness also exert some negative effects on quality of fruits and vegetables and their products as has been reported in many research studies. This section discusses in detail about the traditional and novel processing techniques along with their advantages as well as disadvantages on fruits and vegetables and their products.

13.2.1 THERMAL PROCESSING

Heat treatment remains one of the most important methods used in food processing, not only because of the desirable effect on eating quality, but also because of the preservative effect on foods by the destruction of enzymes, microorganisms, insects, and parasites. Thermal processing or heat processing involves application of heat to food, either in a sealed container or by passing it through a heat exchanger, followed by packaging. It is important to ensure that the food is adequately heat treated to reduce post-processing contamination. Thermal processing can be achieved by techniques using hot water or steam such as cooking, blanching, pasteurization, sterilization, evaporation, and extrusion; hot air in baking, roasting, and drying; hot oil in frying; and irradiated energy in microwave (MW), infrared radiation, and ionizing radiation.[16,39]

13.2.1.1 BLANCHING

Blanching is an unique processing method, commonly employed to soften the product and for destruction of enzymes, which are accountable for browning and leaching of color in fruits and vegetables.[32] Blanching also offers the advantages of preservation of natural color of dried products and reduction of cooking time during reconstitution of dried foods.[41] Generally, effectiveness of blanching depends on the total destruction of peroxidase activity. Conventionally blanching was done by conveying the food through saturated steam or hot water. However, MW blanching and gas blanching have been demonstrated on experimental scale. But MW blanching is very expensive and further research needs to be conducted for gas blanching. Hence, these techniques are not currently used in industries on commercial scale. Generally, the time for blanching using hot water or steam depends on the type and dimensions of produce. The ideal blanching time required for an individual fruit or vegetable is important for retention of its nutritive and health-promoting constituents.[32] In comparison to peas, beans, and carrots; leafy vegetables require less blanching time.[8] Nevertheless, the average blanching time for most of the product is about 6 min.[41]

13.2.1.1.1 Negative Effects of Blanching on Fruits and Vegetables

Blanching can lead to change in sensory qualities and degradation of nutrients such as vitamins and phenolic compounds that are relatively unstable when subjected to heat treatments.[16,35,38] Vitamins B and C soluble in water are sensitive and can be damaged easily during the blanching. Latest research on the influence of processing on the antioxidants level of several vegetables (peas, carrots, spinach, potatoes, and brassicas) revealed that vitamin C loss ranged between 10% and 40% during the blanching process.[24] Steam blanching shadowed by frozen storage ($-20°C$) for a time period of 1 year showed substantial reduction of vitamin C content of various vegetables such as carrots, broccoli, and green beans. It was also found that maximum loss increased in blanching stage and during the initial storage period; afterward the frequency of vitamin C degradation was slowed. Similarly in another study, blanching was also responsible for 20% reduction in pea's ascorbate content. Phenolics and other antioxidant compounds are also highly susceptible to thermal treatment. Puupponen et al.[36] demonstrated that total phenolics and antioxidant activity using 1,1-diphenyl-2-picrylhydrazyl (DPPH) assay were decreased slightly during blanching of cauliflower. Loss of color may also be encountered during blanching of fruits and vegetables. Inyang and Ike[23] and Lin and Brewer[30] demonstrated slight change in color following blanching of okra and peas, respectively.

13.2.1.2   DRYING/DEHYDRATION

Drying/dehydration is one of the oldest methods commonly used for preservation of foods for future use. Both terms "drying" and "dehydration" imply removal of water.[32,41] Drying refers to removal of moisture from foods by using nonconventional energy sources such as sun and air; while dehydration is based on the method of deduction of humidity through synthetic warmness under the organized sets of conditions (temperature, humidity, and air flow). Drying and dehydration involve simultaneous application of heat and removal of moisture from foods (except for osmotic dehydration). The various factors that affect the rate at which foods dry are categorized as (1) factors related to processing conditions and (2) factors related to nature and properties of food.[16]

Various kinds of dryers (such as solar, kiln, cabinet, tray, tunnel, belt trough, fluidized bed, etc.) are commonly used for drying of solid foods; whereas drum and spray dyers have been extensively used for slurry or liquid foods.[41] Examples of commercially important dried foods are sugar, coffee, milk powder, flour, beans, pulses, nuts, breakfast cereals, soup powder, tea, spices, etc.[16] The main reason for drying a food is to extend its shelf life without the need for refrigerated transport and storage. This goal is achieved by reducing the available moisture or water activity (a_w) to a level, which inhibits the growth and development of spoilage and pathogenic microorganisms, reducing the activity of enzymes and the rate at which undesirable chemical changes occur.

Appropriate packaging is necessary to maintain the low a_w during storage and distribution.[9] Drying also reduces the weight and bulk in food providing ease of handling; reduced packaging, transport, and storage costs; and provides greater variety and convenience for the consumer.[16] Besides, dried foods can be considered as source of energy because they contain high concentration of fruit sugars and high amount of vitamins (riboflavin) and minerals (iron). Dried fruits or vegetables can also be added directly to products (such as soups or stews) without presoaking, or they can be drawn on the liquid in the soup or stew for rehydration during cooking.[26]

13.2.1.2.1 Negative Effects of Drying on Fruits and Vegetables

Main disadvantages of drying/dehydration of fruits and vegetables involve physical changes such as structural deformation or crust formation due to case hardening. The removal of moisture causes collapse of structure of the produce contributing to adverse changes in shape when compared to fresh counterparts. Also, faster drying rate due to improper drying conditions may result in case hardening, causing crust formation, which makes the dried fruits or vegetables undesirable.[37] Another, major disadvantage in the dried fruits involves leaching and loss of food nutrients (viz., vitamins B and C) in cooking water during the reconstitution process.[13]

Drying, being a thermal process, also causes reduction of heat degradative nutrients such as vitamin C in the dehydrated fruits and vegetables. Concentration of some nutrients (fiber, sugar) during dehydration may result in products having extra energy, which may contribute to weight gain. Nonenzymatic browning (or Maillard reaction) is another limitation, which may occur when drying fruits and vegetables, where carbonyl groups of reducing sugars combine with amino groups of proteins and amino acids leading to the formation of more complex reactions. Maillard browning can lead to the formation of unnatural color along with the development of bitter taste in fruits and vegetables. Preventive measures by treating vegetables with sulfur dioxide prior to dehydration can prevent vitamin C loss along with significant reduction of nonenzymatic browning.[8]

The common chemical changes in dried products are general discoloration problems and failure of the dried product to fully rehydrate caused by structural changes that can further toughen the rehydrated, cooked products. Exposure to high temperatures or prolonged drying time may also cause discoloration and considerable change in the original color of the dried produce.[37]

13.2.1.3 CANNING

Canning was first invented by Nicolas Appert from France who developed a heat treatment process for sterilization of sealed food products. Hence, canning is also termed as appertization. It is based on the principle of heating the food products inside the can to destroy any hazardous microorganisms and thereby prolong the shelf life of the processed fruits or vegetables.[4] Canning involves the hermetically sealing of fruits and vegetables, dipped in brine (for vegetables) or syrup (for fruits), in containers that is followed by heat sterilization thereby, ensuring the safe storage and prolonged shelf life. The steps involved in canning of fruits and vegetables include preparation of food, filling, exhausting the containers, sealing, thermal processing, and cooling the containers and contents.[44] Canning of fruits and vegetables is carried out during the period of higher availability of raw material. Canned foods are retailed in off-season which provides better revenues to grower.

13.2.1.3.1   Negative Effects of Canning

Exposure to very high temperature during canning may rob the significant amount of nutrients of fruits and vegetables due to thermal degradation process. It was reported that about 10–90% reduction of vitamin C content occurs throughout the canning of several vegetables and there was only a minute alteration observed during storage of canned products. The B-group vitamins (B_1 and B_6) are somewhat sensitive to temperature and light and this may be the reason for their high rate of degradation during canning ranging from 7% to 70% for most of the vegetables.

Canning also resulted in reduction of polyphenols during the storage of fresh fruits and vegetables. Canning of fruit and vegetable products showed minute variation in nutrient content if temperature used for storage is maintained properly.[4] According to a study, it was reported that vitamin C degrades significantly during blanching and thermal processing of foods. Level of vitamin C was reduced by 23–35% during blanching; and further enhancement in this loss also occurred in canning as it permits absorption of oxygen from dissolved gases and from headspace.[44]

13.2.2   LOW-TEMPERATURE PROCESSING

Commonly practiced processes for preservation in the food industry generally involve the application of heat. As discussed earlier, high temperature and longer exposure to heat largely affect the amount of nutrient loss, deterioration of functional properties of the food products, and development of undesirable flavors. Lowering the temperature of foodstuffs reduces microbiological and biochemical spoilage by decreasing microbial growth rates and by removing liquid water that then becomes unavailable to support microbial growth. With development of mechanical refrigeration systems, cold preservation of food, storage and distribution have become widespread. It has also influenced postharvest agricultural practices. It has become possible to transport perishable products for long distances from production to consumption center and make available seasonal foods at all times of the year.

13.2.2.1   FREEZING

Freezing has been successfully employed for long-term preservation of fruits and vegetables. Freezing is a unit operation in which the commodities are subjected to temperature below their freezing point. The process involves lowering the product temperature below $-18°C$ and a proportion of the water changes its state to form ice crystals that immobilizes the water content and lowers the a_w of these produce. This slows down the biochemical and physicochemical reactions that may enhance the deterioration of the produce. Freezing process also slows down both enzymatic and nonenzymatic changes to preserve the quality of the fruits and vegetables. Therefore, preservation of fruits and vegetables is obtained due to combination of low temperature and lowered a_w.

There are different types of freezers available for freezing of fruits and vegetables and their products. Generally, freezers are broadly classified into mechanical and cryogenic freezers. In mechanical freezers, cooled air, cooled liquid, or cooled surfaces are used to remove the heat and freeze the fruits, vegetables, and their products. Examples of these types of freezers are immersion freezer, cabinet freezer, air-blast freezer, fluidized bed freezer, belt freezer, tunnel freezer, and spiral freezer. In cryogenic freezers, cryogens such as liquid nitrogen, solid and liquid carbon dioxide, and liquid Freon are used directly in contact with the fruits, vegetables, and other products for freezing these foods.

13.2.2.1.1   Negative Effects of Freezing

Maintaining the correct freezing process and storage helps in retaining nutritional and sensory qualities of frozen fruits and vegetables without any significant change. Any nutrient losses, if any, might occur during preprocessing before freezing or cooking losses once the frozen food is thawed.

The loss of quality of frozen fruits and vegetables depends on the freezing rate, storage temperature, length of storage time, and thawing procedure. The transformation of water into ice during freezing causes expansion in volume, which greatly causes damage to the cellular membrane, resulting in poor quality of the frozen fruits and vegetables. Weight loss due to dehydration may also occur and must be strictly considered during freezing of unpacked products.

It is very important to maintain almost constant temperature during handling of frozen foods. Temperature fluctuations that occur during storage, retail display, or during transit from retail store to home may cause partial fusion of ice and reforming of irregular large ice crystals that move to the product surface. This affects the cell wall integrity and causes moisture loss resulting in a freeze-dried or freeze-burned product, which subsequently affects the quality and shelf life of final frozen products.[12,37] Exposing of unpacked fruits and vegetables to cold, dry air causes freezer burn with white color spots appearing on the surface of the frozen fruits or vegetables due to direct sublimation of ice to vapor.[26,45] This can be controlled by proper packaging of the product, by decreasing the storage temperature, and by maintaining proper humidification in the freezer.[2]

The ice crystal formation and volume expansion also cause mechanical damage and breakdown of cellular chloroplasts and changes in the natural pigments such as chlorophylls, anthocyanins, and carotenoids. For example, loss of chlorophyll occurs because it slowly degrades to pheophytin to give a dull khaki color. Further, as the water transforms to ice, the solute concentration increases that greatly manipulate the physical properties of the food such as pH which in turn affect the anthocyanin stability in the frozen food.[46]

Furthermore, oxidation process or enzymatic reaction of the enzyme ascorbic acid oxidase may result in loss of water-soluble vitamin C, which is greatly influenced by factors such as pretreatment conditions (blanching), freezing methods, packaging type, time–temperature conditions during storage, and thawing conditions. Vegetables stored at $-24°C$ generally showed better ascorbic acid retention than those stored at lower temperatures. The rate of loss of ascorbic acid is estimated to increase 6-folds in vegetables and 20–30-folds in fruits for every $10°C$ rise in storage temperature. Pretreatment by blanching can greatly improve the retention of ascorbic acid in frozen vegetables. Loss of other vitamins may be attributed to drip loss on thawing.[37]

13.2.3 NOVEL PROCESSING TECHNOLOGIES

Consumers' demand for safe and "fresh-like" product has given impetus to research and has led to studies for development of alternative nonthermal processing methods. A wide range of novel processes have been studies over the last 100 years. Many of these technologies remain much in the research arena; however, others have come onto the brink of commercialization.[28]

This chapter discusses the effects of some of the novel techniques that have found commercial applications in the food industries, namely, cold sterilization (irradiation), MW, ultrasound, and hydrostatic pressure processing. Other processing techniques that have promising applications for the food industry include radio frequency, light pulses, pulsed electric field, etc. Novel process techniques have been observed to produce products having higher freshness, good flavor, color, and nutrient score which might be the reason of increasing food industries attractiveness toward these processes.[19]

13.2.3.1 IRRADIATION

Irradiation or cold sterilization is one of the utmost methods of foodstuff preservation practiced even more than five eras of research and development.[27] The process of food irradiation is done by exposing the food to ionizing radiation, which may be generated through the gamma rays or a high-energy electron beam or powerful X-rays. The waves pass through the foods in the same manner as MWs pass inside the MW oven; however, the heating of food product does not occur to any significant extent. There is no need of radioactive material for producing the electron beams and X-rays as they can be made directly from electricity, which can be switched on or off easily depending on the requirements.[26] Depending on the dosage level, irradiation has been classified into three categories radappertization (analogous to radiation sterilization; dosage level 30–40 kGy; used in canning industry); radurization (used to enhance shelf life; dosage level is 0.75–2.5 kGy; used for fresh meats, poultry, fruits, vegetables, and cereals); and radicidation (analogous to pasteurization of milk; dosage level is 2.5–10 kGy). Irradiation has been successfully used for the inhibition and removal of food allergens and antinutritional factors.[1]

Main advantages of irradiation are (1) prevention of food spoilage by finishing or deactivating food spoilage microorganism; (2) facilitates safe storage devoid of refrigeration conditions[32]; (3) retains sensory qualities of foods irradiated at low doses (citrus fruits and papaya); and (4) juice yield can be improved by using radiation doses of several kGy.[26]

13.2.3.1.1 Negative Impacts of Irradiation on Fruits and Vegetables

The negative impacts of irradiation on food products have been observed by scientists from time to time. Major applications of irradiation around the world include disinfestation of fresh fruits to eliminate pests and reduction of tuber crop losses, such as potato, onion and garlic, by inhibiting sprouting with a maximum permissible dose of 1 kGy. However, in 2008 after stringent reviewing, the US-FDA approved the use of irradiation up to 4.0 kGy to inactivate microorganisms and to extend the shelf life only for fresh lettuce and fresh spinach.[15] The formations of trace levels of furan, a possible carcinogen, have been reported in other types of fruits and vegetables on irradiation. This created obstacles for approval of irradiation processing of other fruits and vegetables for microbial safety and shelf-life extension.[14]

Tissue softening is yet another common problem that occurs during irradiation of fresh fruits and vegetables which result from radiation-induced depolymerization of carbohydrates such as cellulose, hemicellulose, starch, and pectin. Other disorders include discoloration of skin, internal browning, and increased susceptibility to chilling injury. Irradiation of produce having high protein content may also contribute to change in odor and flavor. In comparison with other amino acids, aromatics are more sensitive and undergo a change in the ring structure during ionizing process. However, these effects can be minimized by irradiating at chill or frozen temperatures.[8]

The loss in vitamin C is attributed to transformation of ascorbic acid to *dehydroascorbic acid*. The reduction is, however, insignificant from the nutritional viewpoint. Korkmaz and Polat[27] demonstrated that vitamin C loss increases with radiation dose and time elapsed after radiation. Tocopherol or vitamin E is also found to be very sensitive during the irradiation especially in presence of oxygen. Nevertheless, the negative impact of irradiation on vitamins could be decreased through elimination of oxygen and light, treatment at low temperature, and by applying lowest dose during food irradiation.[26]

Studies have also reported significant reduction in crude lipid and phospholipid concentrations after irradiation process. However, the negative effects largely depend on the type of fruits or vegetables being treated and on the dosage used for treatment. The conditions used for irradiating any fruit or vegetable cultivars are very specific, depending on the intended purpose and the susceptibility of the tissue to irradiation damage. For example, some citrus fruits can withstand dosage of 7.5 kGy, whereas avocados may be sensitive even at a low dosage of 0.1 kGy. Specific examples are discussed in detail by Arvanitoyannis et al.,[1] Salunkhe,[40] Thomas,[47] and Urbain.[50]

13.2.3.2 MW PROCESSING

MWs are electromagnetic waves with frequencies between 300 MHz and 300 GHz. MW technologies have been increasingly used in the food industry. MW generates heat in foods by induced molecular vibration as a result of dipole rotation and/or ionic polarization. Popularly, MW ovens are used at household levels for cooking, heating, baking, etc. At the industrial level, MWs have found plentiful applications in drying, blanching, pasteurization, sterilization, extraction, selective heating, disinfestations, and MW heat treatment of fruits of oil palm bunch. The advantages of MW processing over the conventional processing methods have been reviewed in detail by Marsaioli et al.[31]; Orsat and Raghavan[33]; and Ramaswamy et al.[40]

13.2.3.2.1 Negative Impacts of MW Processing on Fruits and Vegetables

Despite many advantages of MW processing over conventional methods, there are some negative effects that have been observed. These effects are, however, significantly lower when compared with conventional methods of processing. Nevertheless, some of the studies have shown negative effects of MW processing on nutrient content of various fruits and vegetables. MW processing has been reported to cause loss of vitamin A and/or carotenoid. Total carotenoid, namely, β-carotene and lycopene, losses amounting to 57% in papaya puree have been observed after application of various levels of MW power (285–850 W).[11]

MW treatment has also been reported to cause significant reduction in chlorophylls and lycopene in kiwi fruits and cherry tomatoes, respectively.[12,20] Significant losses of vitamin B_1 (32–60%) and vitamin B_2 (9–47%) were demonstrated during MW cooking of Swiss chard and green beans.[54] Studies on the effects of MW processing on ascorbic acid have shown contradictory results where few researchers reported reduction of ascorbic acid with MW treatment whereas others showed increasing effect of MW on ascorbic content. Begum et al.[6] reported 10% loss of vitamin C in tomatoes; whereas Vikram et al.[53] showed 30–50% loss of vitamin C in orange juice after treatment with MW. In contrast, according to Howard et al.[22] and Vina et al.,[55] increasing vitamin C content was demonstrated in carrots after MW processing (120–130%), green beans (117%), and Brussels sprouts (10–15%).

Similar effect was observed with total phenolic content, where the content either was decreased or increased with MW processing depending on the study and commodity. The 15% loss of total anthocyanins was demonstrated in sweet potato,[42] 15% loss of flavonoid was observed in Brussels sprouts[34]; and 57% total phenolic loss was obtained in apple puree after treatment with MW. More elaborated and detailed effects of MW preservation on nutrient quality of specific fruits and vegetables have been discussed in details by Barett et al.[4]

13.2.3.3   ULTRASOUND

The principle for the application of ultrasound processing in the food industry has been described in detail by Bermudez-Aguire et al.[7] With advances in the recent years, ultrasound technology has found commercial applications in the food industry mostly as a processing aid for assisting processes such as drying, emulsification, homogenization, crystallization and freezing, mass and heat transfer, antifouling during filtration and separation, deaeration and defoaming, fermentation, inactivation of enzymes, cutting or particle size reduction, extrusion, viscosity alteration, or for cleaning and decontamination of surfaces.[5]

13.2.3.3.1   Negative Effects of Ultrasound Processing on Fruits and Vegetables

The acoustic energy of ultrasonic waves may induce chemical effects such as generation of free radical and hydrogen peroxide in the final product or physical impacts such as localized high temperature and pressure, shock waves, and microstreaming that can affect the final quality of the product. However, till date there are no documented reports that describe reliable process to measure the cavitation-induced physical and chemical effects and predict the food quality changes.[29] Hence, one can conclude that effects of ultrasound treatment on food quality are strictly limited to specific products as reported by researchers.

Texture is one of the parameters that have been observed to be slightly altered when treating fruits or vegetables with ultrasound process. Ultrasound treatment (47 kHz, 55°C, 0–10,800 s at ambient pressure) induced cell damage in bell pepper was observed, which subsequently reduces the textural integrity of the produce.[18] Flavor alteration on reduction of 1-butanol-3-methyl acetate due to ultrasound treatment (20 kHz, 0.6 W/mL, 60°C, 252 s at ambient pressure) has also been observed in apple juice.[29] Several studies have also demonstrated color changes in orange juice, apple cider, dehydrated rabbiteye blueberry, and blanched watercress, which may affect the overall acceptability of these produce by the consumers.[10,43,49,51]

Very few reports are available on the effects of ultrasound-aided processing on nutritional value of several fruits and vegetables. Reduction of anthocyanins and phenolic content has been observed in osmotic dehydrated rabbiteye blueberry by Stojanovic et al.[43] on treating with ultrasound (850 kHz, 21°C, 3 h at ambient pressure). Vercet et al.[52] demonstrated 100% loss of ascorbic acid and carotenoids in orange juice after ultrasound treatment (20 kHz, 62°C, 15 and 30 s, 200 kPa). Study by Khandpur et al.[25] showed that there was slight reduction in ascorbic acid content, phenolic content, and antioxidant property of orange, sweet lime, carrot, and spinach juices on treating with ultrasound (20 kHz, 100 W, 15 min, duty cycle as 50%). However, the overall retention of these compounds was significantly better when compared to the untreated and thermally treated samples.

13.2.3.4   HIGH PRESSURE PROCESSING

High pressure processing (HPP) is progressive processing method that has been accepted by the food industry as a possible substitute of food pasteurization process. It involves treatment of solid or liquid food with or without the packaging to a pressure of 50–1000 MPa, which resulted in successful destruction of microorganism and enzymes.[21] Fruit juices are most commonly processed by this treatment. Vitamin content, flavor, and color of foods remain unaffected since there is no use of heat during the processing.

Currently, hydrostatic processing is also being considered as sterilization practice at 100 MPa up to 900 MPa and a pressure applied commercially fluctuates between 400 and 700 MPa. The degree of temperature rise during pressure application varies according to food composition but is normally

3–9°C/100 MPa. Examples of successful high-pressure treated foods commercially available are fruit jams and sauces, guacamole, sliced cooked hams, oysters, and meal kits that contain meat, salsa, guacamole peppers, and onions.

13.2.3.4.1 Negative Aspects of HPP on Fruits and Vegetables

During hydrostatic treatments, cell biopolymers of cell (proteins, polysaccharides, and lipids) go through various changes. Functionality and texture of plant foods get affected due to polysaccharide alterations caused by pressure treatment. Fat crystallization along with alteration in protein structure (unfolding, aggregation, and gelation) is major change that takes place after the hydrostatic treatments. Physical disruption of the tissue can induce various changes in plant texture.

According to a recent study, hydrostatic processing at 400 MPa revealed various changes in cellular structure and also caused the folding of membrane, particularly in case of cauliflower and spinach leaves. Microscopic examination of onion epidermis cells displayed severe destruction of vacuoles after 300 MPa treatments at 25°C; and fresh onions odor altered to that of braised or fried onions. In recent studies, it has been reported that high-pressure treatments of lettuce carried out at 150 MPa produced permanent cell damage, and serious changes in the chloroplast membrane integrity.

13.3 SUMMARY

There is no doubt that food processing techniques are necessary for providing the availability of foods beyond the area and time of production. These methods are also necessary for stabilizing the supply and enhancement of food security at national and household levels. However, there are some negative effects associated with these processing techniques which can reduce the quality of fruits and vegetables. By adopting proper processing methods, suitable to particular crop, could minimize the quality and nutritional loss that occurred during processing. It is also helpful to identify the other factors that are associated with the loss of nutrients in fruits and vegetables during processing so that quality of processed produce should remain in acceptable condition.

KEYWORDS

- **ascorbic acid**
- **blanching**
- **canning**
- **drying**
- **freezing**
- **Maillard reaction**
- **vitamins**

REFERENCES

1. Arvanitoyannis, I. S.; Stratakos, A. C.; Tsarouhas, P. I. Irradiation Applications in Vegetables and Fruits: A Review. *Crit. Rev. Food Sci. Nutr.* **2009,** *49*, 427–462.
2. ASHRAE. *Handbook, Refrigeration Systems and Applications*; American Society of Heating, Refrigerating, and Air-conditioning Engineers: Atlanta, GA, 1994; p 310.
3. Barrett, D. M. Maximizing the Nutritional Value of Fruits and Vegetables. *Food Technol.* **2007,** *61* (4), m40–m44.
4. Barrett, D. M.; Lloyd, B. Advanced Preservation Methods and Nutrient Retention in Fruits and Vegetables. *J. Sci. Food Agric.* **2011,** *92* (1), 7–22.
5. Bates, D.; Patist, A. Industrial Applications of High Power Ultrasonics in the Food, Beverage and Wine Industry (Chapter 6). In *Case Studies in Novel Food Processing Technologies: Innovations in Processing, Packaging, and Predictive Modelling*; Doona, C. J., Kustin, K., Feeherry, F. E., Eds.; Woodhead Publishing: Cambridge, UK, 2010; pp 119–137.
6. Begum, S.; Brewer, M. S. Chemical, Nutritive and Sensory Characteristics of Tomatoes Before and After Conventional and Microwave Blanching and During Frozen Storage. *J. Food Qual.* **2001,** *24*, 1–15.
7. Bermudez-Aguire, D.; Mobbs, T., Barbosa-Canovas, G. V. Ultrasound Applications in Food Processing (Chapter 3). In *Ultrasound Technologies for Food and Bioprocessing*; Feng, H., Barbosa-Cánovas, G. V., Weiss, J., Eds.; Springer-Verlag: New York, 2011; pp 65–105.
8. Bhatia, S. C. *Handbook of Food Processing Technology: Biochemical and Microbiological Aspects*; Atlantic Press: New Delhi, 2008; p 312.
9. Brenan, J. G. Evaporation and Dehydration (Chapter 3). In *Food Processing Handbook*; Brenan, J. G., Ed.; Wiley-VCH Verlag GmbH & Co. KGaA: Weinheim, Germany, 2006; pp 71–124.
10. Cruz, R. M. S.; Vieira, M. C.; Silvia, C. L. M. Modeling Kinetics of Watercress (*Nasturtium officinale*) Color Changes due to Heat and Thermosonication Treatments. *Innov. Food Sci. Emerg. Technol.* **2007,** *8*, 244–252.

11. De Ancos, B.; Cano, M. P.; Hernandez, A.; Monreal, M. Effects of Microwave Heating on Pigment Composition and Color of Fruit Purees. *J. Sci. Food Agric.* **1999,** *79,* 663–670.

12. De Ancos, B.; Sanchez-Moreno, C.; De Pascual-Teresa, S.; Cano, M. P. Fruit Freezing Principles (Chapter 4). In *Handbook of Fruits and Fruit Processing*; Hui, Y. H., Barta, J., Cano, M. P., Gusek, T. W., Sidhu, J. S., Sinha, N. K., Eds.; Blackwell Publishing: Iowa, USA, 2006; pp 59–80.

13. Devi, R. Food Processing and Impact on Nutrition. *Sch. J. Agric. Vet. Sci.* **2015,** *2* (4A), 304–331.

14. Fan, X. Irradiation of Fresh Fruits and Vegetables: Principles and Considerations for Further Commercialization (Chapter 8). In *Case Studies in Novel Food Processing Technologies. Innovations in Processing, Packaging, and Predictive Modelling*; Doona, C. J., Kustin, K., Feeherry, F. E., Eds.; Woodhead Publishing: Cambridge, UK, 2010; pp 427–441.

15. FDA (US Food and Drug Administration). *Final Rule (73 FR 49593), Irradiation in the Production, Processing and Handling of Food*; 21 CFR Part 179, Federal Register 73, 2008; pp 49593–49603.

16. Fellows, P. J. *Food Processing Technology: Principles and Practices*; CRC Press: Boca Raton, USA, 2000; p 310.

17. Floros, J. D.; Newsome, R.; Fisher, W.; Barbosa-Cánovas, G. V.; Chen, H., Dunne, C. P.; German, J. B.; Hall, R. L.; Heldman, D. R.; Karwe, M. V.; Knabel, S. J.; Labuza, T. P.; Lund, D. B.; Newell-McGloughlin, M.; Robinson, J. L.;. Sebranek, J. G.; Shewfelt, R. L.; Tracy, W. F.; Weaver, C. M.; Ziegler, G. R. Feeding the World Today and Tomorrow: The Importance of Food Science and Technology. *Compr. Rev. Food Sci. Food Saf.* **2010,** *9,* 572–599.

18. Gabaldón-Leyva, C. A.; Quintero-Ramos, A.; Barnard, J.; Balandrán-Quintana, R. R.; Talamás-Abbud, R. T.; Jiménez-Castro, J. Effect of Ultrasound on the Mass Transfer and Physical Changes in Brine Bell Pepper at Different Temperature. *J. Food Eng.* **2007,** *81,* 374–379.

19. Gonzalez, M. E.; Barrett, D. M. Thermal, High Pressure and Electric Field Processing Effects on Plant Cell Membrane Integrity and Relevance to Fruit and Vegetable Quality. *J. Food Sci.* **2010,** *75* (7), 121–130.

20. Heredia, A.; Peinado, I.; Rosa, E.; Andres, A. Effect of Osmotic Pretreatment and Microwave Heating on Lycopene Degradation and Isomerization in Cherry Tomato. *Food Chem.* **2010,** *123,* 92–98.

21. Hogan, E.; Kelly, A. L.; Sun, D. W. High Pressure Processing of Foods: An Overview. In *Emerging Technologies for Food Processing*; Sun, D. W., Ed.; Elsevier Academic Press: California, USA, 2005; pp 3–32.

22. Howard, L. A.; Wong, A. D.; Perry, A. K.; Klein, B. P. Beta-carotene and Ascorbic Acid Retention in Fresh and Processed Vegetables. *J. Food Sci.* **1999,** *64,* 929–936.

23. Inyang, U. E.; Ike, C. I. Effect of Blanching, Dehydration Method and Temperature on the Ascorbic Acid, Color, Sliminess and Other Constituents of Okra Fruit. *Int. J. Food Sci. Nutr.* **1998,** *49* (2), 125–130.

24. Kalt, W. Effects of Production and Processing Factors on Major Fruit and Vegetable Antioxidants. *J. Food Sci.* **2005,** *70* (1), R11–R19.

25. Khandpur, P.; Gogate, P. R. Effect of Novel Ultrasound Based Processing on the Nutrition Quality of Different Fruit and Vegetable Juices. *Ultrason. Sonochem.* **2015** (Online). DOI: http://dx.doi.org/10.1016/ j.ultsonch.2015.05.008.

26. Khetarpaul, N. *Food Processing and Preservation*; Daya Publishing House: Daryaganj, New Delhi, 2012; pp 269–383.

27. Korkmaz, M.; Polat, M. Irradiation of Fresh Fruit and Vegetables (Chapter 13). In *Improving the Safety of Fresh Fruits and Vegetables*; Jongen, W., Ed.; Woodhead Publishing Ltd.: Cambridge, England, 2005; pp 387–428.

28. Leadley, C. E.; Williams, A. Pulsed Electric Field Processing, Power Ultrasound and Other Emerging Technologies (Chapter 7). In *Food Processing Handbook*; Brennan, J. G., Ed.; Wiley-VCH Verlag GmbH & Co. KGaA: Weinheim, Germany, 2006; pp 201–236.

29. Lee, H.; Feng, H. Effect of Power Ultrasound on Food Quality (Chapter 22). In *Ultrasound Technologies for Food and Bioprocessing*; Feng, H., Barbosa-Cánovas, G. V., Weiss, J., Eds.; Springer-Verlag: New York, NY, 2011; pp 559–582.

30. Lin, S.; Brewer, M. S. Effects of Blanching Method on the Quality Characteristics of Frozen Peas. *J. Food Qual.* **2005**, *28* (4), 350–360.

31. Marsaioli Jr., A.; Berteli, M. N.; Nádia, R.; Pereira, N. R. Applications of Microwave Energy to Postharvest Technology of Fruits and Vegetables. *Stewart Postharvest Rev.* **2009**, *6*, 1–5.

32. Nayak, B.; Liu, R. H.; Tang, J. Effect of Processing on Phenolic Antioxidants of Fruits, Vegetables, and Grains: A Review. *Crit. Rev. Food Sci. Nutr.* **2015**, *55* (7), 887–918.

33. Orsat, V.; Raghavan, G. S. V. Microwave in Postharvest Applications with Fresh Fruits and Vegetables. *Fresh Prod.* **2007**, *1* (1), 16–22.

34. Picouet, P. A.; Landl, A.; Abadias, M.; Castellari, M.; Vinas, I. Minimal Processing of a Granny Smith Apple Puree by Microwave Heating. *Innov. Food Sci. Emerg. Technol.* **2009**, *10*, 545–550.

35. Prochaska, L. J.; Nguyen, X. T.; Donat, N.; Piekutowski, W. V. Effects of food Processing on the Thermodynamic and Nutritive Value of Foods: Literature and Database Survey. *Med. Hypotheses* **2000**, *54* (2), 254–262.

36. Puupponen-Pimia, R.; Hakkinen, S. T.; Aarni, M.; Suortti, T.; Lampi, M.-A.; Eurola, M.; VienoPiironen, V.; Nuutila, A. M.; Oksman-Caldentey, K.-M. Blanching and Long Term Freezing Affect on Various Bioactive Compounds of Vegetables in Different Ways. *J. Sci. Food Agric.* **2003**, *83*, 1389–1402.

37. Rahman, M. S.; Velez-Ruiz, J. Food Preservation by Freezing. In *Handbook of Food Preservation*; Rahman, M. S., Ed.; CRC Press: Boca Raton, USA, 2007; pp 635–666.

38. Rahman, M. S.; Perera, C. O. Drying and Food Preservation. In *Handbook of Food Preservation*; Rahman, M. S., Ed.; CRC Press: Boca Raton, USA, 2007; pp 404–432.

39. Ramaswamy, H.; Tang, J. Microwave and Radio Frequency Heating. *Food Sci. Technol. Int.* **2008**, *14*, 423–427.

40. Salunkhe, D. K. Gamma Radiation Effects on Fruits and Vegetables. *Econ. Bot.* **1961**, *15* (1), 28–56

41. Srivastava, R. P.; Kumar, S. *Fruits and Vegetables Preservation: Principles and Practices*; International Book Distributing Co.: Lucknow, India, 2002; pp 127–169.

42. Steed, L. E.; Truong, V. D.; Simunovic, J.; Sandeep, K. P.; Kumar, P.; Cartwright, G. D.; Swartzel, K. R. Continuous Flow Microwave-assisted Processing and Aseptic Packaging of Purple-fleshed Sweetpotato Purees. *J. Food Sci.* **2008**, *73*, E455–E462.

43. Stojanovic, J.; Silva, J. L. Influence of Osmotic Concentration, Continuous High Frequency Ultrasound and Dehydration on Antioxidants, Color and Chemical Properties of Rabbiteye Blueberries. *Food Chem.* **2007,** *101,* 898–906.

44. Sudheer, K. P.; Indira, V. *Post-harvest Technology of Horticultural Crops*; New India Publishing House: Pitam Pura, New Delhi, India, 2007; pp 101–106.

45. Symons, H. Frozen Foods (Chapter 15). In *Shelf-life Evaluation of Foods*; Man, C. M. D., Jones, A. A., Eds.; Blackie Academic and Professional: London, UK, 1994; pp 296–319.

46. Talens, P.; Escriche, I.; Martinez-Navarret, N.; Chiralt, A. Study of the Influence of Osmotic Dehydration and Freezing on the Volatile Profile of Strawberries. *J. Food Sci.* **2002,** *67* (5), 1648–1653.

47. Thomas, P. Irradiation of Fruits and Vegetables (Chapter 8). In *Food Irradiation: Principles and Applications*; Molins, R., Ed.; Wiley-Blackwell: USA, 1990; pp 213–240.

48. Tiwari, U.; Cummins, E. Factors Influencing Levels of Phytochemicals in Selected Fruit and Vegetables During Pre- and Post-harvest Food Processing Operations. *Food Res. Int.* **2013,** *50,* 497–506.

49. Ugarte-Romero, E.; Feng, H.; Martin, S. E.; Cadwallader, K. R.; Robinson, S. J. Inactivation of *Escherichia coli* with Power Ultrasound in Apple Cider. *J. Food Sci.* **2006,** *71,* E102–E108.

50. Urbain, W. M. *Food Irradiation*; Academic Press: London, UK, 1986; p 413.

51. Valero, M.; Recrosio, N.; Saura, D.; Muñoz, N.; Martí, N., Lizama, V. Effects of Ultrasonic Treatments in Orange Juice Processing. *J. Food Eng.* **2007,** *80,* 509–516.

52. Vercet, A.; Burgos, J.; Lopez-Buesa, P. Manothermosonication of Foods and Food Resembling System: Effect on Nutrient Content and Non-enzymatic Browning. *J. Agric. Food Chem.* **2001,** *49,* 483–489.

53. Vikram, V. B.; Ramesh, M. N.; Prapulla, S. G. Thermal Degradation Kinetics of Nutrients in Orange Juice Heated by Electromagnetic and Conventional Methods. *J. Food Eng.* **2005,** *69,* 31–40.

54. Villanueva, M. T. O.; Marquina, A. D.; Vargas, E. F.; Abellan, G. B. Modification of Vitamins B$_1$ and B$_2$ by Culinary Processes: Traditional Systems and Microwaves. *Food Chem.* **2000,** *71,* 417–421.

55. Vina, S. Z.; Olvera, D. F.; Marani, C. M.; Ferreyra, R. M.; Mugridge, A.; Chaves, A. R.; Mashcharoni, R. H. Quality of Brussels Sprouts (*Brassica oleracea* L. *gemmifera* DC) as Affected by Blanching Method. *J. Food Eng.* **2007,** *80,* 218–225.

56. Wu, Z.; Wang, H. S.; Li, S. J. Advices and Considerations to Actuality of Vegetables Exports in China. *Chin. Agric. Sci. Bull.* **2004,** *20* (3), 277–280.

CHAPTER 14

EFFECTS OF PROCESSING ON VITAMINS IN FRUITS AND VEGETABLES

SINGATHIRULAN BALASUBRAMANIAN,
ABHIMANNYU ARUN KALNE, and KHURSHEED ALAM KHAN

ABSTRACT

Vitamins are essential to provide support to healthy and regular physiologic functions of body. They are responsible in various ways such as to help growth of body, well functioning of various body systems, metabolic processes, proper cell functioning, etc. Long-term vitamin deficiency can cause several diseases such as scurvy, skin disorders, rickets, anemia, and many more. In this chapter, the effects of different processing techniques (viz., freezing, canning, and nonthermal processing), cooking, and storage condition on vitamin housed by fruits and vegetables (FAV) has been discussed. Stored foods are mostly affected by (1) rise in temperature; (2) oxygen level; and (3) increase in concentration of reactants. Additionally this chapter explores the role of nonthermal processing in reducing enzymatic activity and inactivation of microorganism. This chapter also deals with degradation mechanism and precaution measures to be followed during processing of FAV for maximum retention of vitamins.

14.1 INTRODUCTION

Fruits and vegetables (FAV) must be harvested at the proper stage, size, and at the peak quality. Although quality could not be improved after harvest but careful harvesting, proper packaging, storage, and transport can give better quality produce. Storage and processing practices are being used since centuries to develop delicious products for consumption from perishable raw FAV. Refrigeration allows longer shelf life as it retards respiration of harvested FAV. Operations such as freezing, canning of FAV, and drying are being used extensively to process raw perishable FAV and to develop products which will be utilized throughout the year and are suitable for easy transportation to the far away market safely where raw fruits/vegetables are difficult to transport due to perishable nature.

Processing of biological commodities such as FAV inhibits respiration which in turn prevents moisture loss from produce, microorganism growth, as well as loss of nutritious components of FAV. Though developing safe processed product fit for consumption is a prime objective of processing, processors are also worked to provide high-quality processed products. Now, depending on way of processing methods followed, quality parameters are more or less affected. The major changes in texture, color, flavor, and nutritional value of product affect acceptability of processed product and this chapter particularly deals with the changes occurred in nutritional quality of processed products.

FAV are rich source of number of micronutrients and bioactive compounds which are required by human for well-functioning of body systems. Many of these micronutrients are available in FAV at minute level and so it is imperative to retain them during processing chain. Different researchers have studied the effect of processing on loss of such micronutrients and vitamins for different foods during the postharvest handling practices, domestic/industrial processing, or during the preparation.[8,16,22,30] Though losses of these nutritional components during processing are unavoidable to some extent, selection of proper processing practices can retain the major portion of micronutrients in processed foods.

Vitamins are low-molecular-weight organic compounds which are essential in low amounts in daily diet. Vitamins' role is crucial in regulating normal body functions. They helps in growth of body, maintaining normal body system functions, metabolic processes, proper cell functioning, etc. Long-term shortage of vitamins leads to rise of several diseases such as scurvy, skin disorders, reduced bone density, rickets, anemia, etc.[25] According to Modern Nutritional Science, 13 vitamins were grouped in two categories as fat-soluble vitamins and water-soluble vitamins. Out of 13 vitamins, 4 vitamins are considered as fat soluble and 9 vitamins are grouped in water-soluble category (Table 14.1). Some vitamins are working as precursor (also called provitamins) which do not show any biological activity but are needed to produce some active vitamin molecules within body (e.g., β-carotene). Generally, leafy vegetables are good sources of most vitamins. The vitamin content of FAV depends upon variety, growing conditions, and several other factors.

This chapter focuses on the effects of different processing techniques (viz., freezing, canning, and nonthermal processing), cooking, and storage condition on vitamin content of FAV. Additionally degradation mechanism of carotenoids, vitamin C, and folate has been discussed along with precautionary measures to be taken during processing of FAV for maximum retention of vitamins.

TABLE 14.1 Vitamins and Vitamers and Their Commonly Used Synonyms.

Vitamin group	Vitamins	Synonyms	Vitamers
Fat-soluble	Retinol	Vitamin A_1, all-trans-retinol, vitamin A_1 alcohol, axerophtol, or axerol	–
Vitamin A	Retinal	Vitamin A_1 aldehyde, retinene, and retinaldehyde	–
	Retinoic acid	Tretinoin and vitamin A_1 acid	–
	Vitamin A_2	3-dehydroretinol and 3,4-didehydroretinol	–
	Provitamin A	Carotenoids	α-carotene, β-carotene, γ-carotene, β-cryptoxanthin, and echinenon
Vitamin D	Ergocalciferol	Vitamin D_2	–
	Cholecarciferol	Vitamin D_3	–
	Provitamin D_3	7-dehydrocholesterol and 7-procholesterol	–
	Provitamin D_2	Ergosterol	–
Vitamin E	Tocopherol	–	α-tocopherols, β-tocopherols, γ-tocopherols, and δ-tocopherols
	Tocotrienols	–	α-tocotrienol, β-tocotrienol, γ-tocotrienol, and δ-tocotrienol

TABLE 14.1 *(Continued)*

Vitamin group	Vitamins	Synonyms	Vitamers
Vitamin K	Phylloquinone	Vitamin K_1, vitamin $K_{1(20)}$, and phytomenadione	–
	Menaquinone	vitamin K_2, vitamin $K_{2(n)}$, and farnoquinone	–
	Menadione	Vitamin K_3	–
Water-soluble B complex	Thiamin	Vitamin B_1 and aneurin	–
	Riboflavin	Vitamin B_2	–
	Folates	Vitamin Bc and folacin	Folic acid and folate
	Pantothenic acid	Vitamin B_5	–
	Niacin	Vitamin PP and vitamin B_3	Nicotinic acid and nicotinamide
	Pyridoxine	Vitamin B_6	Pyridoxal, pyridoxol, and pyridoxamine
	Cobalamin	Vitamin B_{12}	Cyanocobalamin and hydroxocobalamin
	Biotin	Vitamin H	
Vitamin C		–	Ascorbic acid and dehydroascorbic acid

Source: Reprinted with permission from Ref. [25]. © 2006 Elsevier.

14.2 FRESH VERSUS PROCESSED FAV

In all food services food preparation, minimal processing, and cooking are found to be associated with considerable amount of nutrient loss. In hospitals too,[13] most common food service systems, namely, cook/hot–cold and cook/chill are accountable for almost 30% loss of vitamin C and folate, when food is reheated after storage for 24 h at 3°C.[45] During initial steps of processing (washing and peeling of fruits/vegetables, blanching treatment), some water-soluble vitamins are affected. Oxidation also lowers/damages the nutrients particularly during heat treatment and during storage of produce (Fig. 14.1).

Thermal processing of FAV is also responsible to loss of heat-sensitive vitamins. Vitamin C (ascorbic acid) and thiamin are the example of these.[12] Blanching of fresh fruits/vegetables is performed before freezing in order to inactivate some enzymes. This blanching step may affect the vitamins present in produce to some extent

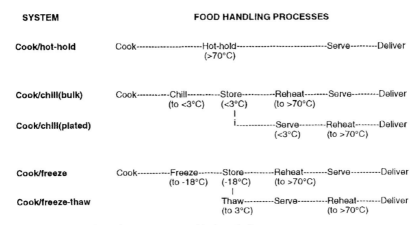

FIGURE 14.1   Food service systems used in hospitals.

14.2.1   EFFECTS OF PRELIMINARY OPERATIONS

Vitamins and bioactive compounds are housed in plant parts naturally. Vitamins and carotenoids are also packed heavily in peel and so the peeling/trimming operation can affect the availability of vitamins in fruit/vegetables. Some operations such as cutting, shredding, and pulping during fruit/vegetable processing destroys peel protection and exposes the fruit/vegetable to surrounding oxygen which is responsible for further degradation. Therefore, thermal processing must be performed immediately after peeling and cutting of fruits/vegetables. Enzymatic degradation in case of minimally processed fruits/vegetables is equally serious as degradation in nutritional values in thermal treatments. Cutting and pulping operations also result in loss of cellular integrity and thus destroy the natural shield of health-promoting components and expose them for oxidation.

Water-soluble ascorbic acid (vitamin C) is highly sensitive to heat, exposure to light, and oxygen and so is lost easily during heat treatment as well as during simple cooking. Loss of ascorbic acid is commonly used as an index of nutrient degradation. Ascorbic acid is considered as GRAS (Generally Recognized As Safe) and so it is commonly used for prevention of browning and other oxidative reactions. Ascorbic acid hinders polyphenol oxidase,[43] which is credited to decrease in enzymatically formed o-quinones to their precursor diphenols. Figure 14.2 shows production of the catecholase enzyme.

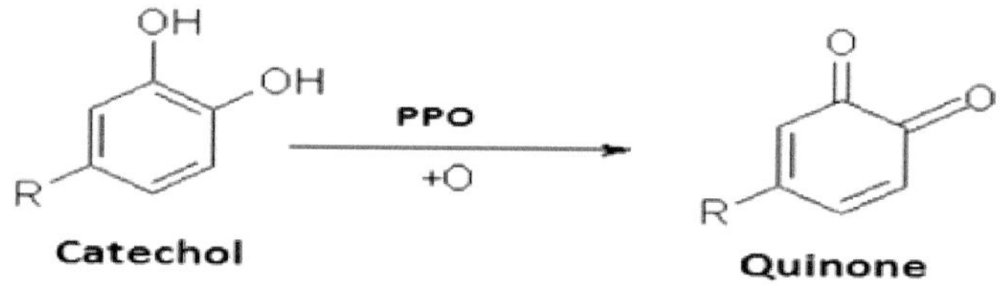

Production of the catecholase enzyme; *Source:* Mishra,
B.B. and S. Gautam (2016). Polyphonel oxidases:
biochemical and molecular characterization, distribution,
role and its control. *Enzyme Engineering*, 5, 141; Open
access online; doi:10.4172/2329-6674.1000141.

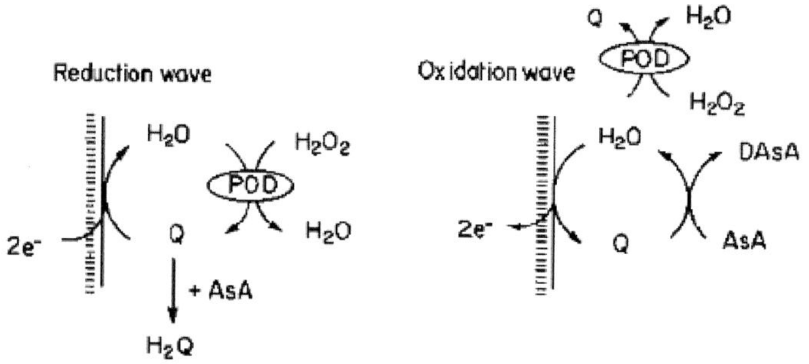

Schematics of mechanism of the enzymatic process: *Source:*
Orlando Fatibello-Filho and Iolanda da C. Vieira (2000). L-
ascorbic acid determination in pharmaceutical formulations
using a biosensor based on carbon paste modified with crude
extract of zucchini (*Cucurbita pepo*). *Journal of the Brazilian
Chemical Society*, *11*(4), online: open access licensed under
a Creative Commons Attribution License;
<http://www.scielo.br/scielo.php?script=sci_arttext&pid=S010
3-50532000000400015>

FIGURE 14.2   Oxidation–reduction mechanism for ascorbic acid.

In order to retain micronutrients and bioactive compounds in minimally processed foods, low-temperature storage and modified atmosphere
packaging was reported to extend the shelf life. During extraction of juice
from fruits/vegetables, it is observed that considerable amount of vitamins,
flavonoids, and carotenoids remain present in pomace (press cake).[41]

Blanching is one of the pretreatments which is usually used to retard
microbial load and to inactivate enzymes that are responsible for quality
deterioration. Blanching is performed by immersion in hot water, exposure
to steam, etc. Hot-water blanching results in leaching of water-soluble
vitamins. If cut or bruised part of fruit/vegetables is exposed to water, water-
soluble vitamins (folate, ascorbic acid, and anthocyanin) are lost.[8] Earlier

studies have reported 40% loss of ascorbic acid (vitamin C) occurred in different vegetables during blanching.[14,20] McKillop et al.[27] also reported 50% reduction in folate content of spinach and broccoli compared to raw foods due to boiling for a minimal time. However, it was observed that steam blanching results in greater retention of folate compared to water blanching.[7,26] Although blanching results in loss of heat-sensitive nutrients, yet the inactivation of oxidative as well as other enzymes such as poly-phenol oxidase for flavonoids, lipoxygenase for carotenoids, and ascorbic acid oxidase for vitamin C avoids further nutrient loss by enzyme-catalyzed degradation during slow processing (such as in drying) and during storage. Blanching is carried out before the canning operation in order to expel air from tissues so that thermal conductivity is increased and to ease better packing into the container. The sole purpose of performing blanching opera-tion before the freezing is to inactivate those enzymes which may remain active even in frozen product. Blanching is an important treatment during canning and freezing of a number of vegetables. Generally, fruits are not introduced to blanching before the freezing due to their delicate nature and characteristics of acidity.

14.2.2 EFFECTS OF PROCESSING

Thermal processing is well known for the fact that it causes substantial loss of vitamins, anthocyanins, and carotenoids. At higher temperature, reactions occur rapidly as compared to ambient temperature. Irrespective of processing methods selected, retention of these compounds will be less with increased processing time and temperature. Sensory and functional properties of processed fruits/vegetables and nutrients availability may get affected due to oxidation and enzymatic destruction. Reduction in vitamin C content is dependent on several factors such as pH, degree of heating, exposed surface area to water and oxygen, leaching into medium, and other related factors which facilitate oxidation.[8] Thermal treatment, freezing, and high pressure processing (HPP) may result in release of carotenoids as well as fat-soluble vitamins such as vitamin A and E.

14.2.2.1 FREEZING

Freezing is generally performed after pretreatment of blanching to inactivate or retard enzymes. Some amount of vitamins is affected by this blanching.

Blanched products usually have greater resistance to decomposition as compared to unblanched food products during the storage practice.

Freezing and frozen storage usually retain nutrients and bioactive compounds in produce, but slow thawing may be damaging. In frozen spinach, peas, green beans, and okra commercially cooked in different stew pans with and without thawing (at room temperature 22.37°C, 3 or 4 h), loss of vitamin C ranged from minimum 3.5% to maximum 19.6%, respectively.[30] Favell[10] had reported negligible loss of ascorbic acid (vitamin C) content in frozen carrots, but 20% and 30% losses in broccoli and green peas, respectively.

14.2.2.2 CANNING

During canning process, FAV are exposed to high temperature levels which significantly degrade vitamin C content and may result in leaching in canning medium. Murcia et al.[28] had reported 84% loss of vitamins C in broccoli as compared to fresh food. Similar reduction in ascorbic acid content has been reported for broccoli, carrot, green pea, and spinach during commercial thermal processing (Table 14.2). Canned products which are unheated are generally compared with cooked frozen foods and/or cooked fresh foods.

Ascorbic acid levels are found superior in cooked frozen products than in canned foods. Some vitamins are sensitive to heat and light. B vitamins such as thiamin and B_6 are particularly sensitive to heat and light. During canning process of various vegetables, losses in these vitamins have been reported from 7% to 70%, respectively.[14] Olunlesi and Lee[31] reported 65% and 56% decrease in original β-carotene content of carrots and carrot pureed for baby during canning. However, in case of canned carrots, Kim and Gerber[18] reported that β-carotene content has increased which might be because of loss of soluble solids.

TABLE 14.2 Loss of Ascorbic Acid (% Wet Weight) due to Canning and Freezing Process.

Commodity	Canning	Blanching and freezing	References
Broccoli	84	50–55	[28]
	–	30	[14]
Carrots	90	0–35	[14]
Green peas	84	63	[11]
Spinach	–	50	[15]
	62	61	[44]

14.2.2.3 NONTHERMAL PROCESSING

Nonthermal processing technologies such as osmotic dehydration (OD), HPP, high-intensity pulsed electric field (PEF) processing, and irradiation have been found to be successful in inactivating enzymes, and in reducing loss of water-soluble nutrients. These technologies are used to inactivate microorganisms and enzymes, without triggering adverse effects on products sensory as well as nutritional properties. Alternative forms of processing such as ultrahigh temperature and irradiation were reported to have greater retention of nutrients in fruit nectars as compared to traditional thermal processing.[35]

Ade-Omowayeet et al.[1] conducted comparative study on red paprika, PEF-pretreated and OD, with osmotic treated at higher temperature. They found much better retention of ascorbic acid and carotenoids in PEF-pretreated and OD red paprika. Shi et al.[36] have reported lower thermal damage in osmotically dehydrated tomatoes due to application of low temperatures, which favors retention of nutrients. Taiwo et al.[39] studied the effect of high-PEF and OD on the rehydration features of apple slices and found higher rehydration capacity values at low temperature in the range of 24–45°C. High hydrostatic pressure (HHP) is effective in reducing enzymatic activity and in inactivation of microorganism. During HHP, food may be preserved without input of large amount of heat, which helps in retention of nutritive value.

Rastogi and Niranjan[32] reported change in tissue architecture, leaving cells more permeable which intern favors OD. Bateet et al.[2] reported advantage of proper blending in retention of quality as well as nutritional value of fruit juices. Rosario et al.[34] had observed that exposure to irradiation results in greater retention of ascorbic acid and β-carotene in fruit nectars obtained by mixing papaya and strawberry. Some of the recent studies have been carried on mango, papaya, strawberry, and peach. All reported about 20% reduction of ascorbic acid (vitamin C) in fresh cut and fruit nectars, respectively.[3,17,37,44,46]

14.2.3 EFFECTS OF COOKING

While cooking of fresh vegetables at home, approximate 15–55% loss in vitamin C was observed depending on the method used for cooking.[33] Table 14.3 shows comparisons of vitamin C in fresh vegetables, frozen vegetables, and canned vegetables which were stored and then utilized for home cooking. During home cooking, retention of bioactive compounds and vitamins were found better in microwave cooking than steaming followed by boiling.[44] Fat-soluble carotenoids are more easily affected in sautéing than in boiling. Some fat-soluble vitamins such as vitamins A and E, and carotenoids are affected by light, heat, oxygen, and pH. Due to their fat-soluble nature, there is less leaching.[14] Deep-frying, extended cooking, as well as combination of numerous preparation methods all do results in substantial loss of nutritive value.

TABLE 14.3   Loss of Ascorbic Acid (% Dry Weight) due to Fresh and Frozen Storage.

Commodity	Loss of ascorbic acid (% dry weight)		
	Fresh (20°C, 7 days)	Fresh (4°C, 7 days)	Frozen (−20°C, 12 months)
Broccoli	56	0	10
Carrots	27	10	–
Green beans	55	77	20
Green peas	60	15	10
Spinach	100	75	30

Folate is sensitive to sunlight, air, and light and being heated in acid solutions. During cooking process, food has lost folate due to breakdown by heating and leaches in cooking water. Due to presence of reducing agents in food, folate retention in cooked food is increased in thermal processing. However, folates of animal origin are fairly stable during cooking methods such as boiling and frying. Cooking methods that minimizes direct contact of food with the cooking water, such as in pressure cooking, microwave cooking, or stir-frying have been found preferable to boiling for folate retention.

McKillop et al.[27] have reported significant reductions of 51% and 56% in the folate content of spinach or broccoli and this reduction could be reduced by applying steam. They found no loss of folate content in spinach or broccoli even after the maximum steaming periods of 4.5 and 15 min, which produced overcooked consistencies. Leichter et al.[24] reported that boiling of broccoli for 10 min causes 62% folate loss. Similar studies based on industrial processing had also reported that steam blanching resulted in greater retention of folate compared to water blanching, pressure cooking, and microwave cooking.[6,7,19,26]

Carotenoids are very sensitive to oxygen, heat, and light and are more susceptible to degradation due to their highly unsaturated structure.[4,5] Nagra and Khan[29] reported variable cooking loss of β-carotene in range from 10% (vegetable sponge) to 59% (carrot), respectively. In a comparative study of different cooking methods (traditional cooking in boiling water, sautéing, and microwave cooking) on the thiamin and riboflavin content of fresh cabbage and cauliflower, loss of thiamin decreases in the following order[42]: cooked in oil > pressure cooked in water > microwave cooked in oil > microwave cooked in small amount of water.

14.2.4 EFFECTS OF STORAGE CONDITIONS

Utmost attention must be given toward processed food otherwise chemical changes will predominate in a processed food that gives undesirable texture, color and flavor. Comparing to losses during processing, losses during storage are less but have considerable effect on amount of vitamin and bioactive compounds. FAV, both raw and processed, have to undergo transport and storage before they are consumed by end user. During this journey, some loss/degradation is expected. But the amount of vitamin C degrades quickly after harvesting of produce and this loss continues during the practices of storage of fruits/vegetables. At lower temperatures, reaction rates are comparatively slow and thus shelf life of fruits/vegetables is extended.[21] Some measures such as elimination of oxygen (by vacuum filling or by using oxygen impermeable packaging), protection from heat and light, and proper storage at low temperature levels protect vitamins and bioactive compounds from decomposition. Stored foods are mostly affected by (1) rise in temperature; (2) oxygen level; and (3) increase in concentration of reactants.

Weits et al.[44] had reported that refrigeration retards the deterioration of vitamin C and variation in storage temperature has considerable effect on storage life (Table 14.3). Vitamin C losses in vegetables which were stored at 4°C for the period of 7 days observed in the range of 15–77% depending on nature of fresh produce stored.[14] Refrigerated storage of carrots and green beans for period of 14–16 days ended in 10% increase in the β-carotene content of carrots and 10% loss in case of green beans. Howard et al.[14] observed no considerable loss of vitamin C in broccoli after 7 days of storage at 0°C but resulted up to 56% loss of vitamin C at 20°C which was observed to be wilted and of faded color.

Howard et al.[14] studied and compared vitamin C loss during storage of fresh produce and processed produce of broccoli, green beans, and carrots. They have observed the maximum loss of vitamin C in canned carrot with the exception of 50% increase in fresh carrot (Table 14.4). Cooked frozen products have often been shown to be equal or superior to cooked fresh products in their ascorbic acid levels (Table 14.4). This is probably because of vitamin C loss during fresh produce storage.[10,15,33,38] Frozen food products are reported to be associated with loss of vitamin C and carotene, while thiamin and riboflavin remained stable during 6 months of storage. However, prolonged storage for 12 months at −4°F showed 22% loss of thiamin.[9]

TABLE 14.4 Cumulative Loss of Vitamin C due to Fresh Storage or Processing Storage, Followed by Home Cooking in All Cases.

Vegetable	Initial concentration (g/kg)	Refrigerated storage time before processing and cooking (days)	Loss after cooking (% wet weight)			References
			Fresh	Frozen	Canned	
Spinach	0.28	0	64	81[c]	67[c]	[44]
Broccoli	1.80	21	38	62[b]	–	
Carrots	0.039	7	50[e]	56[b]	95[b]	[14]
Green beans	0.152	21	37	20[b]	–	
	0.163[a]	0	23	48[c]	68[c]	[44]
Green peas	0.40[a]	0	28	66[c]	77[c]	
	0.354	1–2	61	70[d]	85[d]	[11]

[a]Authors did not provide value. Values are taken from USDA, 2005.[40]
[b]Stored for 12 months prior to cooking.
[c]Stored for 6 months prior to cooking.
[d]Authors did not mention storage time before cooking.
[e]Authors reported increase in vitamin C with fresh storage.

14.3 DEGRADATIVE REACTIONS DURING PROCESSING AND STORAGE

Degradation of vitamins depends on many parameters followed during culinary practices such as light, moisture, pH, oxygen, temperature, and time of exposure of food to these parameters. The losses in vitamins varied the processing practices followed and the type of cooking method employed and also depends on type of food considered. It is not possible that processing methods employed will cause a real surge in nutrient concentration, but the processing methods may cause nutrients more measurable by instruments and may be more chances of available biologically. In the following section, chemical changes during processing as well as during storage of health-supporting food components such as ascorbic acid, folates, carotenoids, and anthocyanins.

Isomerization and enzymatic or nonenzymatic oxidation are considered as major changes in highly unsaturated carotenoids in course of processing and storage. Isomerization of *trans*-carotenoids, the usual configuration in nature, to the *cis*-isomers is promoted by acids, heat, and light. During trimming, slicing, juicing, or pulping organic acids are released in sufficient quantity to cause isomerization; however, in thermal processing isomerization occurs to much more extent than simple slicing and pulping operation. As a result, color of food became paler, reduction in provitamin A occurred, and bioavailability also got affected.

Enzymatic as well as nonenzymatic oxidation is the main reason for destruction of carotenoid during the processing and storage. This also depends on oxygen availability (Fig. 14.2) and is accelerated by heat, light, enzymes, and metals. This oxidation process is inhibited by presence of antioxidants. Enzyme-catalyzed oxidation occurs mainly before heat treatment and during slicing, peeling, or pulping. Nonenzymatic oxidation usually takes place during as well as after thermal treatment. It is augmented by damaging cellular structure, expansion of surface area, time span, and degree of severity of processing operation, storage duration and conditions, permeability of packing material used to oxygen, and also augmented by exposure of light to food samples. Oxidation starts with initiation of oxygen into the carotenoid molecule thus forming carotenoid epoxidase.

During the process of ascorbic acid degradation, dehydroascorbic acid gets oxidized followed by hydrolysis to 2,3-diketogulonic acid[23] and then further oxidation and polymerization produce a broad range of nutritionally inactive products.

Both ascorbic acid as well as dehydroascorbic acid have vitamin C activity. Loss of vitamin C activity happens when dehydroascorbic acid gets hydrolyzed with ring opening and 2,3-diketogulonic acid is formed. This hydrolysis is preferred by alkaline conditions. In the range of pH 2.5–5.5, this dehydroascorbic acid is utmost and its stability reduces as pH increases.

Green leafy vegetables are considered as basic source of folates. During the "folate cycle" tetrahydrofolate (THF) changes to methylene-THF which can be successively involved in the cellular synthesis of nucleic acid (Fig. 14.3). Folate can result in huge losses due to its reactivity as well as solubility in water during the food processing and during preparation methods.[23]

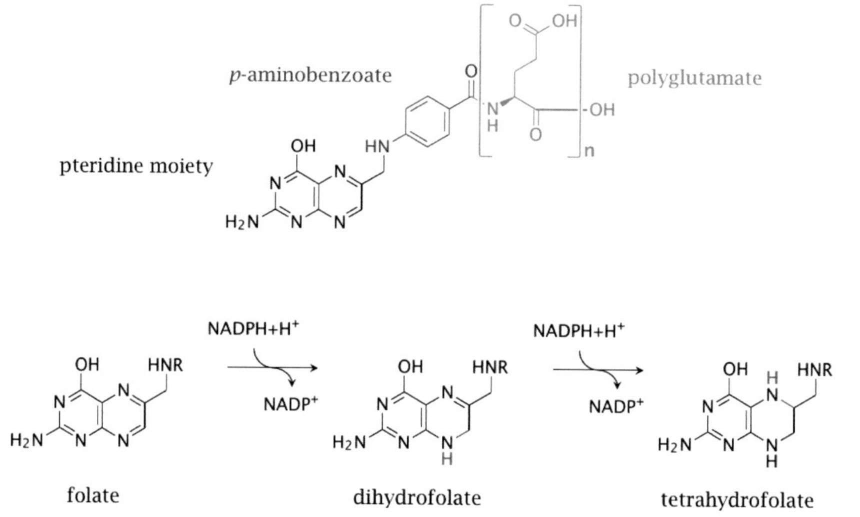

FIGURE 14.3 **(See color insert.)** Folate cycle conversion of tetrahydrofolate to methylene-THF.
Source: Online, open access. http://watcut.uwaterloo.ca/webnotes/Metabolism/FolateB12.html

As a water-soluble vitamin, folate has significantly leached to the water used for various operations such as prewash, washing, and pretreatment such as blanching; canning, cooking, and oxidative degradation can also take place. Literature on processing effects on folates is limited. Natural form of folate tetrahydrofolic acid (THA) is more susceptible to oxidative degradation whereas in its synthetic form, which is being used for food fortification, folic acid is very stable.

Folate metabolism is also influenced by group of chemotherapeutics, the example of which is sulfonamides which are used for treating infections. Human body system does not produce THF, aminobenzoic acid, which is a substrate for folate synthesis. When bacteria attempt to incorporate sulfonamides into the molecule of THF, the consequential molecule possesses a nonfunctional cofactor. Therefore, sulfonamides can be considered as competitive inhibitors of enzyme synthesizing THF.[23]

14.4 RECOMMENDATIONS FOR MAXIMUM RETENTION OF VITAMINS

The stability of individual vitamin varies from the relatively stable niacin and biotin to the relatively unstable such as thiamin and vitamin C. However, loss of nutritive value of processed foods could be minimized by applying following measures:

- Optimized conditions for blanching and processing should be followed for respective food commodity.
- Expelling of oxygen by vacuum filling, use of oxygen permeable packaging can avoid oxidation.
- Where drying is essential, food should be avoided by direct exposure to sunlight. Use of good solar dryer may be good alternative.
- To minimize the loss of water-soluble and oxygen-sensitive vitamins, cooking of fruits/vegetables must be accomplished in such a way that it utilizes optimum amount of water and cooking time.
- Processing time and temperature levels should be optimum. High temperature with short time processing is better approach.
- Processing activities should be instantly initiated after cutting, slicing, etc., so that minimum loss in vitamins and bioactive compounds will occur. Also processed products should be protected from light and should be stored at optimum temperature.
- In case of raw fruits/vegetables, intact fruit/vegetable should be introduced to storage. Temperature during storage should be low and duration for storage should be optimized.
- In case of fat-soluble vitamins, to reduce losses of vitamins A and E, less cooking oil should be used. Grilling and baking method of cooking may prove good alternative.
- FAV should be washed before peeling, cutting, or trimming.

14.5 SUMMARY

During processing of FAV, some losses of vitamins and bioactive compounds are generally occurred and to some extent these are unavoidable. But selection of proper processing and cooking methods, and storage for fruit/vegetables processed products can results in maximum retention of vitamins and bioactive compounds. Knowledge of factors responsible for degradation of bioactive compounds and possible measures to avoid these is a key solution for minimizing the vitamin loss in processed fruit/vegetables. The effect of processing operations, cooking methods employed, and storage conditions are highly correlated with the food commodity. Canning affects water-soluble and thermally liable vitamins of FAV, whereas storage practices and cooking methods used can also considerably affect the nutritional content of commodities. Nonthermal processing methods such as irradiation and high-intensity PEF; HPP would certainly help in maximum retention of nutrients and bioactive compounds.

KEYWORDS

- **ascorbic acid**
- **blanching**
- **canning**
- **carotenoids**
- **drying**
- **enzymes**
- **nutritive value**

REFERENCES

1. Ade-Omowaye, B. I. O.; Taiwo, K. A.; Eshtiaghi, N. M.; Angersbach, A.; Knorr, D. Comparative Evaluation of the Pulsed Electric Field and Freezing on Cell Membrane Permeabilization and Mass Transfer During Dehydration of Red Bell Peppers. *Innov. Food Sci. Emerg. Technol.* **2003,** *4,* 177–188.

2. Bates, R. P.; Morris, J. R.; Crandall, P. G. Principles and Practices of Small and Medium-scale Fruit Juice Processing. *FAO Agric. Serv. Bull.* **2001,** *146,* 59–71.

3. Boylston, T. D.; Reitmeier, C. A.; Moy, J. H.; Mosher, G. A.; Taladriz, L. Sensory Quality and Nutrient Composition of Three Hawaiian Fruits Treated by X-irradiation. *J. Food Qual.* **2002,** *25* (5), 419–433.

4. Britton, G. Carotenoids. In *Natural Food Colorants*; Hendry, G. F., Eds.; Blackie: New York, 1992; pp 141–148.

5. Clydesdale, F. M.; Francis, F. J. Pigments. In *Principles of Food Science: Food Chemistry*; Fennema, O. R., Ed.; Marcel Dekker: New York, NY, 1976; pp 417–430.

6. Dang, J.; Arcot, J.; Shrestha, A. Folate Retention in Selected Processed Legumes. *Food Chem.* **2000,** *68* (3), 295–298.

7. De Souza, S. C.; Eitenmiller, R. R. Effects of Processing and Storage on the Folate Content of Spinach and Broccoli. *J. Food Sci.* **1986,** *51* (3), 626–628.

8. Eitenmiller, R. R.; Laden, W. O. Vitamin A and B-Carotene, Ascorbic Acid, Thiamin, Vitamin B-6, Folate. In *Vitamin Analysis for the Health and Food Science*; Eitenmiller, R. R., Laden, W. O., Eds.; CRC Press: Boca Raton, FL, 1999; pp 15–19, 226–228, 275, 375, 411–465.

9. Farrell, K. T.; Fellers, C R. Vitamin Content of Green Snap Beans-influence of Freezing, Canning, and Dehydration on the Content of Thiamin, Riboflavin, and Ascorbic Acid. *J. Food Sci.* **1942,** *7* (3), 171–177.

10. Favell, D. J. A Comparison of the Vitamin C Content of Fresh and Frozen Vegetables. *Food Chem.* **1998,** *62* (1), 59–64.

11. Fellers, C. R.; Stepat, W. Effect of Shipping, Freezing and Canning on the Ascorbic Acid (Vitamin C) Content of Peas. *Proc. Am. Soc. Hort. Sci.* **1935,** *32,* 627–633.

12. Fennema, O. Effect of Processing on Nutritive Value of Food: Freezing. In *Handbook of Nutritive Value of Processed Food*; Rechcigl, M., Ed.; CRC Press: Boca Raton, FL, 1982; pp 31–43.

13. Glew, G. The Contribution of Large-scale Feeding Operations to Nutrition. *World Rev. Nutr. Diet.* **1980,** *34,* 1–45.

14. Howard, L. A.; Wong, A. D.; Perry, A. K.; Klein, B. P. β-Carotene and Ascorbic Acid Retention in Fresh and Processed Vegetables. *J. Food Sci.* **1999,** *64* (5), 929–936.

15. Hunter, K. J.; Fletcher, J. M. The Antioxidant Activity and Composition of Fresh, Frozen, Jarred and Canned Vegetables. *Innov. Food Sci. Emerg. Technol.* **2002,** *3* (4), 399–406.

16. Ilow, R.; Regulska-Ilow, B.; Szymcak, J. Assessment of Vitamin C Losses in Conventionally Cooked and Microwave-processed Vegetables. *Bromatol. Chem. Toksykol.* **1995,** *28* (4), 317–321.

17. Jeffrey, G. S.; Christopher J. K.; Mohammad A. G.; Nicki J. E. Synergistic Potential of Papaya and Strawberry Nectar Blends Focused on Specific Nutrients and Antioxidants Using Alternative Thermal and Non-thermal Processing Techniques. *Food Chem.* **2016,** *199,* 87–95.

18. Kim, H. Y.; Gerber, L. E. Influence of Processing on Quality of Carrot Juice. *Korean J. Food Sci. Technol.* **1988,** *20,* 683–690.

19. Klein, B. P. Retention of Nutrients in Microwave-cooked Foods. *Boletin de la Asociacion Medica de Puerto Rico* **1989,** *81,* 277–279.

20. Kmiecik, W.; Lisiewska, Z. Effect of Pretreatment and Conditions and Period of Storage on some Quality Indices of Frozen Chive (*Allium schoenoprasum* L.). *Food Chem.* **1999,** *67,* 61–66.

21. Kramer, A. Effect of Storage on Nutritive Value of Food. In *Handbook of Nutritive Value of Processed Food*; Rechcigl, M. Eds.; CRC Press, Boca Raton, FL; 1982; pp 275–299.

22. Krehl, W. A.; Winters, R. W. Effect of Cooking Methods on Retention of Vitamins and Minerals in Vegetables. *J. Am. Diet. Assoc.* **1950,** *26,* 966–1003.

23. Lavrikova, P.; Fontana, J.; Trnka, J. Vitamins and Nutrients (Chapter 6). In *Functions of Cells and Human Body*; World Press, 2015; pp 1–17 (Online). http://fblt.cz/skripta/ix-travici-soustava/7-vitaminy-a-vyziva/ (accessed Sept 28, 2016).

24. Leichter, J.; Switzer, V. P.; Landymore, A. F. Effect of Cooking on Folate Content of Vegetables. *Nutr. Rep. Int.* **1978,** *18,* 475–479.

25. Leskova, E.; Kubikova, J.; Kovacikova, E,; Kosicka, M,; Porubska, J.; Holcikova, K. Vitamin Losses: Retention During Heat Treatment and Continual Changes Expressed by Mathematical Models. *J. Food Comp. Anal.* **2006,** *19,* 252–276.

26. Lund, D. Effects of Heat Processing on Nutrients. In *Nutritional Evaluation of Food Processing*; Karmas, E., Harris, R. S., Eds.; Van Nostrand Reinhold: New York, NY, 1988; pp 319–354.

27. McKillop, D. J.; Pentieva, K.; Daly, D.; McPartlin, J. M.; Hughes, J.; Strain, J. J.; Scott, J. M.; McNulty, H. The Effect of Different Cooking Methods on Folate Retention in Various Foods that are Amongst the Major Contributors to Folate Intake in the UK Diet. *Br. J. Nutr.* **2002,** *88,* 681–688.

28. Murcia, M. A.; Lopez-Ayerra, B.; Martinez-Tome, M.; Vera, A. M.; Garcia-Carmona, F. Evolution of Ascorbic Acid and Peroxidase During Industrial Processing of Broccoli. *J. Sci. Food Agric.* **2000,** *80,* 1882–1886.

29. Nagra, A. S.; Khan, S. Vitamin A (B-Carotene) Losses in Pakistani Cooking. *J. Sci. Food Agric.* **1988,** *46* (2), 249–251.

30. Nursal, B.; Yucecan, S. Vitamin C Losses in some Frozen Vegetables due to Various Cooking Methods. *Nahrung* **2000,** *44,* 451–453.

31. Olunlesi, A. T.; Lee, C. Y. Effect of Thermal Processing on the Stereoisomerization of Major Carotenoids and Vitamin A Value of Carrots. *Food Chem.* **1979,** *4,* 311–318.

32. Rastogi, N. K.; Niranjan, K. Enhanced Mass Transfer During Osmotic Dehydration of High Pressure Treated Pineapple. *J. Food Sci.* **2008,** *63,* 508–511.

33. Rickman, J. C.; Barreet, D. M.; Bruhn, C. M. Nutritional Composition of Fresh, Frozen and Vegetables. Part 1. Vitamins C and B Complex and Phenolic Compounds. *J. Sci. Food Agric.* **2007,** *87,* 930–944.

34. Rosario, R. C.; Julieta, S. G.; Emilia, B. G.; Valdivia-López, M. A. Irradiation Effects on the Chemical Quality of Guavas. *Adv. J. Food Sci. Technol.* **2013,** *5,* 90–98.

35. Satpute, M.; Annapure, U. Approaches for Delivery of Heat Sensitive Nutrients Through Food Systems for Selection of Appropriate Processing Techniques: A Review. *J. Hyg. Eng. Des.* **2013,** *4,* 71–92.

36. Shi, J. X.; Maguer, M.; Wang S. L.; Liptay, A. Application of Osmotic Treatment in Tomato Processing: Effect of Skin Treatments on Mass Transfer in Osmotic Dehydration of Tomatoes. *Food Res. Int.* **1997,** *30,* 669–674.

37. Sogi, D. S.; Siddiq, M.; Roidoung, S.; Dolan, K. D. Total Phenolics, Carotenoids, Ascorbic Acid, and Antioxidant Properties of Fresh-cut Mango (*Mangiferaindica* L., cv. Tommy Atkin) as Affected by Infrared Heat Treatment. *J. Food Sci.* **2012,** *77,* 1197–1202.

38. Smith, J. W.; Kramer, A. Palatability and Nutritive Value of Fresh, Canned, and Frozen Collard Greens. *J. Am. Soc. Hort. Sci.* **1972,** *97,* 161–163.

39. Taiwo, K. A.; Angersbach, A.; Knoor, D. Effects of Pulsed Electric Field on Quality Factors and Mass Transfer During Osmotic Dehydration of Apples. *J. Food Process Eng.* **2003,** *26,* 31–48.

40. USDA-ARS (United States Department of Agriculture—Agricultural Research Service). *USDA Nutrient Database for Standard Reference, Release 18*; USDA-ARS: Beltsville, MD, 2005; Nutrient Data Laboratory Home Page (Online). http://www.nal.usda.gov/fnic/foodcomp (accessed Aug 13, 2016).

41. Valente, A.; Sanches-Silva, A.; Albuquerque, T. G.; Costa, H. S. Development of an Orange Juice In-house Reference Material and Its Application to Guarantee the Quality of Vitamin C Determination in Fruits, Juices and Fruit Pulps. *Food Chem.* **2014,** *154,* 71–77.

42. Vargas, F. E.; Villanueva, O. M. T.; Marquina, D. A. Influences of Cooking on Hydrosoluble Vitamins in Cabbage: Cabbage and Cauliflower I. Thiamin and Riboflavin. *Alimentaria* **1995,** *261,* 111–117.

43. Walker, J. R. L. Enzymatic Browning in Foods. Its Chemistry and Control. *Food Technol. New Zealand* **1977,** *12,* 19–25.

44. Weits, J.; Van der Meer, M. A.; Lassche, J. B.; Meyer, J. C.; Steinbuch, E.; Gersons, L. Nutritive Value and Organoleptic Properties of Three Vegetables Fresh and Preserved in Six Different Ways. *Int. J. Vit. Nutr. Res.* **1970,** *40,* 648–658.

45. Williams, P. G. Vitamin Retention in Cook/Chill and Cook/Hot-hold Hospital Food Services. *J. Am. Diet. Assoc.* **1996,** *96*, 490–498.

46. Zaman, A.; Ihsanullah, I.; Shah, A. A.; Khattak, T. N.; Gul, S.; Muhammadzai, I. U. Combined Effect of Gamma Irradiation and Hot Water Dipping on the Selected Nutrients and Shelf Life of Peach. *J. Radio Anal. Nucl. Chem.* **2013,** *298*, 1665–1672.

INDEX

G

H